To my parents and to Lucia

Note: Readers may require stereoscopic viewers in order to obtain the three-dimensional effects of the stereo images illustrated in this book. These viewers may be ordered from the following companies; please write to request current prices.

Hubbard Scientific Company
P.O. Box 104
Northbrook, Illinois 60062

Edmund Scientific Company
1776 Edscorp Building
Barrington, New Jersey 08007

The reader may be able to obtain the stereoscopic effect without an optical device: Hold the stereo image about ten inches away from the eyes and relax the eyes as if staring into the distance. Eventually the left-hand member of the pair seen by the right eye and the right-hand member of the pair seen by the left eye will merge to produce what will appear to be a three-dimensional image.

It will help to hold a fingertip about halfway between the stereo pair and your eyes. Adjust the position of the finger so that when looking with only your left eye, you see the finger in front of the right edge of the right-hand member of the pair. At the same time, when looking with your right eye only, try to see the finger in front of the right edge of the left-hand member of the pair. When your finger is so positioned, look at the finger with both eyes. This procedure will bring the two members of the stereo pair into registration, but they will be out of focus. Now relax your eyes and try to focus the stereo pair without losing the fixation on your finger. This trick seems to get easier as you get older.

VISION

A Computational Investigation
into the Human Representation
and Processing of Visual Information

David Marr
Late of the Massachusetts Institute of Technology

W. H. Freeman and Company
New York

Project Editor: Judith Wilson
Copy Editor: Paul Monsour
Production Coordinator: Linda Jupiter
Illustration Coordinator: Richard Quiñones
Designer: Ron Newcomer
Artists: Catherine Brandel and Victor Royer
Compositor: Graphic Typesetting Service
Printer and Binder: The Maple-Vail Book Manufacturing Group

Library of Congress Cataloging in Publication Data

Marr, David, 1945-1980.
  Vision.

  Bibliography: p.
  Includes index.
    1. Vision—Data processing.   2. Vision—Mathematical models.   3. Human
information processing.   I. Title.
QP475.M27 1982   152.1'4028'54   81-15076
ISBN 0-7167-1284-9

Contents

Detailed Contents

PART II

VISION

Chapter 2

Chapter 3

Chapter 4

The Immediate Representation of Visible Surfaces   268

Chapter 5

Representing Shapes for Recognition   295

Chapter 6

Synopsis   329

Preface

This book is meant to be enjoyed. It describes the adventures I have had in the years since Marvin Minsky and Seymour Papert invited me to the Artificial Intelligence Laboratory at the Massachusetts Institute of Technology in 1973. Working conditions were ideal, thanks to Patrick Winston's skillful administration, to the generosity of the Advanced Research Projects Agency of the Department of Defense and of the National Science Foundation, and to the freedom arranged for me by Whitman Richards, under the benevolent eye of Richard Held. I was fortunate enough to meet and collaborate with a remarkable collection of people, most especially, Tomaso Poggio. Included among these people were many erstwhile students who became colleagues and from whom I learned much—Keith Nishihara, Shimon Ullman, Ken Forbus, Kent Stevens, Eric Grimson, Ellen Hildreth, Michael Riley, and John Batali. Berthold Horn kept us close to the physics of light, and Whitman Richards, to the abilities and inabilities of people.

In December 1977, certain events occurred that forced me to write this book a few years earlier than I had planned. Although the book has important gaps, which I hope will soon be filled, a new framework for studying vision is already clear and supported by enough solid results to be worth setting down as a coherent whole.

Many people have helped me to live through this somewhat difficult period. Particularly, my parents, my sister, my wife Lucia, and Jennifer, Tomaso, Shimon, Whitman, and Inge gave to me more than I often deserved; although mere thanks are inadequate, I thank them. William Prince steered me to Professor F. G. Hayhoe and Dr. John Rees at Addenbrooke's Hospital in Cambridge, and them I thank for giving me time.

Summer 1979   David Marr

We should like to express our gratitude to those who helped us bring David Marr's *Vision* to fulfillment.

We thank Gunther Stent, whose friendship brought David Marr and W. H. Freeman and Company together and whose sound guidance helped us prepare the book for publication.

We thank David Marr's colleague, Keith Nishihara, for his skill and great effort; the work could not have been finished without him.

We thank David Marr's assistant, Carol Papineau, for attending so well to the needs of the manuscript and the publisher.

We thank the vision group at the MIT Artificial Intelligence Laboratory, especially Ellen Hildreth and Eric Grimson, who participated in ways large and small to bring this book to life.

The Publisher

PART I

Introduction and Philosophical Preliminaries

General
Introduction

What does it mean, to see? The plain man's answer (and Aristotle's, too) would be, to know what is where by looking. In other words, vision is the *process* of discovering from images what is present in the world, and where it is.

Vision is therefore, first and foremost, an information-processing task, but we cannot think of it just as a process. For if we are capable of knowing what is where in the world, our brains must somehow be capable of *representing* this information—in all its profusion of color and form, beauty, motion, and detail. The study of vision must therefore include not only the study of how to extract from images the various aspects of the world that are useful to us, but also an inquiry into the nature of the internal representations by which we capture this information and thus make it available as a basis for decisions about our thoughts and actions. This duality— the representation and the processing of information—lies at the heart of most information-processing tasks and will profoundly shape our investigation of the particular problems posed by vision.

The need to understand information-processing tasks and machines has arisen only quite recently. Until people began to dream of and then to build such machines, there was no very pressing need to think deeply about them. Once people did begin to speculate about such tasks and machines, however, it soon became clear that many aspects of the world around us could benefit from an information-processing point of view. Most of the phenomena that are central to us as human beings—the mysteries of life and evolution, of perception and feeling and thought— are primarily phenomena of information processing, and if we are ever to understand them fully, our thinking about them must include this perspective.

The next point—which has to be made rather quickly to those who inhabit a world in which the local utility's billing computer is still capable of sending a final demand for $0.00—is to emphasize that saying that a job is "only" an information-processing task or that an organism is "only" an information-processing machine is not a limiting or a pejorative description. Even more importantly, I shall in no way use such a description to try to limit the kind of explanations that are necessary. Quite the contrary, in fact. One of the fascinating features of information-processing machines is that in order to understand them completely, one has to be satisfied with one's explanations at many different levels.

For example, let us look at the range of perspectives that must be satisfied before one can be said, from a human and scientific point of view, to have understood visual perception. First, and I think foremost, there is the perspective of the plain man. He knows what it is like to see, and unless the bones of one's arguments and theories roughly correspond to what this person knows to be true at first hand, one will probably be wrong (a point made with force and elegance by Austin, 1962). Second, there is the perspective of the brain scientists, the physiologists and anatomists who know a great deal about how the nervous system is built and how parts of it behave. The issues that concern them—how the cells are connected, why they respond as they do, the neuronal dogmas of Barlow (1972)—must be resolved and addressed in any full account of perception. And the same argument applies to the perspective of the experimental psychologists.

On the other hand, someone who has bought and played with a small home computer may make quite different demands. "If," he might say, "vision really is an information-processing task, then I should be able to make my computer do it, provided that it has sufficient power, memory, and some way of being connected to a home television camera." The explanation he wants is therefore a rather abstract one, telling him what to program and, if possible, a hint about the best algorithms for doing so. He doesn't want to know about rhodopsin, or the lateral geniculate nucleus, or inhibitory interneurons. He wants to know how to program vision.

The fundamental point is that in order to understand a device that performs an information-processing task, one needs many different kinds of explanations. Part I of this book is concerned with this point, and it plays a prominent role because one of the keystones of the book is the realization that we have had to be more careful about what constitutes an explanation than has been necessary in other recent scientific developments, like those in molecular biology. For the subject of vision, there *is* no single equation or view that explains everything. Each problem has to be addressed from several points of view—as a problem in representing information, as a computation capable of deriving that representation, and as a problem in the architecture of a computer capable of carrying out both things quickly and reliably.

If one keeps strongly in mind this necessarily rather broad aspect of the nature of explanation, one can avoid a number of pitfalls. One consequence of an emphasis on information processing might be, for example, to introduce a comparison between the human brain and a computer. In a sense, of course, the brain is a computer, but to say this without qualification is misleading, because the essence of the brain is not simply that it is a computer but that it is a computer which is in the habit of performing some rather particular computations. The term *computer* usually refers to a machine with a rather standard type of instruction set that usually runs serially but nowadays sometimes in parallel, under the control of programs that have been stored in a memory. In order to understand such a computer, one needs to understand what it is made of, how it is put together, what its instruction set is, how much memory it has and how it is accessed, and how the machine may be made to run. But this forms only a small part of understanding a computer that is performing an information-processing task.

This point bears reflection, because it is central to why most analogies between brains and computers are too superficial to be useful. Think, for example, of the international network of airline reservation computers, which performs the task of assigning flights for millions of passengers all over the world. To understand this system it is not enough to know how a modern computer works. One also has to understand a little about what aircraft are and what they do; about geography, time zones, fares, exchange rates, and connections; and something about politics, diets, and the various other aspects of human nature that happen to be relevant to this particular task.

Thus the critical point is that understanding computers is different from understanding computations. To understand a computer, one has to study that computer. To understand an information-processing task, one has to study that information-processing task. To understand fully a particular machine carrying out a particular information-processing task, one has to do both things. Neither alone will suffice.

From a philosophical point of view, the approach that I describe is an extension of what have sometimes been called representational theories of mind. On the whole, it rejects the more recent excursions into the philosophy of perception, with their arguments about sense-data, the molecules of perception, and the validity of what the senses tell us; instead, this approach looks back to an older view, according to which the senses are for the most part concerned with telling one what is there. Modern representational theories conceive of the mind as having access to systems of internal representations; mental states are characterized by asserting what the internal representations currently specify, and mental processes by how such internal representations are obtained and how they interact.

This scheme affords a comfortable framework for our study of visual perception, and I am content to let it form the point of departure for our inquiry. As we shall see, pursuing this approach will lead us away from traditional avenues into what is almost a new intellectual landscape. Some of the things we find will seem strange, and it will be hard to reconcile subjectively some of the ideas and theories that are forced on us with what actually goes on inside ourselves when we open our eyes and look at things. Even the basic notion of what constitutes an explanation will have to be developed and broadened a little, to ensure that we do not leave anything out and that every important perspective on the problem is satisfied or satisfiable.

The book itself is divided into three parts. In the first are contained the philosophical preliminaries, a description of the approach, the representational framework that is proposed for the overall process of visual perception, and the way that led to it. I have adopted a fairly personal style in the hope that if the reader understands why particular directions were taken at each point, the reasons for the overall approach will be clearer.

The second part of the book, Chapters 2 to 6, contains the real analysis. It describes informally, but in some detail, how the approach and framework are actually realized, and the results that have been achieved.

The third part is somewhat unorthodox and consists of a set of questions and answers that are designed to help the reader to understand the way of thinking behind the approach—to help him acquire the right prejudices, if you like—and to relate these explanations to his personal experience of seeing. I have often found that one or two of the remarks set out in Part III have helped a person to see the point of part of the theory or to circumvent some private difficulty with it, and I hope they may serve a similar purpose here. The reader may find this section means more after having read the first two parts of the book, but an early glance at it may provide the motivation to take the trouble.

The detailed exposition comes, then, in Part II. Of course, the subject of human visual perception is not solved here by a long way. But over the last six years, my colleagues and I have been fortunate enough to see the establishment of an overall theoretical framework as well as the solution of several rather central problems in visual perception. We feel that the combination amounts to a reasonably strong case that the representational approach is a useful one, and the point of this book is to make that case. How far this approach can be pursued, of course, remains to be seen.

CHAPTER 1

The Philosophy
and the Approach

1.1   BACKGROUND

The problems of visual perception have attracted the curiosity of scientists for many centuries. Important early contributions were made by Newton (1704), who laid the foundations for modern work on color vision, and Helmholtz (1910), whose treatise on physiological optics generates interest even today. Early in this century, Wertheimer (1912, 1923) noticed the apparent motion not of individual dots but of wholes, or "fields," in images presented sequentially as in a movie. In much the same way we perceive the migration across the sky of a flock of geese: the flock somehow constitutes a single entity, and is not seen as individual birds. This observation started the Gestalt school of psychology, which was concerned with describing the qualities of wholes by using terms like *solidarity* and *distinctness,* and with trying to formulate the "laws" that governed the creation of these wholes. The attempt failed for various reasons, and the Gestalt school dissolved into the fog of subjectivism. With the death of the school, many

8

Figure 1–1. A random-dot stereogram of the type used extensively by Bela Julesz. The left and right images are identical except for a central square region that is displaced slightly in one image. When fused binocularly, the images yield the impression of the central square floating in front of the background.

of its early and genuine insights were unfortunately lost to the mainstream of experimental psychology.

Since then, students of the psychology of perception have made no serious attempts at an overall understanding of what perception is, concentrating instead on the analysis of properties and performance. The trichromatism of color vision was firmly established (see Brindley, 1970), and the preoccupation with motion continued, with the most interesting developments perhaps being the experiments of Miles (1931) and of Wallach and O'Connell (1953), which established that under suitable conditions an unfamiliar three-dimensional shape can be correctly perceived from only its changing monocular projection.*

The development of the digital electronic computer made possible a similar discovery for binocular vision. In 1960 Bela Julesz devised computer-generated random-dot stereograms, which are image pairs constructed of dot patterns that appear random when viewed monocularly but fuse when viewed one through each eye to give a percept of shapes and surfaces with a clear three-dimensional structure. An example is shown in Figure 1–1. Here the image for the left eye is a matrix of black and white squares generated at random by a computer program. The image for the

---

*The two dimensional image seen by a single eye.

right eye is made by copying the left image, shifting a square-shaped region at its center slightly to the left, and then providing a new random pattern to fill the gap that the shift creates. If each of the eyes sees only one matrix, as if the matrices were both in the same physical place, the result is the sensation of a square floating in space. Plainly, such percepts are caused solely by the stereo disparity between matching elements in the images presented to each eye; from such experiments, we know that the analysis of stereoscopic information, like the analysis of motion, can proceed independently in the absence of other information. Such findings are of critical importance because they help us to subdivide our study of perception into more specialized parts which can be treated separately. I shall refer to these as independent modules of perception.

The most recent contribution of psychophysics has been of a different kind but of equal importance. It arose from a combination of adaptation and threshold detection studies and originated from the demonstration by Campbell and Robson (1968) of the existence of independent, spatial-frequency-tuned channels—that is, channels sensitive to intensity variations in the image occurring at a particular scale or spatial interval—in the early stages of our perceptual apparatus. This paper led to an explosion of articles on various aspects of these channels, which culminated ten years later with quite satisfactory quantitative accounts of the characteristics of the first stages of visual perception (Wilson and Bergen, 1979). I shall discuss this in detail later on.

Recently a rather different approach has attracted considerable attention. In 1971, Roger N. Shepard and Jacqueline Metzler made line drawings of simple objects that differed from one another either by a three-dimensional rotation or by a rotation plus a reflection (see Figure 1–2). They asked how long it took to decide whether two depicted objects differed by a rotation and a reflection or merely a rotation. They found that the time taken depended on the three-dimensional angle of rotation necessary to bring the two objects into correspondence. Indeed, the time varied linearly with this angle. One is led thereby to the notion that a mental rotation of sorts is actually being performed—that a mental description of the first shape in a pair is being adjusted incrementally in orientation until it matches the second, such adjustment requiring greater time when greater angles are involved.

The significance of this approach lies not so much in its results, whose interpretation is controversial, as in the type of questions it raised. For until then, the notion of a representation was not one that visual psychologists took seriously. This type of experiment meant that the notion had to be considered. Although the early thoughts of visual psychologists were naive compared with those of the computer vision community, which had had

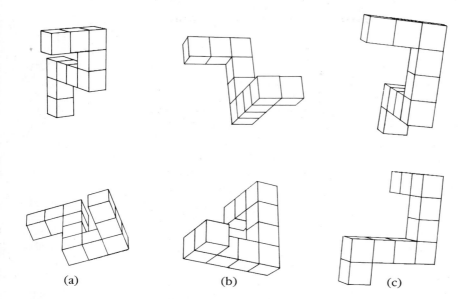

(a)   (b)   (c)

Figure 1–2. Some drawings similar to those used in Shepard and Metzler's experiments on mental rotation. The ones shown in (a) are identical, as a clockwise turning of this page by 80° will readily prove. Those in (b) are also identical, and again the relative angle between the two is 80°. Here, however, a rotation in depth will make the first coincide with the second. Finally, those in (c) are not at all identical, for no rotation will bring them into congruence. The time taken to decide whether a pair is the same was found to vary linearly with the angle through which one figure must be rotated to be brought into correspondence with the other. This suggested to the investigators that a stepwise mental rotation was in fact being performed by the subjects of their experiments.

to face the problem of representation from the beginning, it was not long before the thinking of psychologists became more sophisticated (see Shepard, 1979).

But what of explanation? For a long time, the best hope seemed to lie along another line of investigation, that of electrophysiology. The development of amplifiers allowed Adrian (1928) and his colleagues to record the minute voltage changes that accompanied the transmission of nerve signals. Their investigations showed that the character of the sensation so produced depended on which fiber carried the message, not how the fiber was stimulated—as one might have expected from anatomical studies. This led to the view that the peripheral nerve fibers could be thought of as a simple mapping supplying the sensorium with a copy of the physical events at the body surface (Adrian, 1947). The rest of the explanation, it was thought, could safely be left to the psychologists.

The next development was the technical improvement in amplification that made possible the recording of single neurons (Granit and Svaetichin, 1939; Hartline, 1938; Galambos and Davis, 1943). This led to the notion of a cell's "receptive field" (Hartline, 1940) and to the Harvard School's famous series of studies of the behavior of neurons at successively deeper levels of the visual pathway (Kuffler, 1953; Hubel and Wiesel, 1962, 1968). But perhaps the most exciting development was the new view that questions of psychological interest could be illuminated and perhaps even explained by neurophysiological experiments. The clearest early example of this was Barlow's (1953) study of ganglion cells in the frog retina, and I cannot put it better than he did:

> If one explores the responsiveness of single ganglion cells in the frog's retina using handheld targets, one finds that one particular type of ganglion cell is most effectively driven by something like a black disc subtending a degree or so moved rapidly to and fro within the unit's receptive field. This causes a vigorous discharge which can be maintained without much decrement as long as the movement is continued. Now, if the stimulus which is optimal for this class of cells is presented to intact frogs, the behavioural response is often dramatic; they turn towards the target and make repeated feeding responses consisting of a jump and snap. The selectivity of the retinal neurons and the frog's reaction when they are selectively stimulated, suggest that they are "bug detectors" (Barlow 1953) performing a primitive but vitally important form of recognition.
>
> The result makes one suddenly realize that a large part of the sensory machinery involved in a frog's feeding responses may actually reside in the retina rather than in mysterious "centres" that would be too difficult to understand by physiological methods. The essential lock-like property resides in each member of a whole class of neurons and allows the cell to discharge only to the appropriate key pattern of sensory stimulation. Lettvin *et al.* (1959) suggested that there were five different classes of cell in the frog, and Barlow, Hill and Levick (1964) found an even larger number of categories in the rabbit. [Barlow *et al.*] called these key patterns "trigger features," and Maturana *et al.* (1960) emphasized another important aspect of the behaviour of these ganglion cells; a cell continues to respond to the same trigger feature in spite of changes in light intensity over many decades. The properties of the retina are such that a ganglion cell can, figuratively speaking, reach out and determine that something specific is happening in front of the eye. Light is the agent by

which it does this, but it is the detailed pattern of the light that carries the information, and the overall level of illumination prevailing at the time is almost totally disregarded. (p. 373)

Barlow (1972) then goes on to summarize these findings in the following way:

The cumulative effect of all the changes I have tried to outline above has been to make us realise that each *single neuron can perform a much more complex and subtle task than had previously been thought* (emphasis added). Neurons do not loosely and unreliably remap the luminous intensities of the visual image onto our sensorium, but instead they detect pattern elements, discriminate the depth of objects, ignore irrelevant causes of variation and are arranged in an intriguing hierarchy. Furthermore, there is evidence that they give prominence to what is informationally important, can respond with great reliability, and can have their pattern selectivity permanently modified by early visual experience. This amounts to a revolution in our outlook. It is now quite inappropriate to regard unit activity as a noisy indication of more basic and reliable processes involved in mental operations: instead, we must regard single neurons as the prime movers of these mechanisms. Thinking is brought about by neurons and we should not use phrases like "unit activity reflects, reveals, or monitors thought processes," because the activities of neurons, quite simply, are thought processes.

This revolution stemmed from physiological work and makes us realize that the activity of each single neuron may play a significant role in perception. (p. 380)

This aspect of his thinking led Barlow to formulate the first and most important of his five dogmas: 'A description of that activity of a single nerve cell which is transmitted to and influences other nerve cells and of a nerve cell's response to such influences from other cells, is a complete enough description for functional understanding of the nervous system. There is nothing else "looking at" or controlling this activity, which must therefore provide a basis for understanding how the brain controls behaviour' (Barlow, 1972, p. 380).

I shall return later on to more carefully examine the validity of this point of view, but for now let us just enjoy it. The vigor and excitement of these ideas need no emphasis. At the time the eventual success of a reductionist approach seemed likely. Hubel and Wiesel's (1962, 1968) pioneering studies had shown the way; single-unit studies on stereopsis (Barlow, Blakemore, and Pettigrew, 1967) and on color (DeValois, Abramov, and Mead, 1967; Gouras, 1968) seemed to confirm the close links between perception and single-cell recordings, and the intriguing results of Gross,

Rocha-Miranda, and Bender (1972), who found "hand-detectors" in the inferotemporal cortex, seemed to show that the application of the reductionist approach would not be limited just to the early parts of the visual pathway.

It was, of course, recognized that physiologists had been lucky: If one probes around in a conventional electronic computer and records the behavior of single elements within it, one is unlikely to be able to discern what a given element is doing. But the brain, thanks to Barlow's first dogma, seemed to be built along more accommodating lines—people *were* able to determine the functions of single elements of the brain. There seemed no reason why the reductionist approach could not be taken all the way.

I was myself fully caught up in this excitement. Truth, I also believed, was basically neural, and the central aim of all research was a thorough functional analysis of the structure of the central nervous system. My enthusiasm found expression in a theory of the cerebellar cortex (Marr, 1969). According to this theory, the simple and regular cortical structure is interpreted as a simple but powerful memorizing device for learning motor skills; because of a simple combinatorial trick, each of the 15 million Purkinje cells in the cerebellum is capable of learning over 200 different patterns and discriminating them from unlearned patterns. Evidence is gradually accumulating that the cerebellum is involved in learning motor skills (Ito, 1978), so that something like this theory may in fact be correct.

The way seemed clear. On the one hand we had new experimental techniques of proven power, and on the other, the beginnings of a theoretical approach that could back them up with a fine analysis of cortical structure. Psychophysics could tell us what needed explaining, and the recent advances in anatomy—the Fink-Heimer technique from Nauta's laboratory and the recent successful deployment by Szentagothai and others of the electron microscope—could provide the necessary information about the structure of the cerebral cortex.

But somewhere underneath, something was going wrong. The initial discoveries of the 1950s and 1960s were not being followed by equally dramatic discoveries in the 1970s. No neurophysiologists had recorded new and clear high-level correlates of perception. The leaders of the 1960s had turned away from what they had been doing—Hubel and Wiesel concentrated on anatomy, Barlow turned to psychophysics, and the mainstream of neurophysiology concentrated on development and plasticity (the concept that neural connections are not fixed) or on a more thorough analysis of the cells that had already been discovered (for example, Bishop, Coombs, and Henry, 1971; Schiller, Finlay, and Volman, 1976a, 1976b), or on cells in species like the owl (for example, Pettigrew and Konishi, 1976).

None of the new studies succeeded in elucidating the *function* of the visual cortex.

It is difficult to say precisely why this happened, because the reasoning was never made explicit and was probably largely unconscious. However, various factors are identifiable. In my own case, the cerebellar study had two effects. On the one hand, it suggested that one could eventually hope to understand cortical structure in functional terms, and this was exciting. But at the same time the study has disappointed me, because even if the theory was correct, it did not much enlighten one about the motor system—it did not, for example, tell one how to go about programming a mechanical arm. It suggested that if one wishes to program a mechanical arm so that it operates in a versatile way, then at some point a very large and rather simple type of memory will prove indispensable. But it did not say why, nor what that memory should contain.

The discoveries of the visual neurophysiologists left one in a similar situation. Suppose, for example, that one actually found the apocryphal grandmother cell.* Would that really tell us anything much at all? It would tell us that it existed—Gross's hand-detectors tell us almost that—but not *why* or even *how* such a thing may be constructed from the outputs of previously discovered cells. Do the single-unit recordings—the simple and complex cells—tell us much about how to detect edges or why one would want to, except in a rather general way through arguments based on economy and redundancy? If we really knew the answers, for example, we should be able to program them on a computer. But finding a hand-detector certainly did not allow us to program one.

As one reflected on these sorts of issues in the early 1970s, it gradually became clear that something important was missing that was not present in either of the disciplines of neurophysiology or psychophysics. The key observation is that neurophysiology and psychophysics have as their business to *describe* the behavior of cells or of subjects but not to *explain* such behavior. What are the visual areas of the cerebral cortex actually doing? What are the problems in doing it that need explaining, and at what level of description should such explanations be sought?

The best way of finding out the difficulties of doing something is to try to do it, so at this point I moved to the Artificial Intelligence Laboratory at MIT, where Marvin Minsky had collected a group of people and a powerful computer for the express purpose of addressing these questions.

---

*A cell that fires only when one's grandmother comes into view.

The first great revelation was that the problems are difficult. Of course, these days this fact is a commonplace. But in the 1960s almost no one realized that machine vision was difficult. The field had to go through the same experience as the machine translation field did in its fiascoes of the 1950s before it was at last realized that here were some problems that had to be taken seriously. The reason for this misperception is that we humans are ourselves so good at vision. The notion of a feature detector was well established by Barlow and by Hubel and Wiesel, and the idea that extracting edges and lines from images might be at all difficult simply did not occur to those who had not tried to do it. It turned out to be an elusive problem: Edges that are of critical importance from a three-dimensional point of view often cannot be found at all by looking at the intensity changes in an image. Any kind of textured image gives a multitude of noisy edge segments; variations in reflectance and illumination cause no end of trouble; and even if an edge has a clear existence at one point, it is as likely as not to fade out quite soon, appearing only in patches along its length in the image. The common and almost despairing feeling of the early investigators like B.K.P. Horn and T.O. Binford was that practically anything could happen in an image and furthermore that practically everything did.

Three types of approach were taken to try to come to grips with these phenomena. The first was unashamedly empirical, associated most with Azriel Rosenfeld. His style was to take some new trick for edge detection, texture discrimination, or something similar, run it on images, and observe the result. Although several interesting ideas emerged in this way, including the simultaneous use of operators* of different sizes as an approach to increasing sensitivity and reducing noise (Rosenfeld and Thurston, 1971), these studies were not as useful as they could have been because they were never accompanied by any serious assessment of how well the different algorithms performed. Few attempts were made to compare the merits of different operators (although Fram and Deutsch, 1975, did try), and an approach like trying to prove mathematically which operator was optimal was not even attempted. Indeed, it could not be, because no one had yet formulated precisely what these operators should be trying to do. Nevertheless, considerable ingenuity was shown. The most clever was probably Hueckel's (1973) operator, which solved in an ingenious way the problem of finding the edge orientation that best fit a given intensity change in a small neighborhood of an image.

Operator refers to a local calculation to be applied at each location in the image, making use of the intensity there and in the immediate vicinity.

The second approach was to try for depth of analysis by restricting the scope to a world of single, illuminated, matte white toy blocks set against a black background. The blocks could occur in any shapes provided only that all faces were planar and all edges were straight. This restriction allowed more specialized techniques to be used, but it still did not make the problem easy. The Binford–Horn line finder (Horn, 1973) was used to find edges, and both it and its sequel (described in Shirai, 1973) made use of the special circumstances of the environment, such as the fact that all edges there were straight.

These techniques did work reasonably well, however, and they allowed a preliminary analysis of later problems to emerge—roughly, what does one do once a complete line drawing has been extracted from a scene? Studies of this had begun sometime before with Roberts (1965) and Guzman (1968), and they culminated in the works of Waltz (1975) and Mackworth (1973), which essentially solved the interpretation problem for line drawings derived from images of prismatic solids. Waltz's work had a particularly dramatic impact, because it was the first to show explicitly that an exhaustive analysis of all possible local physical arrangements of surfaces, edges, and shadows could lead to an effective and efficient algorithm for interpreting an actual image. Figure 1–3 and its legend convey the main ideas behind Waltz's theory.

The hope that lay behind this work was, of course, that once the toy world of white blocks had been understood, the solutions found there could be generalized, providing the basis for attacking the more complex problems posed by a richer visual environment. Unfortunately, this turned out not to be so. For the roots of the approach that was eventually successful, we have to look at the third kind of development that was taking place then.

Two pieces of work were important here. Neither is probably of very great significance to human perception for what it actually accomplished— in the end, it is likely that neither will particularly reflect human visual processes—but they are both of importance because of the way in which they were formulated. The first was Land and McCann's (1971) work on the retinex theory of color vision, as developed by them and subsequently by Horn (1974). The starting point is the traditional one of regarding color as a perceptual approximation to reflectance. This allows the formulation of a clear computational question, namely, How can the effects of reflectance changes be separated from the vagaries of the prevailing illumination? Land and McCann suggested using the fact that changes in illumination are usually gradual, whereas changes in reflectance of a surface or of an object boundary are often quite sharp. Hence by filtering out slow changes, those changes due to the reflectance alone could be isolated. Horn devised a

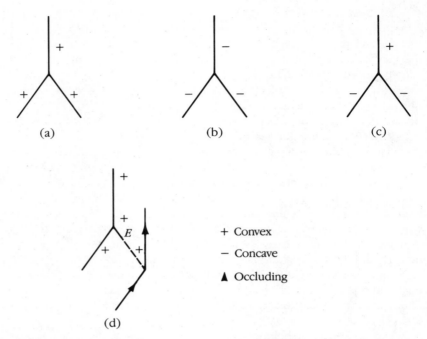

Figure 1–3. Some configurations of edges are physically realizable, and some are not. The trihedral junctions of three convex edges (a) or of three concave edges (b) are realizable, whereas the configuration (c) is impossible. Waltz cataloged all the possible junctions, including shadow edges, for up to four coincident edges. He then found that by using this catalog to implement consistency relations [requiring, for example, that an edge be of the same type all along its length like edge *E* in (d)], the solution to the labeling of a line drawing that included shadows was often uniquely determined.

clever parallel algorithm for this, and I suggested how it might be implemented by neurons in the retina (Marr, 1974a).

I do not now believe that this is at all a correct analysis of color vision or of the retina, but it showed the possible style of a correct analysis. Gone are the ad hoc programs of computer vision; gone is the restriction to a special visual miniworld; gone is any explanation *in terms of* neurons— except as a way of implementing a method. And present is a clear understanding of what is to be computed, how it is to be done, the physical assumptions on which the method is based, and some kind of analysis of algorithms that are capable of carrying it out.

The other piece of work was Horn's (1975) analysis of shape from shading, which was the first in what was to become a distinguished series of articles on the formation of images. By carefully analyzing the way in which the illumination, surface geometry, surface reflectance, and viewpoint conspired to create the measured intensity values in an image, Horn formulated a differential equation that related the image intensity values to the surface geometry. If the surface reflectance and illumination are known, one can solve for the surface geometry (see also Horn, 1977). Thus from shading one can derive shape.

The message was plain. There must exist an additional level of understanding at which the character of the information-processing tasks carried out during perception are analyzed and understood in a way that is independent of the particular mechanisms and structures that implement them in our heads. This was what was missing—the analysis of the problem as an information-processing task. Such analysis does not usurp an understanding at the other levels—of neurons or of computer programs—but it is a necessary complement to them, since without it there can be no real understanding of the function of all those neurons.

This realization was arrived at independently and formulated together by Tomaso Poggio in Tübingen and myself (Marr and Poggio, 1977; Marr, 1977b). It was not even quite new—Leon D. Harmon was saying something similar at about the same time, and others had paid lip service to a similar distinction. But the important point is that if the notion of different types of understanding is taken very seriously, it allows the study of the information-processing basis of perception to be made *rigorous*. It becomes possible, by separating explanations into different levels, to make explicit statements about what is being computed and why and to construct theories stating that what is being computed is optimal in some sense or is guaranteed to function correctly. The ad hoc element is removed, and heuristic computer programs are replaced by solid foundations on which a real subject can be built. This realization—the formulation of what was missing, together with a clear idea of how to supply it—formed the basic foundation for a new integrated approach, which it is the purpose of this book to describe.

1.2   UNDERSTANDING COMPLEX INFORMATION-PROCESSING SYSTEMS

Almost never can a complex system of any kind be understood as a simple extrapolation from the properties of its elementary components. Consider, for example, some gas in a bottle. A description of thermodynamic effects—

temperature, pressure, density, and the relationships among these factors—is not formulated by using a large set of equations, one for each of the particles involved. Such effects are described at their own level, that of an enormous collection of particles; the effort is to show that in principle the microscopic and macroscopic descriptions are consistent with one another. If one hopes to achieve a full understanding of a system as complicated as a nervous system, a developing embryo, a set of metabolic pathways, a bottle of gas, or even a large computer program, then one must be prepared to contemplate different kinds of explanation at different levels of description that are linked, at least in principle, into a cohesive whole, even if linking the levels in complete detail is impractical. For the specific case of a system that solves an information-processing problem, there are in addition the twin strands of process and representation, and both these ideas need some discussion.

Representation and Description

A *representation* is a formal system for making explicit certain entities or types of information, together with a specification of how the system does this. And I shall call the result of using a representation to describe a given entity a *description* of the entity in that representation (Marr and Nishihara, 1978).

For example, the Arabic, Roman, and binary numeral systems are all formal systems for representing numbers. The Arabic representation consists of a string of symbols drawn from the set (0, 1, 2, 3, 4, 5, 6, 7, 8, 9), and the rule for constructing the description of a particular integer n is that one decomposes n into a sum of multiples of powers of 10 and unites these multiples into a string with the largest powers on the left and the smallest on the right. Thus, thirty-seven equals $3 \times 10^1 + 7 \times 10^0$, which becomes 37, the Arabic numeral system's description of the number. What this description makes explicit is the number's decomposition into powers of 10. The binary numeral system's description of the number thirty-seven is 100101, and this description makes explicit the number's decomposition into powers of 2. In the Roman numeral system, thirty-seven is represented as XXXVII.

This definition of a representation is quite general. For example, a representation for shape would be a formal scheme for describing some aspects of shape, together with rules that specify how the scheme is applied to any particular shape. A musical score provides a way of representing a symphony; the alphabet allows the construction of a written representation of words; and so forth. The phrase "formal scheme" is critical to the definition, but the reader should not be frightened by it. The reason is simply that we are dealing with information-processing machines, and the way such machines work is by using symbols to stand for things—to represent things, in our terminology. To say that something is a formal scheme means only that it is a set of symbols with rules for putting them together—no more and no less.

A representation, therefore, is not a foreign idea at all—we all use representations all the time. However, the notion that one can capture some aspect of reality by making a description of it using a symbol and that to do so can be useful seems to me a fascinating and powerful idea. But even the simple examples we have discussed introduce some rather general and important issues that arise whenever one chooses to use one particular representation. For example, if one chooses the Arabic numeral representation, it is easy to discover whether a number is a power of 10 but difficult to discover whether it is a power of 2. If one chooses the binary representation, the situation is reversed. Thus, there is a trade-off; any particular representation makes certain information explicit at the expense of information that is pushed into the background and may be quite hard to recover.

This issue is important, because how information is represented can greatly affect how easy it is to do different things with it. This is evident even from our numbers example: It is easy to add, to subtract, and even to multiply if the Arabic or binary representations are used, but it is not at all easy to do these things—especially multiplication—with Roman numerals. This is a key reason why the Roman culture failed to develop mathematics in the way the earlier Arabic cultures had.

An analogous problem faces computer engineers today. Electronic technology is much more suited to a binary number system than to the conventional base 10 system, yet humans supply their data and require the results in base 10. The design decision facing the engineer, therefore, is, Should one pay the cost of conversion into base 2, carry out the arithmetic in a binary representation, and then convert back into decimal numbers on output; or should one sacrifice efficiency of circuitry to carry out operations directly in a decimal representation? On the whole, business computers and pocket calculators take the second approach, and general purpose computers take the first. But even though one is not restricted to using just one representation system for a given type of information, the choice of which to use is important and cannot be taken lightly. It determines what information is made explicit and hence what is pushed further into the background, and it has a far-reaching effect on the ease and difficulty with which operations may subsequently be carried out on that information.

Process

The term *process* is very broad. For example, addition is a process, and so is taking a Fourier transform. But so is making a cup of tea, or going shopping. For the purposes of this book, I want to restrict our attention to the meanings associated with machines that are carrying out information-processing tasks. So let us examine in depth the notions behind one simple such device, a cash register at the checkout counter of a supermarket.

There are several levels at which one needs to understand such a device, and it is perhaps most useful to think in terms of three of them. The most abstract is the level of *what* the device does and *why*. What it does is arithmetic, so our first task is to master the theory of addition. Addition is a mapping, usually denoted by $+$, from pairs of numbers into single numbers; for example, $+$ maps the pair $(3, 4)$ to 7, and I shall write this in the form $(3 + 4) \rightarrow 7$. Addition has a number of abstract properties, however. It is commutative: both $(3 + 4)$ and $(4 + 3)$ are equal to 7; and associative: the sum of $3 + (4 + 5)$ is the same as the sum of $(3 + 4) + 5$. Then there is the unique distinguished element, zero, the adding of which has no effect: $(4 + 0) \rightarrow 4$. Also, for every number there is a unique "inverse," written (-4) in the case of 4, which when added to the number gives zero: $[4 + (-4)] \rightarrow 0$.

Notice that these properties are part of the fundamental *theory* of addition. They are true no matter how the numbers are written—whether in binary, Arabic, or Roman representation—and no matter how the addition is executed. Thus part of this first level is something that might be characterized as *what* is being computed.

The other half of this level of explanation has to do with the question of *why* the cash register performs addition and not, for instance, multiplication when combining the prices of the purchased items to arrive at a final bill. The reason is that the rules we intuitively feel to be appropriate for combining the individual prices in fact define the mathematical operation of addition. These can be formulated as *constraints* in the following way:

1. If you buy nothing, it should cost you nothing; and buying nothing and something should cost the same as buying just the something. (The rules for zero.)

2. The order in which goods are presented to the cashier should not affect the total. (Commutativity.)

3. Arranging the goods into two piles and paying for each pile separately should not affect the total amount you pay. (Associativity; the basic operation for combining prices.)

4. If you buy an item and then return it for a refund, your total expenditure should be zero. (Inverses.)

It is a mathematical theorem that these conditions define the operation of addition, which is therefore the appropriate computation to use.

This whole argument is what I call the *computational theory* of the cash register. Its important features are (1) that it contains separate arguments about what is computed and why and (2) that the resulting operation is defined uniquely by the constraints it has to satisfy. In the theory of visual processes, the underlying task is to reliably derive properties of the world from images of it; the business of isolating constraints that are both powerful enough to allow a process to be defined and generally true of the world is a central theme of our inquiry.

In order that a process shall actually run, however, one has to realize it in some way and therefore choose a representation for the entities that the process manipulates. The second level of the analysis of a process, therefore, involves choosing two things: (1) a *representation* for the input and for the output of the process and (2) an *algorithm* by which the transformation may actually be accomplished. For addition, of course, the input and output representations can both be the same, because they both consist of numbers. However this is not true in general. In the case of a Fourier transform, for example, the input representation may be the time domain, and the output, the frequency domain. If the first of our levels specifies what and why, this second level specifies *how*. For addition, we might choose Arabic numerals for the representations, and for the algorithm we could follow the usual rules about adding the least significant digits first and "carrying" if the sum exceeds 9. Cash registers, whether mechanical or electronic, usually use this type of representation and algorithm.

There are three important points here. First, there is usually a wide choice of representation. Second, the choice of algorithm often depends rather critically on the particular representation that is employed. And third, even for a given fixed representation, there are often several possible algorithms for carrying out the same process. Which one is chosen will usually depend on any particularly desirable or undesirable characteristics that the algorithms may have; for example, one algorithm may be much more efficient than another, or another may be slightly less efficient but more robust (that is, less sensitive to slight inaccuracies in the data on which it must run). Or again, one algorithm may be parallel, and another, serial. The choice, then, may depend on the type of hardware or machinery in which the algorithm is to be embodied physically.

This brings us to the third level, that of the device in which the process is to be realized physically. The important point here is that, once again, the same algorithm may be implemented in quite different technologies. The child who methodically adds two numbers from right to left, carrying a digit when necessary, may be using the same algorithm that is implemented by the wires and transistors of the cash register in the neighborhood supermarket, but the physical realization of the algorithm is quite different in these two cases. Another example: Many people have written computer programs to play tic-tac-toe, and there is a more or less standard algorithm that cannot lose. This algorithm has in fact been implemented by W. D. Hillis and B. Silverman in a quite different technology, in a computer made out of Tinkertoys, a children's wooden building set. The whole monstrously ungainly engine, which actually works, currently resides in a museum at the University of Missouri in St. Louis.

Some styles of algorithm will suit some physical substrates better than others. For example, in conventional digital computers, the number of connections is comparable to the number of gates, while in a brain, the number of connections is much larger ($\times 10^4$) than the number of nerve cells. The underlying reason is that wires are rather cheap in biological architecture, because they can grow individually and in three dimensions. In conventional technology, wire laying is more or less restricted to two dimensions, which quite severely restricts the scope for using parallel techniques and algorithms; the same operations are often better carried out serially.

The Three Levels

We can summarize our discussion in something like the manner shown in Figure 1–4, which illustrates the different levels at which an information-processing device must be understood before one can be said to have understood it completely. At one extreme, the top level, is the abstract computational theory of the device, in which the performance of the device is characterized as a mapping from one kind of information to another, the abstract properties of this mapping are defined precisely, and its appropriateness and adequacy for the task at hand are demonstrated. In the center is the choice of representation for the input and output and the

Computational theory	Representation and algorithm	Hardware implementation
What is the goal of the computation, why is it appropriate, and what is the logic of the strategy by which it can be carried out?	How can this computational theory be implemented? In particular, what is the representation for the input and output, and what is the algorithm for the transformation?	How can the representation and algorithm be realized physically?

Figure 1–4. The three levels at which any machine carrying out an information-processing task must be understood.

algorithm to be used to transform one into the other. And at the other extreme are the details of how the algorithm and representation are realized physically—the detailed computer architecture, so to speak. These three levels are coupled, but only loosely. The choice of an algorithm is influenced for example, by what it has to do and by the hardware in which it must run. But there is a wide choice available at each level, and the explication of each level involves issues that are rather independent of the other two.

Each of the three levels of description will have its place in the eventual understanding of perceptual information processing, and of course they are logically and causally related. But an important point to note is that since the three levels are only rather loosely related, some phenomena may be explained at only one or two of them. This means, for example, that a correct explanation of some psychophysical observation must be formulated at the appropriate level. In attempts to relate psychophysical problems to physiology, too often there is confusion about the level at which problems should be addressed. For instance, some are related mainly to the physical mechanisms of vision—such as afterimages (for example, the one you see after staring at a light bulb) or such as the fact that any color can be matched by a suitable mixture of the three primaries (a consequence principally of the fact that we humans have three types of cones). On the other hand, the ambiguity of the Necker cube (Figure 1–5) seems to demand a different kind of explanation. To be sure, part of the explanation of its perceptual reversal must have to do with a bistable neural network (that is, one with two distinct stable states) somewhere inside the

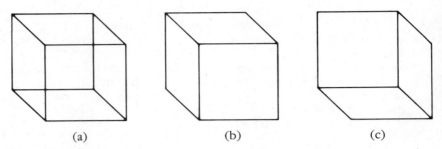

Figure 1–5. The so-called Necker illusion, named after L. A. Necker, the Swiss naturalist who developed it in 1832. The essence of the matter is that the two-dimensional representation (a) has collapsed the depth out of a cube and that a certain aspect of human vision is to recover this missing third dimension. The depth of the cube can indeed be perceived, but two interpretations are possible, (b) and (c). A person's perception characteristically flips from one to the other.

brain, but few would feel satisfied by an account that failed to mention the existence of two different but perfectly plausible three-dimensional interpretations of this two-dimensional image.

For some phenomena, the type of explanation required is fairly obvious. Neuroanatomy, for example, is clearly tied principally to the third level, the physical realization of the computation. The same holds for synaptic mechanisms, action potentials, inhibitory interactions, and so forth. Neurophysiology, too, is related mostly to this level, but it can also help us to understand the type of representations being used, particularly if one accepts something along the lines of Barlow's views that I quoted earlier. But one has to exercise extreme caution in making inferences from neurophysiological findings about the algorithms and representations being used, particularly until one has a clear idea about what information needs to be represented and what processes need to be implemented.

Psychophysics, on the other hand, is related more directly to the level of algorithm and representation. Different algorithms tend to fail in radically different ways as they are pushed to the limits of their performance or are deprived of critical information. As we shall see, primarily psychophysical evidence proved to Poggio and myself that our first stereo-matching algorithm (Marr and Poggio, 1976) was not the one that is used by the brain, and the best evidence that our second algorithm (Marr and Poggio, 1979) *is* roughly the one that is used also comes from psychophysics. Of course, the underlying computational theory remained the same in both cases, only the algorithms were different.

Psychophysics can also help to determine the nature of a representation. The work of Roger Shepard (1975), Eleanor Rosch (1978), or Elizabeth Warrington (1975) provides some interesting hints in this direction. More specifically, Stevens (1979) argued from psychophysical experiments that surface orientation is represented by the coordinates of slant and tilt, rather than (for example) the more traditional (p, q) of gradient space (see Chapter 3). He also deduced from the uniformity of the size of errors made by subjects judging surface orientation over a wide range of orientations that the representational quantities used for slant and tilt are pure angles and not, for example, their cosines, sines, or tangents.

More generally, if the idea that different phenomena need to be explained at different levels is kept clearly in mind, it often helps in the assessment of the validity of the different kinds of objections that are raised from time to time. For example, one favorite is that the brain is quite different from a computer because one is parallel and the other serial. The answer to this, of course, is that the distinction between serial and parallel is a distinction at the level of algorithm; it is not fundamental at all— anything programmed in parallel can be rewritten serially (though not necessarily vice versa). The distinction, therefore, provides no grounds for arguing that the brain operates so differently from a computer that a computer could not be programmed to perform the same tasks.

Importance of Computational Theory

Although algorithms and mechanisms are empirically more accessible, it is the top level, the level of computational theory, which is critically important from an information-processing point of view. The reason for this is that the nature of the computations that underlie perception depends more upon the computational problems that have to be solved than upon the particular hardware in which their solutions are implemented. To phrase the matter another way, an algorithm is likely to be understood more readily by understanding the nature of the problem being solved than by examining the mechanism (and the hardware) in which it is embodied.

In a similar vein, trying to understand perception by studying only neurons is like trying to understand bird flight by studying only feathers: It just cannot be done. In order to understand bird flight, we have to understand aerodynamics; only then do the structure of feathers and the different shapes of birds' wings make sense. More to the point, as we shall see, we cannot understand why retinal ganglion cells and lateral geniculate neurons have the receptive fields they do just by studying their anatomy and physiology. We can understand how these cells and neurons behave as they do by studying their wiring and interactions, but in order to understand *why* the receptive fields are as they are—why they are circularly symmetrical and why their excitatory and inhibitory regions have characteristic shapes and distributions—we have to know a little of the theory of differential operators, band-pass channels, and the mathematics of the uncertainty principle (see Chapter 2).

Perhaps it is not surprising that the very specialized empirical disciplines of the neurosciences failed to appreciate fully the absence of computational theory; but it is surprising that this level of approach did not play a more forceful role in the early development of artificial intelligence. For far too long, a heuristic program for carrying out some task was held to be a theory of that task, and the distinction between what a program did and how it did it was not taken seriously. As a result, (1) a style of explanation evolved that invoked the use of special mechanisms to solve particular problems, (2) particular data structures, such as the lists of attribute value pairs called property lists in the LISP programing language, were held to amount to theories of the representation of knowledge, and (3) there was frequently no way to determine whether a program would deal with a particular case other than by running the program.

Failure to recognize this theoretical distinction between *what* and *how* also greatly hampered communication between the fields of artificial intelligence and linguistics. Chomsky's (1965) theory of transformational grammar is a true computational theory in the sense defined earlier. It is concerned solely with specifying what the syntactic decomposition of an English sentence should be, and not at all with how that decomposition should be achieved. Chomsky himself was very clear about this—it is roughly his distinction between competence and performance, though his idea of performance did include other factors, like stopping in midutterance—but the fact that his theory was defined by transformations, which look like computations, seems to have confused many people. Winograd (1972), for example, felt able to criticize Chomsky's theory on the grounds that it cannot be inverted and so cannot be made to run on a computer; I had heard reflections of the same argument made by Chomsky's colleagues in linguistics as they turn their attention to how grammatical structure might actually be computed from a real English sentence.

The explanation is simply that finding algorithms by which Chomsky's theory may be implemented is a completely different endeavor from formulating the theory itself. In our terms, it is a study at a different level, and both tasks have to be done. This point was appreciated by Marcus (1980), who was concerned precisely with how Chomsky's theory can be realized and with the kinds of constraints on the power of the human grammatical processor that might give rise to the structural constraints in syntax that

Chomsky found. It even appears that the emerging "trace" theory of grammar (Chomsky and Lasnik, 1977) may provide a way of synthesizing the two approaches—showing that, for example, some of the rather ad hoc restrictions that form part of the computational theory may be consequences of weaknesses in the computational power that is available for implementing syntactical decoding.

The Approach of J. J. Gibson

In perception, perhaps the nearest anyone came to the level of computational theory was Gibson (1966). However, although some aspects of his thinking were on the right lines, he did not understand properly what information processing was, which led him to seriously underestimate the complexity of the information-processing problems involved in vision and the consequent subtlety that is necessary in approaching them.

Gibson's important contribution was to take the debate away from the philosophical considerations of sense-data and the affective qualities of sensation and to note instead that the important thing about the senses is that they are channels for perception of the real world outside or, in the case of vision, of the visible surfaces. He therefore asked the critically important question, How does one obtain constant perceptions in everyday life on the basis of continually changing sensations? This is exactly the right question, showing that Gibson correctly regarded the problem of perception as that of recovering from sensory information "valid" properties of the external world. His problem was that he had a much oversimplified view of how this should be done. His approach led him to consider higher-order variables—stimulus energy, ratios, proportions, and so on—as "invariants" of the movement of an observer and of changes in stimulation intensity.

"These invariants," he wrote, "correspond to permanent properties of the environment. They constitute, therefore, information about the permanent environment." This led him to a view in which the function of the brain was to "detect invariants" despite changes in "sensations" of light, pressure, or loudness of sound. Thus, he says that the "function of the brain, when looped with its perceptual organs, is not to decode signals, nor to interpret messages, nor to accept images, nor to *organize* the sensory input or to *process* the data, in modern terminology. It is to seek and extract information about the environment from the flowing array of ambient energy," and he thought of the nervous system as in some way "resonating" to these invariants. He then embarked on a broad study of animals in their environments, looking for invariants to which they might resonate. This was the basic idea behind the notion of ecological optics (Gibson, 1966, 1979).

Although one can criticize certain shortcomings in the quality of Gibson's analysis, its major and, in my view, fatal shortcoming lies at a deeper level and results from a failure to realize two things. First, the detection of physical invariants, like image surfaces, is exactly and precisely an information-processing problem, in modern terminology. And second, he vastly underrated the sheer difficulty of such detection. In discussing the recovery of three-dimensional information from the movement of an observer, he says that "in motion, perspective information alone can be used" (Gibson, 1966, p. 202). And perhaps the key to Gibson is the following:

> The detection of non-change when an object moves in the world is not as difficult as it might appear. It is only made to seem difficult when we assume that the perception of constant dimensions of the object must depend on the correcting of sensations of inconstant form and size. The information for the constant dimension of an object is normally carried by invariant relations in an optic array. Rigidity is *specified*. (emphasis added)

Yes, to be sure, but *how?* Detecting physical invariants is just as difficult as Gibson feared, but nevertheless we can do it. And the only way to understand how is to treat it as an information-processing problem.

The underlying point is that visual information processing is actually very complicated, and Gibson was not the only thinker who was misled by the apparent simplicity of the act of seeing. The whole tradition of philosophical inquiry into the nature of perception seems not to have taken seriously enough the complexity of the information processing involved. For example, Austin's (1962) *Sense and Sensibilia* entertainingly demolishes the argument, apparently favored by earlier philosophers, that since we are sometimes deluded by illusions (for example, a straight stick appears bent if it is partly submerged in water), we see sense-data rather than material things. The answer is simply that usually our perceptual processing does run correctly (it delivers a true description of what is there), but although evolution has seen to it that our processing allows for many changes (like inconstant illumination), the perturbation due to the refraction of light by water is not one of them. And incidentally, although the example of the bent stick has been discussed since Aristotle, I have seen no philosphical inquiry into the nature of the perceptions of, for instance, a heron, which is a bird that feeds by pecking up fish first seen from above the water surface. For such birds the visual correction might be present.

Anyway, my main point here is another one. Austin (1962) spends much time on the idea that perception tells one about real properties of the external world, and one thing he considers is "real shape," (p. 66), a notion which had cropped up earlier in his discussion of a coin that "looked elliptical" from some points of view. Even so,

> it had a real shape which remained unchanged. But coins in fact are rather special cases. For one thing their outlines are well defined and very highly stable, and for another they have a known and a nameable shape. But there are plenty of things of which this is not true. What is the real shape of a cloud? . . . or of a cat? Does its real shape change whenever it moves? If not, in what posture *is* its real shape on display? Furthermore, is its real shape such as to be fairly smooth outlines, or must it be finely enough serrated to take account of each hair? *It is pretty obvious that there is no answer to these questions—no rules according to which, no procedure by which, answers are to be determined.* (emphasis added), (p. 67)

But there *are* answers to these questions. There are ways of describing the shape of a cat to an arbitrary level of precision (see Chapter 5), and there are rules and procedures for arriving at such descriptions. That is exactly what vision is about, and precisely what makes it complicated.

1.3   A REPRESENTATIONAL FRAMEWORK FOR VISION

Vision is a process that produces from images of the external world a description that is useful to the viewer and not cluttered with irrelevant information (Marr, 1976; Marr and Nishihara, 1978). We have already seen that a process may be thought of as a mapping from one representation to another, and in the case of human vision, the initial representation is in no doubt—it consists of arrays of image intensity values as detected by the photoreceptors in the retina.

It is quite proper to think of an image as a representation; the items that are made explicit are the image intensity values at each point in the array, which we can conveniently denote by $I(x,y)$ at coordinate (x,y). In order to simplify our discussion, we shall neglect for the moment the fact that there are several different types of receptor, and imagine instead that there is just one, so that the image is black-and-white. Each value of $I(x,y)$ thus specifies a particular level of gray; we shall refer to each detector as a picture element or *pixel* and to the whole array I as an image.

But what of the output of the process of vision? We have already agreed that it must consist of a useful description of the world, but that requirement is rather nebulous. Can we not do better? Well, it is perfectly true that, unlike the input, the result of vision is much harder to discern, let alone specify precisely, and an important aspect of this new approach is that it makes quite concrete proposals about what that end is. But before we begin that discussion, let us step back a little and spend a little time formulating the more general issues that are raised by these questions.

The Purpose of Vision

The usefulness of a representation depends upon how well suited it is to the purpose for which it is used. A pigeon uses vision to help it navigate, fly, and seek out food. Many types of jumping spider use vision to tell the difference between a potential meal and a potential mate. One type, for example, has a curious retina formed of two diagonal strips arranged in a V. If it detects a red V on the back of an object lying in front of it, the spider has found a mate. Otherwise, maybe a meal. The frog, as we have seen, detects bugs with its retina; and the rabbit retina is full of special gadgets, including what is apparently a hawk detector, since it responds well to the pattern made by a preying hawk hovering overhead. Human vision, on the other hand, seems to be very much more general, although it clearly contains a variety of special-purpose mechanisms that can, for example, direct the eye toward an unexpected movement in the visual field or cause one to blink or otherwise avoid something that approaches one's head too quickly.

Vision, in short, is used in such a bewildering variety of ways that the visual systems of different animals must differ significantly from one another. Can the type of formulation that I have been advocating, in terms of representations and processes, possibly prove adequate for them all? I think so. The general point here is that because vision is used by different animals for such a wide variety of purposes, it is inconceivable that all seeing animals use the same representations; each can confidently be expected to use one or more representations that are nicely tailored to the owner's purposes.

As an example, let us consider briefly a primitive but highly efficient visual system that has the added virtue of being well understood. Werner Reichardt's group in Tübingen has spent the last 14 years patiently unraveling the visual flight-control system of the housefly, and in a famous collaboration, Reichardt and Tomaso Poggio have gone far toward solving the problem (Reichardt and Poggio, 1976, 1979; Poggio and Reichardt, 1976). Roughly speaking, the fly's visual apparatus controls its flight through a collection of about five independent, rigidly inflexible, very fast responding systems (the time from visual stimulus to change of torque is only 21 ms). For example, one of these systems is the landing system; if the visual field "explodes" fast enough (because a surface looms nearby), the fly automatically "lands" toward its center. If this center is above the fly, the fly automatically inverts to land upside down. When the feet touch, power to the wings is cut off. Conversely, to take off, the fly jumps; when the feet no longer touch the ground, power is restored to the wings, and the insect flies again.

In-flight control is achieved by independent systems controlling the fly's vertical velocity (through control of the lift generated by the wings) and horizontal direction (determined by the torque produced by the asymmetry of the horizontal thrust from the left and right wings). The visual input to the horizontal control system, for example, is completely described by the two terms

$$r(\psi)\dot{\psi} + D(\psi)$$

where r and D have the form illustrated in Figure 1–6. This input describes how the fly tracks an object that is present at angle ψ in the visual field and has angular velocity $\dot{\psi}$. This system is triggered to track objects of a certain angular dimension in the visual field, and the motor strategy is such that if the visible object was another fly a few inches away, then it would be

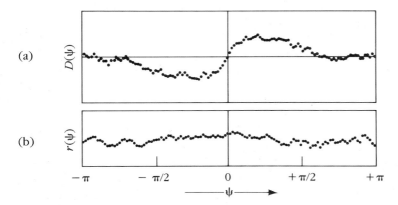

(a)

$D(\psi)$

(b)

$r(\psi)$

$-\pi$ $-\pi/2$ 0 $+\pi/2$ $+\pi$

ψ

Figure 1–6. The horizontal component of the visual input R to the fly's flight system is described by the formula $R = D(\psi) - r(\psi)\,\dot{\psi}$, where ψ is the direction of the stimulus and $\dot{\psi}$ is its angular velocity in the fly's visual field. $D(\psi)$ is an odd function, as shown in (a), which has the effect of keeping the target centered in the fly's visual field; $r(\psi)$ is essentially constant as shown in (b).

intercepted successfully. If the target was an elephant 100 yd away, interception would fail because the fly's built-in parameters are for another fly nearby, not an elephant far away.

Thus, fly vision delivers a representation in which at least these three things are specified: (1) whether the visual field is looming sufficiently fast that the fly should contemplate landing; (2) whether there is a small patch—it could be a black speck or, it turns out, a textured figure in front of a textured ground—having some kind of motion relative to its background; and if there is such a patch, (3) ψ and $\dot{\psi}$ for this patch are delivered to the motor system. And that is probably about 60% of fly vision. In particular, it is extremely unlikely that the fly has any explicit representation of the visual world around him—no true conception of a surface, for example, but just a few triggers and some specifically fly-centered parameters like ψ and $\dot{\psi}$.

It is clear that human vision is much more complex than this, although it may well incorporate subsystems not unlike the fly's to help with specific and rather low-level tasks like the control of pursuit eye movements. Nevertheless, as Poggio and Reichardt have shown, even these simple systems can be understood in the same sort of way, as information-processing tasks. And one of the fascinating aspects of their work is how they have managed not only to formulate the differential equations that accurately describe the visual control system of the fly but also to express these equations, using the Volterra series expansion, in a way that gives direct information about the minimum possible complexity of connections of the underlying neuronal networks.

Advanced Vision

Visual systems like the fly's serve adequately and with speed and precision the needs of their owners, but they are not very complicated; very little objective information about the world is obtained. The information is all very much subjective—the angular size of the stimulus as the fly sees it rather than the objective size of the object out there, the angle that the object has in the fly's visual field rather than its position relative to the fly or to some external reference, and the object's angular velocity, again in the fly's visual field, rather than any assessment of its true velocity relative to the fly or to some stationary reference point.

One reason for this simplicity must be that these facts provide the fly with sufficient information for it to survive. Of course, the information is not optimal and from time to time the fly will fritter away its energy chasing a falling leaf a medium distance away or an elephant a long way away as a direct consequence of the inadequacies of its perceptual system. But this apparently does not matter very much—the fly has sufficient excess energy for it to be able to absorb these extra costs. Another reason is certainly that translating these rather subjective measurements into more objective qualities involves much more computation. How, then, should one think about more advanced visual systems—human vision, for example. What are the issues? What kind of information is vision really delivering, and what are the representational issues involved?

My approach to these problems was very much influenced by the fascinating accounts of clinical neurology, such as Critchley (1953) and Warrington and Taylor (1973). Particularly important was a lecture that Elizabeth Warrington gave at MIT in October 1973, in which she described the capacities and limitations of patients who had suffered left or right parietal lesions. For me, the most important thing that she did was to draw a distinction between the two classes of patient (see Warrington and Taylor, 1978). For those with lesions on the right side, recognition of a common object was possible *provided* that the patient's view of it was in some sense straightforward. She used the words *conventional* and *unconventional*— a water pail or a clarinet seen from the side gave "conventional" views but seen end-on gave "unconventional" views. If these patients recognized the object at all, they knew its name and its semantics—that is, its use and purpose, how big it was, how much it weighed, what it was made of, and so forth. If their view was unconventional—a pail seen from above, for example—not only would the patients fail to recognize it, but they would vehemently deny that it *could* be a view of a pail. Patients with left parietal lesions behaved completely differently. Often these patients had no language, so they were unable to name the viewed object or state its purpose and semantics. But they could convey that they correctly perceived its geometry—that is, its shape—even from the unconventional view.

Warrington's talk suggested two things. First, the representation of the shape of an object is stored in a different place and is therefore a quite different kind of thing from the representation of its use and purpose. And second, vision alone can deliver an internal description of the shape of a viewed object, even when the object was not recognized in the conventional sense of understanding its use and purpose.

This was an important moment for me for two reasons. The general trend in the computer vision community was to believe that recognition was so difficult that it required every possible kind of information. The results of this point of view duly appeared a few years later in programs like Freuder's (1974) and Tenenbaum and Barrow's (1976). In the latter program, knowledge about offices—for example, that desks have telephones on them and that telephones are black—was used to help "segment" out a black blob halfway up an image and "recognize" it as a telephone. Freuder's program used a similar approach to "segment" and "recognize" a hammer in a scene. Clearly, we do use such knowledge in real life; I once saw a brown blob quivering amongst the lettuce in my garden and correctly identified it as a rabbit, even though the visual information alone was inadequate. And yet here was this young woman calmly telling us not only that her patients could convey to her that they had grasped the shapes of things that she had shown them, even though they could not name the objects or say how they were used, but also that they could happily continue to do so even if she made the task extremely difficult visually by showing them peculiar views or by illuminating the objects in peculiar ways. It seemed clear that the intuitions of the computer vision people were completely wrong and that even in difficult circumstances shapes could be determined by vision alone.

The second important thing, I thought, was that Elizabeth Warrington had put her finger on what was somehow the quintessential fact of human vision—that it tells about shape and space and spatial arrangement. Here lay a way to formulate its purpose—building a description of the shapes and positions of things from images. Of course, that is by no means all that vision can do; it also tells about the illumination and about the reflectances of the surfaces that make the shapes—their brightnesses and colors and visual textures—and about their motion. But these things seemed secondary; they could be hung off a theory in which the main job of vision was to derive a representation of shape.

To the Desirable via the Possible

Finally, one has to come to terms with cold reality. Desirable as it may be to have vision deliver a completely invariant shape description from an image (whatever that may mean in detail), it is almost certainly impossible in only one step. We can only do what is possible and proceed from there toward what is desirable. Thus we arrived at the idea of a sequence of representations, starting with descriptions that could be obtained straight from an image but that are carefully designed to facilitate the subsequent recovery of gradually more objective, physical properties about an object's shape. The main stepping stone toward this goal is describing the geometry of the visible surfaces, since the information encoded in images, for example by stereopsis, shading, texture, contours, or visual motion, is due to a shape's local surface properties. The objective of many early visual computations is to extract this information.

However, this description of the visible surfaces turns out to be unsuitable for recognition tasks. There are several reasons why, perhaps the most prominent being that like all early visual processes, it depends critically on the vantage point. The final step therefore consists of transforming the viewer-centered surface description into a representation of the three-dimensional shape and spatial arrangement of an object that does not depend upon the direction from which the object is being viewed. This final description is object centered rather than viewer centered.

   The overall framework described here therefore divides the derivation of shape information from images into three representational stages: (Table 1–1): (1) the representation of properties of the two-dimensional image,

Table 1–1. Representational framework for deriving shape information from images.

Name	Purpose	Primitives
Image(s)	Represents intensity.	Intensity value at each point in the image
Primal sketch	Makes explicit important information about the two-dimensional image, primarily the intensity changes there and their geometrical distribution and organization.	Zero-crossings Blobs Terminations and discontinuities Edge segments Virtual lines Groups Curvilinear organization Boundaries
2½-D sketch	Makes explicit the orientation and rough depth of the visible surfaces, and contours of discontinuities in these quantities in a viewer-centered coordinate frame.	Local surface orientation (the "needles" primitives) Distance from viewer Discontinuities in depth Discontinuities in surface orientation
3-D model representation	Describes shapes and their spatial organization in an object-centered coordinate frame, using a modular hierarchical representation that includes volumetric primitives (i.e., primitives that represent the volume of space that a shape occupies) as well as surface primitives.	3-D models arranged hierarchically, each one based on a spatial configuration of a few sticks or axes, to which volumetric or surface shape primitives are attached

such as intensity changes and local two-dimensional geometry; (2) the representation of properties of the visible surfaces in a viewer-centered coordinate system, such as surface orientation, distance from the viewer, and discontinuities in these quantities; surface reflectance; and some coarse description of the prevailing illumination; and (3) an object-centered representation of the three-dimensional structure and of the organization of the viewed shape, together with some description of its surface properties.

This framework is summarized in Table 1–1. Chapters 2 through 5 give a more detailed account.

PART II

Vision

CHAPTER 2

Representing
the Image

2.1   PHYSICAL BACKGROUND
OF EARLY VISION

We cannot develop a rigorous theory of early vision—the first stages of the vision process—unless we know what the theory is for. We have already seen that, in general terms, the aim is to develop useful canonical descriptions of the shapes and surfaces that form the image. It is now time to state the goals more boldly (Marr 1976, 1978).

There are four main factors responsible for the intensity values in an image. They are (1) the geometry and (2) the reflectances of the visible surfaces, (3) the illumination of the scene, and (4) the viewpoint. In an image, all these factors are muddled up, some intensity changes being due to one cause, others to another, and some to a combination. The purpose of early visual processing is to sort out which changes are due to what factors and hence to create representations in which the four factors are separated.

Roughly speaking, it is proposed that this goal is reached in two stages. First, suitable representations are obtained of the changes and structures in the image. This involves things like the detection of intensity changes, the representation and analysis of local geometrical structure, and the detection of illumination effects like light sources, highlights, and transparency. The result of this first stage is a representation called the *primal sketch*. Second, a number of processes operate on the primal sketch to derive a representation—still retinocentric—of the geometry of the visible surfaces. This second representation, that of the visible surfaces, is called the *2½-dimensional (2½-D) sketch*. Both the primal sketch and the 2½-D sketch are constructed in a viewer-centered coordinate frame, and this is the aspect of their structures denoted by the term *sketch*.

The necessity for representing spatial relations, with its attendant complexities of how much should be made explicit and how much can safely be left implicit, raises problems that are typical of and rather special to vision. For example, the reader, especially if from a nonmathematical background, should not be put off by the notion of a coordinate frame, because it is probably a much more general notion than the reader thinks. To say that early visual representations are retinocentric does not literally imply that a Cartesian coordinate system, marked out in minutes of arc, is somehow laid out across the striate cortex, and that whenever some line or edge is noticed it is somehow associated with its particular x- and y-coordinates, whose values are somehow carried around by the neural machinery. This process would be one way of making the representations, to be sure, but no one would seriously propose it for human vision. There are many other ways in which this scheme can be realized in humans—for example, an (implicit) anatomical mapping that roughly preserves the spatial organization of the retina together with a representation that makes local relations explicit (point A is 5′ from point B in direction 35°) would seem plausible.

The important point about a retinocentric frame is that the spatial relations represented refer to two-dimensional relations on the viewer's retina, not three-dimensional relations relative to the viewer in the world around him, nor two-dimensional relations on another viewer's retina, nor three-dimensional relations relative to an external reference point like the top of a mountain. To say that image point A is below image point B is a remark in a retinocentric frame. To say one's hand is to the left of and below one's chest is a remark in one's own three-dimensional, viewer-centered frame. To say that the tip of a certain cat's tail is above and to the left of its body is a remark in a coordinate frame that is centered on the cat. They are all perfectly good ways of specifying rough spatial relationships, yet none uses sets of numbers. One can speak of each of these frames in terms of numbers—as if one was using (x, y, z), for example—but that does not mean that they have to be implemented this way, and it is important to bear this in mind.

Although it helps a great deal to formulate the purpose of early vision in the rather straightforward terms of separating out the four factors of geometry, reflectance, illumination, and viewpoint, it is important to be aware of the simplifications that are involved in doing so. Perhaps the most important simplification is the rather rigid distinction between surface reflectance and surface geometry. In fact, these two notions are linked, and the distinction between them can be rather imprecise, so that one must be a little cautious when using them. A field of ripening wheat provides a convenient illustration of some of the difficulties. When seen from close by, the individual wheat stems form the reflecting surfaces, and the situation is relatively straightforward. When viewed from afar, however, image resolution is insufficient to distinguish the stems; the field as a whole forms the visible surface, and its reflectance function may now be very complex, since it incorporates considerable variation that should more properly be viewed as spatial (see, for example, Bouguer, 1957; Trowbridge and Reitz, 1975). Thinking of a distant wheat field or the coat of a cat as a surface is probably not too unrealistic an approximation for the theory of perception. We do see surfaces smoothed out. Tyler (1973), for example, found that we cannot see surface corrugations in stereograms if their spatial frequency is higher than about 4 cycles per degree.

In addition to these complexities, the illumination of a scene can only rarely be described in simple terms: Diffuse illumination, reflections, multiple light sources (only some of which are visible), and illumination between surfaces often conspire to create very complex illumination conditions, which will probably never be solved analytically. Nevertheless, our crude division into four categories has its uses. Provided that the variation in depth from the viewer of the surface from which light is reflected is small compared with the viewing distance, I shall assume that what is viewed can be regarded as a reflecting surface, and that the relation between its incident and reflected light may be described by a reflectance function ρ that, for a given illumination and viewpoint, may have a complex spatial structure.

Finally, a general point about the exposition. The purpose of these representations is to provide useful descriptions of aspects of the real world. The structure of the real world therefore plays an important role in determining both the nature of the representations that are used and the nature of the processes that derive and maintain them. An important part of the theoretical analysis is to make explicit the physical constraints and assumptions that have been used in the design of the representations and processes, and I shall be quite careful to do this.

Representing the Image

From an information-processing point of view, our primary purpose now is to define a representation of the image of reflectance changes on a surface that is suitable for detecting changes in the image's geometrical organization that are due to changes in the reflectance of the surface itself or to changes in the surface's orientation or distance from the viewer. If one thinks for a minute about a smooth surface, then changes in orientation and perhaps also in distance are likely to give rise to a change in image intensity. If the surface is textured, then quantities like the orientation or size of tiny elements on the surface—perhaps rough length and width—and measures taken over a small area reflecting the density and spacing of these elements yield the important clues in an image.

Hence we can see in a general way what our representation should contain. It should include some type of "tokens" that can be derived reliably and repeatedly from images and to which can be assigned values of attributes like orientation, brightness, size (length and width), and position (for density and spacing measurements). It is of critical importance that the tokens one obtains correspond to real physical changes on the viewed surface; the blobs, lines, edges, groups, and so forth that we shall use must not be artifacts of the imaging process, or else inferences made from their structure backwards to the structure of the surface will be meaningless. Let us therefore take a look at the general nature of surface reflectance functions, for this will give us important clues as to how we should structure our early representations.

Underlying Physical Assumptions

Existence of surfaces

Our first assumption is that it is proper to speak of surfaces at all, and it refers to the discussion that we had earlier about wheat fields and cats' coats. Stated precisely, it is *that the visible world can be regarded as being composed of smooth surfaces having reflectance functions whose spatial structure may be elaborate.*

Hierarchical organization

The second assumption has to do with the organization of this spatial structure, and it may help to introduce the topic with some examples. As

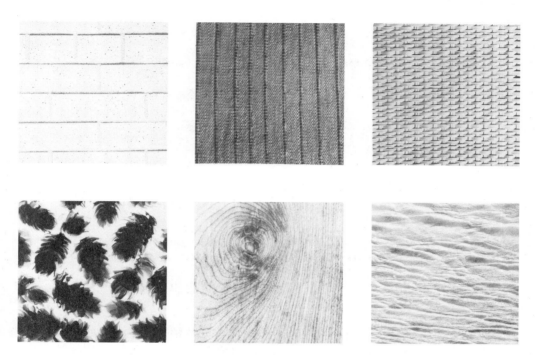

Figure 2–1.   Some images of surfaces. Notice how different types of spatial organization occur almost independently at different scales. An important aspect of early vision is concerned with capturing these different organizations. (Reprinted by permission from Phil Brodatz, *Textures: A Photographic Album for Artists and Designers,* Dover, 1966, pl. D11.)

we have already seen, the coat of a cat is composed at the finest level of single hairs, each of which has its own reflectance function. At the next level up, these are organized into a surface by being placed close and parallel to one another. Then, over the coat so formed is the still higher-level organization of surface markings and coloration. The surface of a river has an analogous organization. At the basic level there is the flat water, randomly perturbed by protrusions like rocks or prominences. Superimposed on this surface are ripples oriented by gusts of wind and patches of weed and vegetation oriented by the flow of the river. There are analogous levels of structure in many surfaces—a hedgerow, a fabric, a rush weave, the bark of a tree, the grain of wood, a rock face, and so on (examine for a moment the surfaces illustrated in Figure 2–1).

Figure 2–2. In a herringbone pattern such as this, a clear part of the spatial organization consists of the vertical stripes. These cannot be recovered by Fourier techniques such as band-pass filtering the images, but yield easily to grouping processes. (Reprinted by permission from Phil Brodatz, *Textures: A Photographic Album for Artists and Designers,* Dover, 1966, pl. 16, 17.)

From these examples, we see that the attributes carrying the valuable information may emerge at any of a range of scales in the real world, and hence even more so in images because of the additional transformations introduced by the imaging process. Whatever tokens are, we must therefore expect them to be capable of making image features explicit over a wide range of sizes. Furthermore, it is important to realize that these different levels of organization do not correspond simply to what would be seen through medium band-pass spatial-frequency filters* centered on different frequencies. Although several types of organization can be detected in this way, many cannot—for example, the vertical stripes in the pattern of Figure 2–2.

We can therefore formulate our second physical assumption: *The spatial organization of a surface's reflectance function is often generated by a number of different processes, each operating at a different scale.* Consequently, a representation that uses changes in the image of such surfaces to find changes in depth and surface orientation must be capable of capturing changes in attribute values applied to tokens that span a wide range of sizes in the image. In other words, the primitives of our representation must work at a number of different scales.

---

*Such filters eliminate all spatial frequency components in the image outside a fixed range of frequencies.

Similarity

Our third assumption is of a rather different kind. Suppose that we already had a representation containing primitives of various sizes. It seems intuitively obvious that they should be kept separate in some way—that a given large-scale descriptor should be compared with other large-scale descriptors much more readily than with small-scale ones. And perhaps it also seems obvious that tokens or descriptors having other extreme dissimilarities—very different or even opposite-signed contrasts, for example—should somehow be kept rather separate.

We can, in fact, find a physical basis for why this should be so, and it is apparent in our earlier examples. Recall that among the various levels of organization present in an animal's coat, on the surface of a river, on the bark of a tree, in woven fabric, and so forth, the processes that operated to generate the reflectance function are relatively independent at each scale, but the items for which each process is responsible are visually much more similar to one another than to other things on the same surface. For example, a given hair in a cat's coat is much more similar to neighboring hairs than to the stripes formed by the arrangement of thousands of hairs. Similarity here may be measured in several ways, but a straightforward measure based on local contrast, size (length and width), orientation, and color would suffice (compare Jardine and Sibson, 1971, for a general discussion of dissimilarity measures).

This observation gives us the means for selecting items from an image during the assignment of primitives in its representation. It is important, and may be formulated as our third physical assumption that *the items generated on a given surface by a reflectance-generating process acting at a given scale tend to be more similar to one another in their size, local contrast, color, and spatial organization than to other items on that surface.*

The importance of this type of similarity is illustrated by Figure 2–3. Following Glass (1969), these patterns are created by superimposing on a set of random dots the same set of dots but rotated or expanded a little (Figure 2–3a). The effect works for tokens made of squares (Figure 2–3b) or for pairs of tokens made in quite different ways (Figure 2–3c). If the tokens are too different (Figure 2–3d), however, no pattern is seen. Glass and Switkes (1976) showed that the effect fails if the dots have opposite contrast or opponent colors. Stevens (1978, fig 51a) showed that if three sets of dots are superimposed—the original, a rotated, and an expanded set—no organization is visible. If, say, the rotated set is made much brighter than the other two, then one sees the organization present in the dimmer pairs. This proves that the effect is based on a symbolic comparison of the

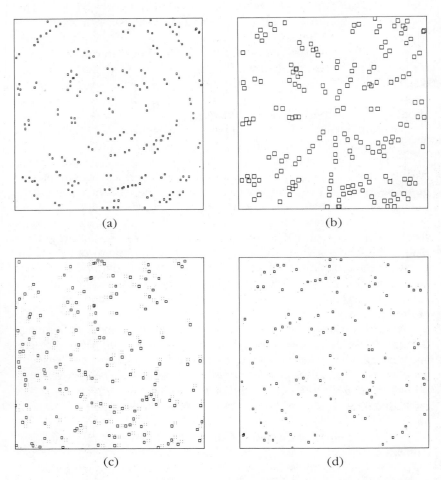

Figure 2–3. These displays are made by superimposing a random pattern of tokens on a slightly rotated or expanded copy of the same pattern. The tokens can be points or small squares (a) or larger squares (b). They do not have to be the same—in (c) one set consists of squares and the other set of four dots—but they do have to be similar. In (d), one set consists of quite large squares, and the other of small dots. These are apparently too dissimilar for us to discern the expanding structure there.

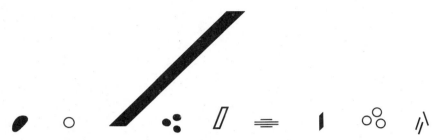

Figure 2–4. More evidence for place tokens. In this diagram every subgroup is defined differently, yet the collinearity of all of them is immediately apparent. This suggests that each group causes a place token to be created, whose collinearity is detected almost independently of the way the token is defined, provided that the tokens represent sufficiently similar items (compare Fig. 2–3d). (Reprinted by permission from D. Marr "Early processing of visual information," *Phil. Trans. R. Soc. Lond. B 275* 1976, fig. 10.)

properties of the local tokens and not, for example, on Hubel and Wiesel simple-cell-like measurements acting directly on the images.

Spatial continuity

In addition to their intrinsic similarity, *markings generated on a surface by a single process are often spatially organized—they are arranged in curves or lines and possibly create more complex patterns.* The basic feature is that markings often form smooth contours on a surface, and hence tokens will do so in an image. We are ourselves very sensitive to spatial continuity. We immediately see the items in Figure 2–4 (from Marr, 1976, fig. 10) as being collinear, despite the fact that every item along the line is defined in a different way: One is a blob, one is a small group of dots, one is the end of a bar, and so forth. They are, however, all about the same size. Figure 2–5 (from Marroquin, 1976, fig. 7) provides another fascinating example. There are very many continuous organizations buried in this pattern, and each one seems to be trying to jump out and dominate the others.

Continuity of discontinuities

One consequence of the cohesiveness of matter is that objects exist in the world and have boundaries. These give rise to the discontinuities in depth or surface orientation with whose detection we are concerned, and an important feature of such boundaries is that they often progress smoothly

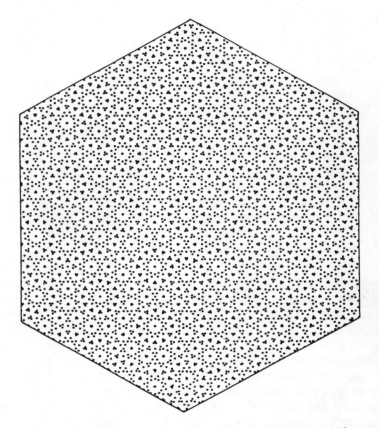

Figure 2–5. Evidence for the existence of active grouping processes. This pattern apparently seethes with activity as the rival organizations seem to compete with one another. (Reprinted by permission from J. L. Marroquin, "Human visual perception of structure," Master's thesis, Department of Electrical Engineering and Computer Science, Massachusetts Institute of Technology, 1976.)

across an image. We can assume, in fact, that *the loci of discontinuities in depth or in surface orientation are smooth almost everywhere.* This is probably the physical constraint that makes the mechanism of smooth subjective contours a useful one (see Figure 2–6 and Section 4.8).

Continuity of flow

Finally, we must not forget that motion is extremely important for vision— it is ubiquitous. Motion of the viewer or of a physical object can cause

Figure 2–6. Subjective contours. The visual system apparently regards changes in depth as so important that they must be made explicit everywhere, including places where there is no direct visual evidence for them.

movements in the images of that object. If the object is rigid, the motions of the images of nearby portions of the object's surface are similar. Hence, the motions of portions of the object that are close to one another in the image are usually similar. In particular, the velocity field of motion in the image varies continuously almost everywhere, and if it is ever discontinuous at more than an isolated point, then a failure of rigidity (like an object boundary) is present in the outside world. In particular, *if direction of motion is ever discontinuous at more than one point—along a line, for example,—then an object boundary is present.*

General Nature of the Representation

The important message of these physical constraints is that although the basic elements in our image are the intensity changes, the physical world imposes on these raw intensity changes a wide variety of spatial organizations, roughly independently at different scales. This organization is reflected in the structure of images, and since it yields important clues about the structure of the visible surfaces, it needs to be captured by the early representations of the image. Specifically, I propose doing this by a set of "place tokens" that roughly correspond to oriented *edge* or *boundary* segments or to points of *discontinuity* in their orientations, to *bars* (roughly parallel edge pairs) or to their *terminations*; or to *blobs*—roughly, doubly terminated bars. These primitives can be defined in very concrete ways— from pure discontinuities in intensity—or in rather abstract ways. A blob can be defined from a cloud of dots, for example, or a boundary from certain (but not all) kinds of texture change or from the lining up of a set of tokens that are themselves defined in quite complex ways, as in the example of Figure 2–4.

A rough illustration of the general idea appears in Figure 2–7; this representational scheme is called the *primal sketch* (Marr, 1976). The critical ideas behind it are the following:

1. The primal sketch consists of primitives of the same general kind at different scales—a blob has a rough position, length, width, and orientation at whatever scale it is defined—but the primitives can be defined from an image in a variety of ways, from the very concrete (a black ink mark) to the very abstract (a cloud of dots).

2. These primitives are built up in stages in a constructive way, first by analyzing and representing the intensity changes and forming tokens directly from them, then by adding representations of the local geometrical structure of their arrangement, and then by operating on these things with active selection and grouping processes to form larger-scale tokens that reflect larger-scale structures in the image, and so forth.

3. On the whole, the primitives that are obtained, the parameters associated with them, and the accuracy with which they are measured are designed to capture and to match the structure in an image so as to facilitate the recovery of information about the underlying geometry of the visible surfaces. This gives rise to a complex trade-off between the accuracy of the discriminations that can be made and the value of making them. For example, projected orientations in the image do change if the surface orientation changes, but on the whole by only a rather small amount and probably usually less than the typical variation in orientation to be found in the objective distribution of markings on a surface. This means that except in special situations, it is not worth having a very powerful apparatus for making subtle orientation discriminations. On the other hand, because only a very small relative movement is compelling evidence that two surfaces are separate, it is worth being very sensitive to relative movement.

The three main stages in the processes that derive the primal sketch are (1) the detection of zero-crossings (Marr and Poggio, 1979; Marr, Poggio, and Ullman, 1979; Marr and Hildreth, 1980); (2) the formation of the raw primal sketch (Marr, 1976; Marr and Hildreth, 1980; Hildreth 1980); and (3) the creation of the full primal sketch (Marr, 1976).

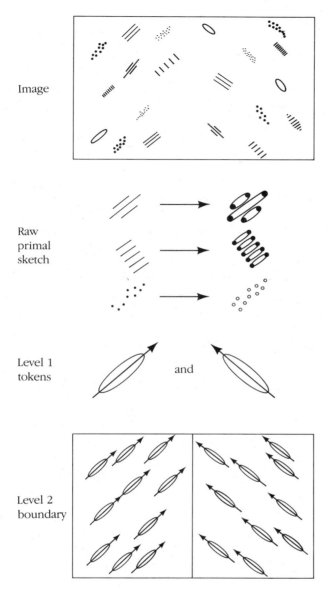

Figure 2–7. A diagrammatic representation of the descriptions of an image at different scales which together constitute the primal sketch. At the lowest level, the raw primal sketch faithfully follows the intensity changes and also represents terminations, denoted here by filled circles. At the next level, oriented tokens are formed for the groups in the image. At the next level, the difference in orientations of the groups in the two halves of the image causes a boundary to be constructed between them. The complexity of the primal sketch depends upon the degree to which the image is organized at the different scales.

2.2   ZERO-CROSSINGS AND
THE RAW PRIMAL SKETCH

Zero-Crossings

The first of the three stages described above concerns the detection of intensity changes. The two ideas underlying their detection are (1) that intensity changes occur at different scales in an image, and so their optimal detection requires the use of operators of different sizes; and (2) that a sudden intensity change will give rise to a peak or trough in the first derivative or, equivalently, to a *zero-crossing* in the second derivative, as illustrated in Figure 2–8. (A zero-crossing is a place where the value of a function passes from positive to negative).

These ideas suggest that in order to detect intensity changes efficiently, one should search for a filter that has two salient characteristics. First and foremost, it should be a differential operator, taking either a first or second spatial derivative of the image. Second, it should be capable of being tuned to act at any desired scale, so that large filters can be used to detect blurry shadow edges, and small ones to detect sharply focused fine detail in the image.

Marr and Hildreth (1980) argued that the most satisfactory operator fulfilling these conditions is the filter $\nabla^2 G$, where ∇^2 is the Laplacian operator $(\partial^2/\partial x^2 + \partial^2/\partial y^2)$ and G stands for the two-dimensional Gaussian distribution

$$G(x,y) = e^{-\frac{x^2+y^2}{2\pi\sigma^2}}$$

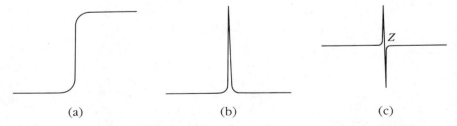

(a)   (b)   (c)

Figure 2–8.   The notion of a zero-crossing. The intensity change (a) gives rise to a peak (b) in its first derivative and to a (steep) zero-crossing Z (c) in its second derivative.

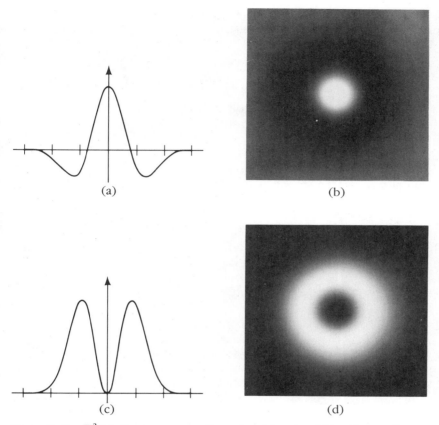

Figure 2–9. $\nabla^2 G$ is shown as a one-dimensional function (a) and in two-dimensions (b) using intensity to indicate the value of the function at each point. (c) and (d) show the Fourier transforms for the one- and two-dimensional cases respectively. (Reprinted by permission from D. Marr and E. Hildreth, "Theory of edge detection," *Proc. R. Soc. Lond. B 207*, pp. 187-217.)

which has standard deviation σ. $\nabla^2 G$ is a circularly symmetric Mexican-hat-shaped operator whose distribution in two dimensions may be expressed in terms of the radial distance r from the origin by the formula

$$\nabla^2 G(r) = \frac{-1}{\pi \sigma^4}\left(1 - \frac{r^2}{2\sigma^2}\right)e^{\frac{-r^2}{2\sigma^2}}$$

Figure 2–9 illustrates the one- and two-dimensional forms of this operator, as well as their Fourier transforms.

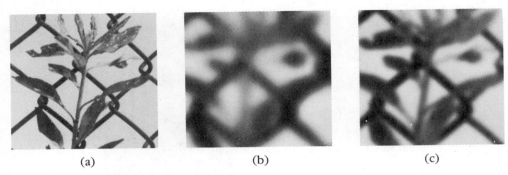

(a)   (b)   (c)

Figure 2–10. Blurring images is the first step in detecting intensity changes in them. (a) In the original image, intensity changes can take place over a wide range of scales, and no single operator will be very efficient at detecting all of them. The problem is much simplified in an image that has been blurred with a Gaussian filter, because there is, in effect, an upper limit to the rate at which changes can take place. The first part of the edge detection process can be thought of as decomposing the original image into a set of copies, each filtered with a different-sized Gaussian, and then detecting the intensity changes separately in each. (b) The image filtered with a Gaussian having $\sigma = 8$ pixels; in (c), $\sigma = 4$. The image is 320 by 320 elements. (Reprinted by permission from D. Marr and E. Hildreth, "Theory of edge detection," *Proc. R. Soc. Lond. B 207,* pp. 187-217.)

There are two basic ideas behind the choice of the filter $\nabla^2 G$. The first is that the Gaussian part of it, G, blurs the image, effectively wiping out all structure at scales much smaller than the space constant σ of the Gaussian. To illustrate this, Figure 2–10 shows an image that has been convolved with two different-sized Gaussians whose space constants σ were 8 pixels (Figure 2–10b) and 4 pixels (Figure 2–10c). The reason why one chooses the Gaussian for this purpose, rather than blurring with a cylindrical pillbox function (for instance), is that the Gaussian distribution has the desirable characteristic of being smooth and localized in both the spatial and frequency domains and, in a strict sense, being the unique distribution that is simultaneously optimally localized in both domains. And the reason, in turn, why this should be a desirable property of our blurring function is that if the blurring is as smooth as possible, both spatially and in the frequency domain, it is least likely to introduce any changes that were not present in the original image.

The second idea concerns the derivative part of the filter, ∇^2. The great advantage of using it is economy of computation. First-order directional derivatives, like $\partial/\partial x$ or $\partial/\partial y$, could be used, in which case one would subsequently have to search for their peaks or troughs at each orientation (as illustrated in Figure 2–8b); or, second-order directional derivatives, like $\partial^2/\partial x^2$ or $\partial^2/\partial y^2$, could be used, in which case intensity changes would

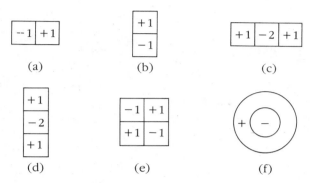

Figure 2–11. The spatial configuration of low-order differential operators. Operators like $\partial/\partial x$ can be roughly realized by filters with the receptive fields illustrated in the figure. (a) $\partial/\partial x$ can be thought of as measuring the difference between the values at two neighboring points along the x-axis. Similarly, (b) shows $\partial/\partial y$. The operator $\partial^2/\partial x^2$ can be thought of as the difference between two neighboring values of $\partial/\partial x$, and so it takes the form shown in (c). The other two second-order operators, $\partial^2/\partial y^2$ and $\partial^2/\partial x\partial y$, appear in (d) and (e), respectively. Finally, the lowest-order isotropic operator, the Laplacian ($\partial^2/\partial x^2 + \partial^2/\partial y^2$), which we denote by ∇^2, has the circularly symmetric form shown in (f).

correspond to their zero-crossings (see Figure 2–8c). However, the disadvantage of all these operators is that they are directional; they all involve an orientation (see Figure 2–11, which illustrates the spatial organizations, or "receptive fields," in neurophysiological terms of the various first- and second-order differential operators). In order to use the first derivatives, for example, both $\partial I/\partial x$ and $\partial I/\partial y$ have to be measured, and the peaks and troughs in the overall amplitude have to be found. This means that the signed quantity $[(\partial I/\partial x)^2 + (\partial I/\partial y)^2]^{-\frac{1}{2}}$ must also be computed.

Using second-order directional derivative operators involves problems that are even worse than the ones involved in using first-order derivatives. The only way of avoiding these extra computational burdens is to try to choose an orientation-independent operator. The lowest-order isotropic differential operator is the Laplacian ∇^2, and fortunately it so happens that this operator can be used to detect intensity changes provided the blurred image satisfies some quite weak requirements (Marr and Hildreth, 1980).* Images on the whole do satisfy these requirements locally,

---

*The mathematical notation for blurring an image intensity function $I(x, y)$ with a Gaussian function G is $G * I$ which is read G convolved with I. The Laplacian of this is denoted by $\nabla^2 (G * I)$ and a mathematical identity allows us to move the ∇^2 operator inside the convolution giving $\nabla^2 (G * I) = (\nabla^2 G) * I$.

(a)   (b)

(c)   (d)

Figures 2–12, 13, 14.   These three figures show examples of zero-crossing detection using $\nabla^2 G$. In each figure, (a) shows the image (320 × 320 pixels); (b) shows the image's convolution with $\nabla^2 G$, with $w_{2-D} = 8$ (zero is represented by gray); (c) shows the positive values in white and the negative in black; (d) shows only the zero-crossings.

so in practice one can use the Laplacian. Hence, in practice, the most satisfactory way of finding the intensity changes at a given scale in an image is first to filter it with the operator $\nabla^2 G$, where the space constant of G is chosen to reflect the scale at which the changes are to be detected, and then to locate the zero-crossings in the filtered image.

      Figures 2–12 to 2–14 show what an image looks like when processed in this way. The numerical values in the $\nabla^2 G$-filtered image are both positive

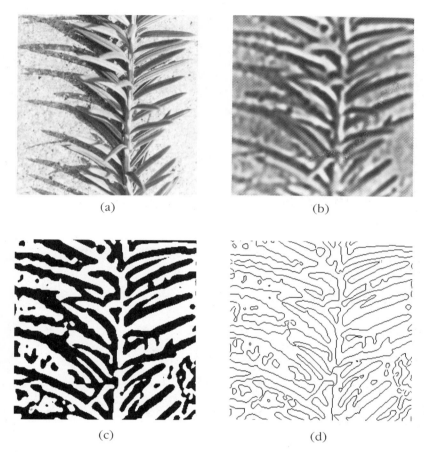

(a) (b)

(c) (d)

Figure 2–13.

and negative, the overall average being zero. Positive values are represented here by whites, negative by blacks, and the value zero by an intermediate gray. As we have seen, the critical fact about the operator $\nabla^2 G$ is that its zero-crossings mark the intensity changes, as seen at the Gaussian's particular scale. The figures show this well. In Figure 2–12(c), for instance, the filtered image has been "binarized"—that is, positive values were all set to $+1$ and negative values to -1, and in Figure 2–12(d) the zero-crossings alone are shown. The advantage of the binarized representation is that it also shows the sign of the zero-crossing—which side in the image is the darker.

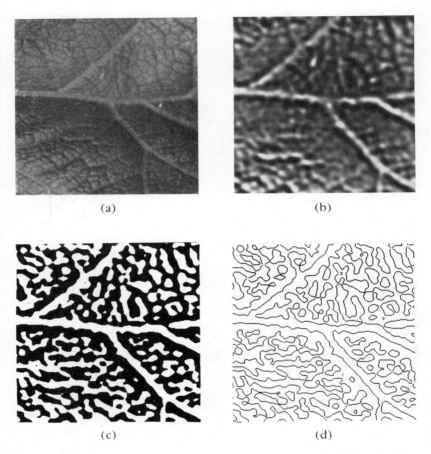

(a)   (b)

(c)   (d)

Figure 2–14.

In addition, the slope of the zero-crossing depends on the contrast of the intensity change, though not in a very straightforward way. This is illustrated by Figure 2–15, which shows an original image together with zero-crossings that have been marked with curves of varying intensity. The more contrasty the curve, the greater the slope of the zero-crossing at that point, measured perpendicularly to its local orientation.

Zero-crossings like those of Figures 2–12 to 2–15 can be represented symbolically in various ways. I choose to represent them by a set of oriented primitives called *zero-crossing segments,* each describing a piece of the contour whose intensity slope (rate at which the convolution changes across the segment) and local orientation are roughly uniform. Because of their eventual physical significance, it is also important to make explicit

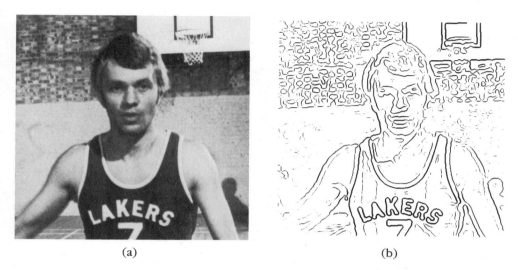

(a)   (b)

Figure 2–15. Another example of zero-crossings; here, the intensity of the lines has been made to vary with the slope of the zero-crossing, so that it is easier to see which lines correspond to the greater contrast. (Courtesy BBC Horizon.)

those places at which the orientation of a zero-crossing changes "discontinuously." The quotation marks are necessary because one can in fact prove that the zero-crossings of $\nabla^2 G * I$ can never change orientation discontinuously, but one can nevertheless construct a practical definition of discontinuity. In addition, small, closed contours are represented as blobs, each also with an associated orientation, average intensity slope, and size defined by its extent along a major and minor axis. Finally, in keeping with the overall plan, several sizes of operator will be needed to cover the range of scales over which intensity changes occur.

Biological Implications

This computational scheme for the very first stages in visual processing leads to an interpretation of many results from the psychophysical and neurophysiological investigations into early vision and to a proposal for the overall strategy behind the design of the first part of the visual pathway.

The psychophysics of early vision

In 1968, Campbell and Robson carried out some adaptation experiments. They found that the sensitivity of subjects to high-contrast gratings was temporarily reduced after exposure to such gratings and this desensitization was specific to the orientation and spatial frequency of the gratings. The experimenters concluded that the visual pathway included a set of "channels" that are orientation and spatial frequency selective.

This finding provided an explosion of articles investigating various aspects of the detailed structure of these channels, culminating recently in an elegant quantitative model for their structure in humans, constructed on the basis of data from threshold detection studies by Wilson and Giese (1977) and Wilson and Bergen (1979). The model is quite easy to understand. The basic idea is that at each point in the visual field, there are four size-tuned filters or masks analyzing the image. The spatial receptive fields of these filters all have approximately the shape of a DOG, that is, of the difference of two Gaussian distributions, but the smaller two filters exhibit relatively sustained temporal properties, whereas the larger two are relatively transient. The channels are labeled N, S, T, and U, in order of increasing size, and their dimensions scale linearly with increasing eccentricity (angular distance from the fovea). The S channel is the most sensitive under both sustained and transient stimulation; the U channel is the least, having only one-fourth to one-eleventh the sensitivity of the S channel. Wilson himself made no statement about whether the filters were oriented, but he measured their dimensions using light and dark lines. With these one-dimensional stimuli, the widths of the central part of the receptive fields, which I shall denote by the symbol w_{1-D}, had the following values: N channel, 3.1'; S channel, 6.2'; T channel, 11.7'; and U channel, 21'. The receptive field sizes increase linearly with eccentricity, being about doubled at 4° eccentricity. Essentially all of the psychophysical data on the detection of spatial patterns below 16 cycles per degree at contrast threshold can be explained by this model, together with the hypothesis that the detection process is based on a form of spatial probability summation in the channels.

It is the $\nabla^2 G$ filters, I think, that form the basis for these psychophysically determined channels. The $\nabla^2 G$ operator approximates a band-pass filter with a bandwidth at half power of 1.25 octaves. It can be approximated closely by a DOG, the best approximation from an engineering point of view being achieved when the two Gaussians that form the DOG have space constants in the ratio 1:1.6. Figure 2–16 shows how good this approximation is. Wilson's estimate of the ratio for his sustained channels was 1:1.75.

In order to relate the numerical values of w_{1-D} measured by Wilson and Bergen to the values of the diameter w_{2-D} of the central part of the receptive fields of the underlying $\nabla^2 G$ operators, one must remember to multiply their values by $\sqrt{2}$, since all the measurements Wilson made correspond to a linear projection of the circularly symmetric receptive

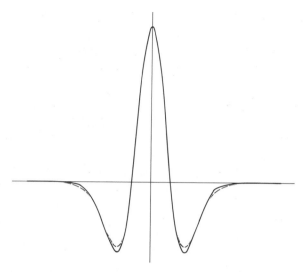

Figure 2–16. The best engineering approximation to $\nabla^2 G$ (shown by the continuous line), obtained by using the difference of two Gaussians (DOG), occurs when the ratio of the inhibitory to excitatory space constraints is about 1:1.6. The DOG is shown here dotted. The two profiles are very similar. (Reprinted by permission from D. Marr and E. Hildreth, "Theory of edge detection," *Proc. R. Soc. Lond. B 207*, pp. 187-217.)

fields. Hence Wilson's N channel would correspond to a $\nabla^2 G$ filter with $w_{2-D} = 3.1 \sqrt{2} = 4.38'$, which corresponds to the diameter of about nine foveal cones. This seems rather large for the smallest channel, and arguments based on a theoretical analysis of acuity and resolution suggest that a smaller one exists. The diameter w_{2-D} of the central part of its receptive field should be about $1' \ 20''$, and because of diffraction in the eye, it could correspond to the midget ganglion cells, whose receptive field centers are driven by a single cone (see Marr, Poggio, and Hildreth, 1980).

Thus if Wilson's figures are correct, they tell us the sizes that the initial center–surround operators should have in order to produce the observed psychophysical adaptation and other effects. These numbers can then in principle be related to the measurements made by physiologists, in the manner that we shall derive in the next section. The final point to note here is that Campbell also found the adaptation to be orientation specific (and it may also be specific for the direction of movement). This we attribute to the stage at which zero-crossings are detected, which is best explained by looking at the neurophysiology.

The physiological realization of the $\nabla^2 G$ filters

It has been known since Kuffler (1953) that the spatial organization of the receptive fields of the retinal ganglion cells is circularly symmetric, with a central excitatory region and an inhibitory surround. Some cells, called on-center cells, are excited by a small spot of light shone on the center of their receptive fields, and others are inhibited. Rodieck and Stone (1965) suggested that this organization was the result of superimposing a small central excitatory region on a larger inhibitory "dome" that extends over the entire receptive field. Enroth-Cugell and Robson (1966) described the two domes as Gaussians, thus describing the receptive field as a difference of two Gaussians (a DOG). In addition, Enroth-Cugell and Robson divided the larger retinal ganglion cells into two classes, X and Y, on the basis of their temporal response properties. X cells show a fairly sustained response, whereas the Y cells show a relatively transient one—a distinction that is preserved at the lateral geniculate nucleus. Wilson's sustained channels probably correspond to the physiological X cells, and the transient, to the Y cells (Tolhurst, 1975).

Thus it is not too unreasonable to propose that the $\nabla^2 G$ function is what is carried by the X cells of the retina and lateral geniculate body, positive values being carried by the on-center X cells, and negative values by the off-center X cells. To illustrate the physiological point, Figure 2–17 compares the predicted X-cell responses, using $\nabla^2 G$, against actual published records of retinal and lateral geniculate cells, which we identified as X cells, for three stimuli—an edge, a thin bar, and a wide bar. As we can see, the qualitative agreement is very good. I shall discuss the function of the Y cells in Section 3.4.

The physiological detection of zero-crossings

From a physiological point of view, zero-crossing segments are easy to detect without relying on the detection of zero values, which would be a physiologically implausible idea. The reason is that just to one side of the zero-crossing will lie a peak positive value of the filtered image $\nabla^2 G * I$, and just to the other side, a peak negative value. These peaks will be roughly $w_{2-D}/\sqrt{2}$ apart, where w_{2-D} is the width of the receptive field center of the underlying filter $\nabla^2 G$. Hence, just to one side, an on-center X cell will be firing strongly, and just to the other side, an off-center X cell will be firing strongly; the sum of their firings will correspond to the slope of the zero-crossing—a high-contrast intensity change producing stronger firing than a low-contrast change. The existence of a zero-crossing can therefore

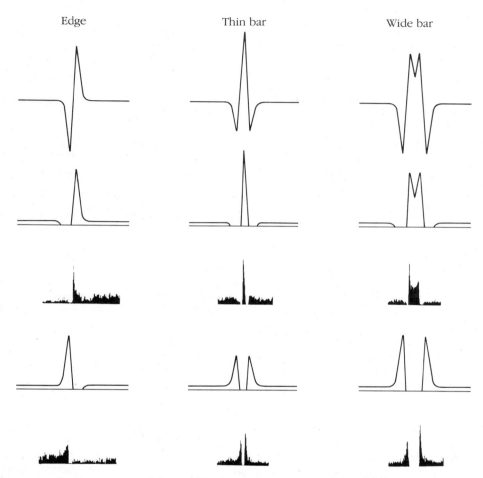

Figure 2–17. Comparison of the predicted responses of on- and off-center X cells with electro-physiological recordings. The first row shows the response to $\nabla^2 G * I$ for an isolated edge, a thin bar (bar width $= 0.5w_{1-D}$, where $_{1-D}$ is the width of the central excitatory region of the receptive field projected onto a line), and a wide bar (bar width $= 2.5w_{1-D}$). The predicted traces are calculated by superimposing the positive (in the second row) or the negative (in the fourth row) parts of $\nabla^2 G * I$ on a small resting or background discharge. The corresponding physiological responses (third and fifth rows) are taken from Dreher and Sanderson (1973, figs. 6d and 6e) for the responses to an edge and from Rodieck and Stone (1965, figs. 1 and 2), using traces from bars 1° and 5° wide. (Reprinted by permission from D. Marr and S. Ullman, "Directional selectivity and its use in early visual processing," *Phil. Trans. R. Soc. B 275*, pp. 483–524.)

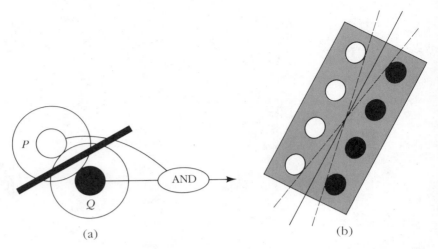

Figure 2–18. A mechanism for detecting oriented zero-crossing segments. In (a), if *P* represents an on-center geniculate X-cell receptive field, and *Q* an off-center, then a zero-crossing must pass between them if both are active. Hence, if they are connected to a logical AND gate as shown, the gate will detect the presence of the zero-crossing. If several are arranged in tandem as in (b) and are also connected by logical AND's, the resulting mechanism will detect an oriented zero-crossing segment within the orientation bounds given roughly by the dotted lines. Ideally, we would use gates that responded by signaling their sum only when all their *P* and *Q* inputs were active. (Reprinted, by permission, by D. Marr and E. Hildreth, "Theory of edge detection," *Proc. R. Soc. Lond. B 207,* pp. 187-217.)

be detected by a mechanism that connects an on-center cell and an off-center cell to an AND gate,* as illustrated in Figure 2–18(a).

It is a simple matter to adapt this idea to create an oriented zero-crossing segment detector: simply arrange on- and off-center X cells into two columns, as illustrated in Figure 2–18(b). If these units are all connected by AND gates or some suitable approximation to them, the result will be a unit that detects a zero-crossing segment whose orientation lies roughly between the dotted lines of Figure 2–18(b). This idea provides the basis for the model of cortical simple cells, which we shall derive in Section 3.4. It is enough to note here that such units would be orientation dependent and spatial-frequency-tuned (as well as directionally selective, after the modifications of Section 3.4). These are the units, I believe, that Campbell and Robson found that they could adapt in their 1968 experiments.

---

*A simple logical device that produces a positive output only when all of its inputs are positive.

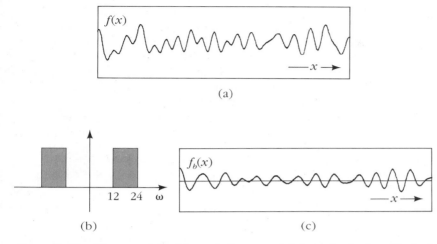

(a)

(b)   (c)

Figure 2–19. The meaning of Logan's theorem. (a) A stochastic, band-limited Gaussian signal $f(x)$. (b) The passband—in the frequency domain—of an ideal one-octave band-pass filter. (c) The result $f_b(x)$ of filtering (a) with the filter described by (b). Provided that (c) has no zeros in common with its Hilbert transform, Logan's theorem tells us that (c) is determined, up to a multiplicative constant, by the positions of its zero-crossings alone. The aspect of Logan's result that is important for early visual processing is that, under the right conditions, the zero-crossings alone are very rich in information. (Reprinted by permission from D. Marr, T. Poggio, and S. Ullman, "Bandpass channels, zero-crossings, and early visual information processing," *J. Opt. Soc. Am. 69,* 1979, fig. 1.)

The first complete symbolic representation of the image

Zero-crossings provide a natural way of moving from an analogue or continuous representation like the two-dimensional image intensity values $I(x,y)$ to a discrete, symbolic representation. A fascinating thing about this transformation is that it probably incurs no loss of information. The arguments supporting this are not yet secure (Marr, Poggio, and Ullman, 1979) and rest on a recent theorem of B. F. Logan (1977). This theorem states that provided certain technical conditions are satisfied, a one-octave band-pass signal can be completely reconstructed (up to an overall multiplicative constant) from its zero-crossings. Figure 2–19 illustrates the idea; the proof of the theorem is difficult, but consists essentially of showing that if the signal is less than an octave in bandwidth, then it must cross the x-axis at least as often as the standard sampling theorem requires.

Unfortunately, Logan's theorem is not quite strong enough for us to be able to make any direct claims about vision from it. The problems are twofold. First, the zero-crossings in the visual application lie in two dimensions and not one, and it is often difficult to extend sampling arguments from one dimension to two. Second, the operator $\nabla^2 G$ is not a pure one-octave band-pass filter; its bandwidth at half power is 1.25 octaves, and at half sensitivity, 1.8 octaves. On the other hand, we do have extra information, namely, the values of the slopes of the curves as they cross zero, since this corresponds roughly to the contrast of the underlying edge in the image. An analytical approach to the problem seems to be very difficult, but in an empirical investigation, Nishihara (1981) found encouraging evidence supporting the view that a two-dimensional filtered image can be reconstructed from its zero-crossings and their slopes.

Figure 2–20 summarizes pictorially the point we have now reached. It shows the image, of a sculpture by Henry Moore, as seen through three different-sized channels; that is, it shows the zero-crossings of the image after filtering it through $\nabla^2 G$ filters where the Gaussians, G, have three different space constants. The next question is, What should we do with this information?

The Raw Primal Sketch

Up to now I have studiously avoided using the word *edge*, preferring instead to discuss the detection of intensity changes and their representation by using oriented zero-crossing segments. The reason for this is that the term *edge* has a partly physical meaning—it makes us think of a real physical boundary, for example—and all we have discussed so far are the zero values of a set of roughly band-pass second-derivative filters. We have no right to call these edges, or, if we do have a right, then we must say so and why. This kind of distinction is vital to the theory of vision and probably to the theories of other perceptual systems, because the true heart of visual perception is the inference from the structure of an image about the structure of the real world outside. The theory of vision is exactly the theory of how to do this, and its central concern is with the physical constraints and assumptions that make this inference possible.

We meet this for the first time now, as we address the problem posed by Figure 2–20—namely, How do we combine information from the different channels? The $\nabla^2 G$ filters that are actually used by the visual system are an octave or more apart, so there is no priori reason why the zero-crossings obtained from the different-sized filters should be related. There is, however, a physical reason why they often should be. It is a consequence of the first of our physical assumptions of the last chapter, and it is called the *constraint of spatial localization* (Marr and Hildreth, 1980). The things

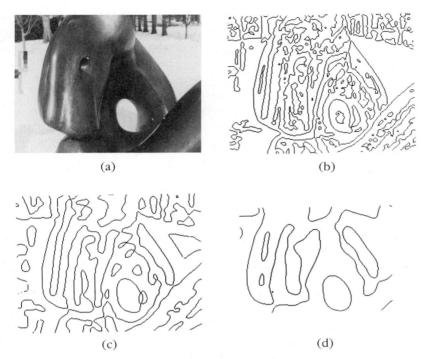

(a)   (b)

(c)   (d)

Figure 2–20. The image (a) has been convolved with $\nabla^2 G$ having $w_{2-\text{D}} = 2\sqrt{2}\,\sigma = 6, 12,$ and 24 pixels. These filters span approximately the range of filters that operate in the human fovea. (b), (c), and (d) show the zero-crossings thus obtained. Notice the fine detail picked up by the smallest. This set of figures neatly poses the next problem—How do we combine all this information into a single description? (Reprinted by permission from D. Marr and E. Hildreth, "Theory of edge detection," *Proc. R. Soc. Lond. B 204,* pp. 301–328.)

in the world that give rise to intensity changes in an image are (1) illumination changes, which include shadows, visible light sources, and illumination gradients; (2) changes in the orientation or distance from the viewer of the visible surfaces; and (3) changes in surface reflectance.

The critical observation here is that, at their own scale, these things can all be thought of as spatially localized. Apart from the occasional diffraction pattern, the visual world is not constructed of ripply, wavelike primitives that extend over an area and that add together over it (compare Marr, 1970, p. 169). By and large, the visual world is made of contours, creases, scratches, marks, shadows, and shading, and these are spatially localized. Hence, it follows that if a discernable zero-crossing is present in an image filtered through $\nabla^2 G$ at one size, then it should be present at the same location for all larger sizes. If this ceases to be so at some larger size, it will be for one of two reasons: Either two or more local intensity changes are interfering—being averaged together—in the larger channel, or two independent physical phenomena are operating to produce intensity changes in the same region of the image but at different scales. An example of the first situation is a thin bar, whose edges would be accurately located by small channels but not by large ones. Situations of this kind can be recognized by the presence of two nearby zero-crossings in the small channels. An example of the second situation is a shadow superimposed on a sharp reflectance change, which can be recognized if the zero-crossings in the large channels are displaced relative to those in the smaller ones. If the shadow has exactly the correct position and orientation, the locations of the zero-crossings may not contain enough information to separate the two physical phenomena, but in practice this situation will be rare.

Thus, the physical world constrains the geometry of the zero-crossings from the different-sized channels. We can exploit this by using it to formulate the *spatial coincidence assumption:*

If a zero-crossing segment is present in a set of independent $\nabla^2 G$ channels over a contiguous range of sizes, and the segment has the same position and orientation in each channel, then the set of such zero-crossing segments indicates the presence of an intensity change in the image that is due to a single physical phenomenon (a change in reflectance, illumination, depth, or surface orientation).

In other words, provided that the zero-crossings from independent channels of adjacent sizes coincide, they can be taken together. If the zero-crossings do not coincide, they probably arise from distinct surfaces or physical phenomena. It follows (1) that the minimum number of $\nabla^2 G$ channels required to establish physical reality is two and (2) that if there is a range of channel sizes, reasonably well separated in the frequency domain and covering an adequate range of the frequency spectrum, rules can be derived for combining their zero-crossings into a description whose primitives are physically meaningful (Marr and Hildreth, 1980).

The actual details of the rules are quite complicated because a number of special cases have to be taken into account, but the general idea is straightforward. Provided the zero-crossings in the larger channels are "accounted for" by what the smaller channels are seeing, either because they are in one-to-one correspondence with the zero-crossings in the smaller channels or because they are blurred, averaged copies of them, then all the evidence points to a physical reality that is roughly what the smaller channels are seeing, perhaps modified and smoothed a little by the noise-reducing, averaging effects of the larger ones. In order to determine whether this accountability holds, configurations in which the zero-crossings of the smaller channels lie close to one another have to be detected explicitly, because it is these situations that can "fool" the larger channels. Hence the need for the explicit detection of spatial configurations such as thin bars and blobs.

If the larger channels' zero-crossings cannot be accounted for by what the smaller channels are seeing, then new descriptive elements must be developed, because the larger channels are recording different physical phenomena. This can happen in many ways: There may be a soft shadow, for example, or a wire grid in focus with an out-of-focus landscape behind; or a water beetle scurrying along the ripply surface of a pond with the weeds at the bottom forming a defocused background.

The description of the image to which these ideas lead is called the *raw primal sketch* (Marr and Hildreth 1980; Hildreth, 1980). Its primitives are edges, bars, blobs, and terminations, and these have attributes of orientation, contrast, length, width, and position. An example appears in Figure 2–21. It can be thought of as a binary map (Figure 2–21a) specifying the precise positions of the edge segments, together with the specifications at each point along them of the local orientation and of the type and extent of the intensity change (Figure 2–21d). Blob (Figure 2–21c), bar (Figure 2–21e), and discontinuity (termination) primitives can be made explicit in the same way. The representation of a long straight line, for example, consists of a termination, several segments having the same orientation, followed by a termination at the other end, as shown in Figure 2–22(a). The width, contrast, and orientation are in principle specified all along the way, although in practice it would be enough to provide this information at an adequate sampling interval. If the line is thicker than about the value of w for the smallest available channel, independent edge descriptions for its two sides would also be available. If the line curves, the orientation would gradually change along it (Figure 2–22b). If a discontinuity in orientation exists at some point along the line, then its location will be identified with a termination or discontinuity assertion (Figure 2–22c).

The raw primal sketch is a very rich description of an image, since it contains virtually all the information in the zero-crossings from several channels (two in the example of Figure 2–21). Its importance is that it is the first representation derived from an image whose primitives have a high probability of reflecting physical reality directly.

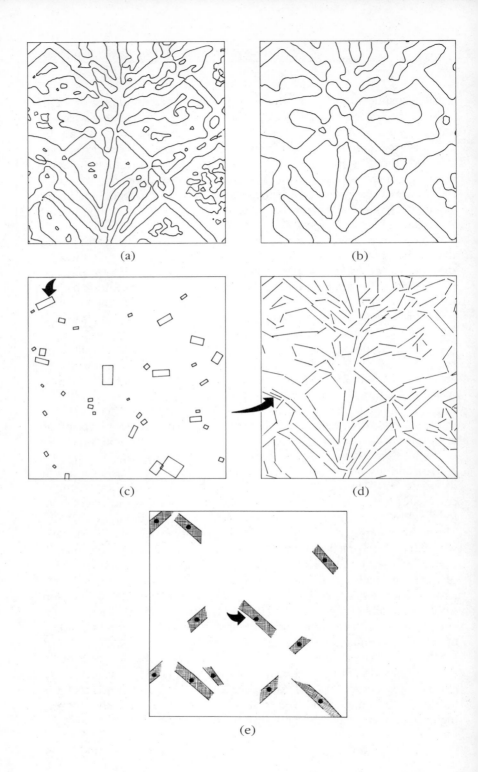

(a)

(b)

(c)

(d)

(e)

Figure 2–21. *(opposite)* The raw primal sketch as computed from two channels. (a), (b) The zero-crossings obtained from the image of Figure 2–12 by using masks with $w_{2-D} = 9$ and 18 pixels. Because there are no zero-crossings in the larger channel that do not correspond to zero-crossings in the smaller channel, the locations of the edges in the combined description also correspond to (a). (c), (d), and (e) Symbolic representations of the descriptors attached to the zero-crossing locations shown in (a). (c) Blobs. (d) Local orientations assigned to the edge segments. (e) Bars. The diagrams show only the spatial information contained in the descriptors. Typical examples of the full descriptors are:

BLOB	EDGE	BAR
(POSITION 146 21)	(POSITION 184 23)	(POSITION 118 134)
(ORIENTATION 105)	(ORIENTATION 128)	(ORIENTATION 128)
(CONTRAST 76)	(CONTRAST −25)	(CONTRAST −25)
(LENGTH 16)	(LENGTH 25)	(LENGTH 25)
(WIDTH 6)	(WIDTH 4)	(WIDTH 4)

The descriptors to which these correspond are marked with arrows. The resolution of this analysis of the image of Figure 2–12 roughly corresponds to what a human would see when viewing it from a distance of about 6 ft. (Reprinted, by permission, from D. Marr and E. Hildreth, "Theory of edge detection," *Proc. R. Soc. Lond. B 204,* pp. 301–328.)

Subjectively, you are aware of the raw primal sketch—and of the full primal sketch of Section 2.5—but you are not aware of the zero-crossings from which it is made. In order to see what the larger channels are telling your brain, you have to screw up your eyes or defocus the image somehow. Only by doing so, for example, can you see Abraham Lincoln in L. D. Harman's discretely sampled and quantized picture of him (Figure 2–23), and only by doing so can you see lines running diagonally down a chessboard (Figure 2–24). Although the larger channels are "seeing" these things all the time, as shown in Figure 2–23, what they see is adequately accounted for by the zero-crossings that occur in the smaller channels. If the middle spectral frequencies are removed from the picture of Lincoln, this is no longer the case. The processes that combine zero-crossings now find no relation between what the smaller channels see and what the larger ones see, so they both give rise to primitives in the raw primal sketch. The result, as Harmon and Julesz (1973) found, is that one sees Abraham Lincoln behind a visible graticule. The primal sketch machinery assumes that the two different kinds of information are due to different physical phenomena, so we see both.

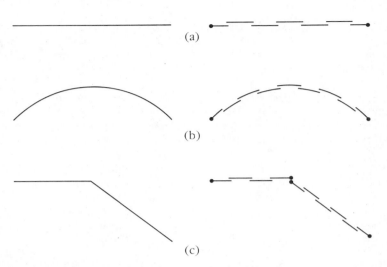

(a)

(b)

(c)

Figure 2–22. The raw primal sketch represents a straight line as a termination, several oriented segments, and a second termination (a). If the line is replaced by a smooth curve, the orientations of the inner segments will gradually change (b). If the line changes its orientation suddenly in the middle (c), its representation will include an explicit pointer to this discontinuity. Thus in this representation, smoothness and continuity are assumed to hold unless explicitly negated by an assertion.

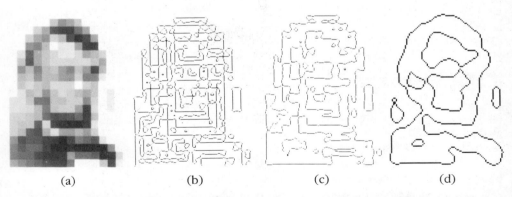

(a)   (b)   (c)   (d)

Figure 2–23. We cannot sense the primitive zero-crossings, only the description to which they give rise in the raw primal sketch. This can be seen in L. D. Harmon's discretely sampled and quantized image of Abraham Lincoln (a). No amount of voluntary effort allows us to see Lincoln without defocusing the image or squinting the eyes, despite the fact that the zero-crossings in the larger channels are producing an approximate representation of Lincoln's face. (b), (c), (d) The zero-crossings from the three sizes of the $\nabla^2 G$ operator used in Figure 2–20.

Philosophical Aside

It is interesting that the visual system takes this spatial, physical approach so seriously. It apparently does not allow the perception of a raw zero-crossing just on its own. Additional evidence, like a coincident zero-crossing from another channel seems to be required. Zero-crossings are also thought to form the input for the stereo matching process (see Chapter 3). Here again the input from two channels is combined, but this time from different eyes. Similar arguments hold for analyses based on directional selectivity, which is probably detected at the level of zero-crossings (see Section 3.4). However, once more, additional information is probably required before it is used, in this case an analysis of the coherence of the local motions in the visual field. The conclusion is that zero-crossings alone are insufficient, and this has a deep message for the whole approach, namely, that the visual system tries to deal only with physical things, using rules based on constraints supplied by the physical structure of the world to build up other descriptions that again have physical meanings.

This means that extreme care is required in the formulation of theories because nature seems to have been very careful and exact in evolving our visual systems. In this respect it is a great help to have the framework of the three levels explicitly available. Having to formulate the computational theory of a process introduces a great and useful discipline into the subject. No longer are we allowed to invoke a mechanism that seems to have some features in common with the problem and to assert that the mechanism works *like* the process. Instead, we have to analyze exactly what will work and be prepared to prove it. Stereo matching, for example, is like a lot of other things, but it is not the same as any of them. It is like a correlation, but it is not a correlation, and if it is treated like a correlation, the methods chosen will be unreliable. The job of stereo fusion is to match items that have definite physical correlates, because the laws of physics can guarantee only that items will be matchable if they correspond to things in space that have a well-defined physical location. Gray-level pixel values do not. Hence, gray-level correlation fails.

Again, the enterprise of looking for structure at different scales in an image, as illustrated by Figure 2–7 and developed in the next section, is reminiscent of ideas like filtering the image with different band-pass filters. Campbell (1977), for example, explicitly suggested that the fine details of a tank, like its registration number, might be explored using a high-pass filter, whereas the overall outline, which indicates that it is a tank, may be derived from a low-pass-filtered image. The point is once again that, just as for gray-level correlation and stereopsis, these ideas based on Fourier theory are *like* what is wanted, but they are not *what* is wanted; the structure of the physical world does not allow us to deduce, for example, that a low-pass-filtered image contains the important information about how the world is physically and spatially arranged at that scale. We can see how this could be so from the chessboard of Figure 2–24. One important aspect of the organization of this image is that the black and white squares line up horizontally and vertically as well as diagonally. To be sure, the approach of low-pass spectral filtering can tell us about the diagonal organization but not about the horizontal and vertical, and mechanisms for detecting the horizontal and vertical arrangements (making tokens for the squares and noticing how they group) will also find the diagonal organization. So the filtering approach is both unnecessary and deficient.

Another example is provided by the herringbone pattern of Figure 2–2. The vertical organization of the stripes is a clear example of an important spatial organization, yet it cannot be detected by Fourier methods because there is no power in the vertical orientation. However, such organization is easily detectable by methods that take a spatial, physical approach, starting with a representation of the basic intensity changes and then using grouping procedures based on similarity, spatial proximity, and arrangement to work up from there (Marr, 1976). Mayhew and Frisby (1978b) were among the first to appreciate the importance of this point, and they adduced further evidence in its support in experiments that explored our ability to perform texture discrimination tasks. I shall return to their work later on.

Finally, let us consider some evidence that terminations are made explicit at this stage and that they are important. I feel that it is a good thing

---

Figure 2–24.   (opposite) The Fourier spectrum of a chessboard pattern (of infinite size) has all its power in the diagonal directions, and none in the horizontal or vertical. Yet in (a) we can see that the vertical, horizontal, and diagonal spatial organizations are all equally visible while in (b) the diagonal organizations are slightly more prominent. (c), (d), and (e) show the analyses of zero-crossings from $\nabla^2 G$ operators of sizes w_{2-D} = 12, 24, and 48 pixels, respectively, on a pattern whose block size is 24 pixels, thus giving a range of w values from half to twice the size of the squares. In the first column are the convolution outputs. The second column shows the zero-crossings, with slope displayed as intensity (light and dark intensities representing positive and negative contrasts). In the third column, all the zero-crossings are displayed at uniform intensity; finally, the fourth column provides a cross-section of the convolution output near the zero-crossing contours. (f) and (g) illustrate symbolically the description obtained by channels much smaller and much larger, respectively, than the block size and should be compared with the perceptions one obtains from the chessboards in (a) and (b)—notice, for example, the roughly diagonal organization we see in looking at (b).

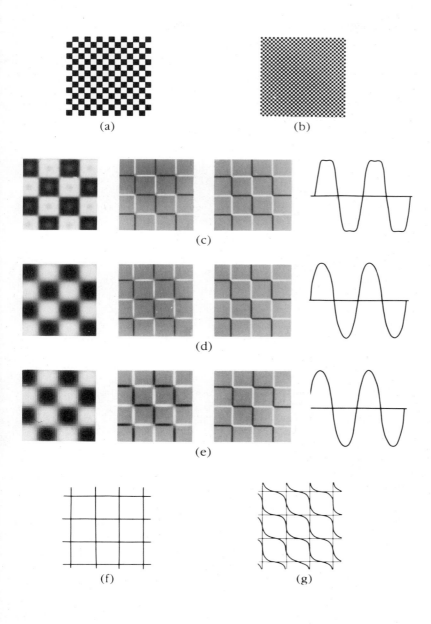

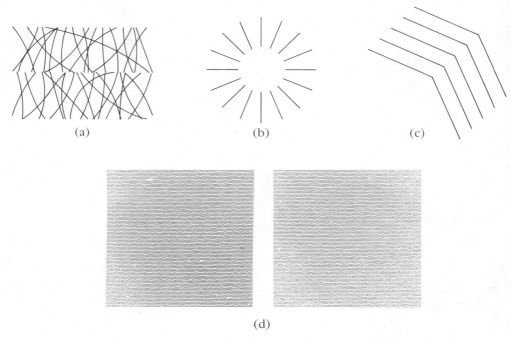

(a)   (b)   (c)

(d)

Figure 2–25.   Examples of terminations being made explicit. In (a) and (b) subjective contours are constructed by joining termination points. In (c), points of discontinuity in orientation are seen to have a linear arrangement. In the stereogram (d), terminations or discontinuities in the small horizontal lines are probably being matched between the images to yield a square in depth. (Figs. (a), (b) reprinted by permission from D. Marr, "Early processing of visual information," *Phil. Trans. R. Soc. Lond. B 275,* 1976, figs. 9(a)(d). Fig. (d) reprinted by permission from B. Julesz, *Foundations of cyclopean perception,* University of Chicago Press, 1971, fig. 3.6-3.)

to give this information here because although edges, bars, and blobs are rather obvious things, terminations are much more symbolic and abstract. The reader may therefore need some additional persuasion that these things are indeed created and at a rather low level.

Figure 2–25 provides some examples on this point. We have defined a termination as a discontinuity in the zero-crossing orientation or as the termination point of a bar. Figures 2–25(a)–(c) show clear examples where such terminations line up and where it is difficult to think of methods for detecting this fact that do not make the actual positions of the discontinuities explicit. Figure 2–25(d), from Julesz (1971, fig. 3.6–3), is even more interesting, because the things that are being matched in this stereo pair are probably the small discontinuities in the horizontal lines, and these images can be seen in stereo even when the discontinuities are tiny—less than 20 seconds of arc. Thus not only are such terminations used by stereopsis (as well as our being subjectively aware of them), but they are apparently used quite routinely even when the discontinuities are in the range of hyperacuity (smaller than a retinal receptor). The human visual system is an amazing machine!

2.3 SPATIAL ARRANGEMENT OF AN IMAGE

We come now to the question of representing spatial relations. Up to now, I have been content to assume that each item—each zero-crossing or each descriptive element of the raw primal sketch—has a coordinate in the image that determines its position there. This is reflected in our computer implementation by our use of a bit map of the image to represent basic positional information. That is, as in Figure 2–21(a), whenever there is a descriptive element, a two-dimensional array the size of the image has a 1 at the corresponding position. This 1 is also associated with a pointer to the element's actual description, which has the form shown in the legend to Figure 2–21. Like others before me, I have found that this rather literal representation, which is reminiscent of the topographically organized projections found in the early visual pathways, provides the most convenient starting point for examining geometrical relations in the image.

The reason for this is that there is quite a wide range of spatial relationships that needs to be made explicit in order to get at the useful information in an image. Once again we have the general point that these spatial relationships—things like density, collinearity, and local parallelism—are all implicit in the positions of each item, just as the binary decomposition of thirty-seven is implicit in its representation as XXXVII. But if that number's binary coefficients are necessary for some purpose, they must be made explicit at some point, so it would be advantageous to use the representation 100101.

A bit map is a good representation from which to start because it makes it relatively easy to limit the search of, for example, the raw primal sketch to just those elements in the local neighborhood of interest. Thus if we wish to know the density of certain elements in a circular neighborhood, we simply search that neighborhood in the bit map. When looking for collinear arrangements, we take a pair and search outward in the bit map along the two directions at roughly the specified orientation. The important point is that the bit map saves the trouble of searching through the whole list of primal sketch descriptors checking each coordinate to see whether it falls within the specified neighborhood. The underlying reason why using a literal bit map representation of an image is more efficient is that most of the spatial relationships that must be examined early on are rather local. If we had to examine arbitrary, scattered, pepper-and-salt-like configurations, then a bit map would probably be no more efficient than a list.

It is not too hard to see the consequences of the bit map representation in terms of nerve cells. If a neuron is to measure the density of a particular type of token in a neighborhood of some fixed size, then provided that the neurons representing the tokens are roughly topographically organized, all our density neuron has to do is count how many of the token neurons are active. Similarly, if a neuron is to measure how much local activity is present at a particular orientation, then provided that the neural representation has a roughly topographical organization, the "oriented-activity neuron" need only count how many neurons tuned to approximately the orientation in question are active within a particular physical neighborhood of the cortex. Of course, if this physical neighborhood is circular, then the neighborhood in image coordinates will not be exactly circular, but it will be roughly so, which is usually good enough.

The reason for laboring this point is that many people have difficulty relating the idea of an x,y-coordinate system of the type that might be used in a computer program to the sort of thinking that must be employed for neurons. I suggested earlier that relating this idea need not be too much of a problem, and I hope it is now clear that at least for certain aspects of local geometry, notions based on rough topographical representation and locally connected receptive fields can provide machinery of adequate power. The other half of the game, the rather precise representation of particular local geometrical relations, is something we turn to now.

The critical question is, What spatial relations are important to make explicit now, and why? The answer to this, of course, depends on the purpose for which the representation is to be used. For us, the purpose is to infer the geometry of the underlying surfaces, and we can use the physical assumptions formulated in Section 2.1, together with the natural consequences for an image of changes in depth and surface orientation. This leads us to the following list of image properties, whose detection will aid the task of decoding surface geometry:

1. Average local *intensity,* from the first physical assumption (changes in average intensity can be caused by changes in illumination, perhaps due to changes in depth, and by changes in surface orientation or surface reflectance).

2. Average *size* of items on a surface that are similar to one another, in the sense of the second and third physical assumptions (the term *size* includes the concepts of length and width).

3. Local *density* of the items defined in image property 2.

4. Local *orientation,* if such exists, of the items defined in image property 2.

5. Local *distances* associated with the spatial arrangement of similar items (the third and fourth physical assumptions), that is, the distance between neighboring pairs of similar items.

6. Local *orientation* associated with the spatial arrangement of similar items (the third, fourth, and fifth physical assumptions), that is, the orientation of the line joining neighboring pairs of similar items.

From a representational point of view, the three broad ideas that we need here are (1) tokens to represent items, and we have already seen that they form one of the pillars of the primal sketch; (2) the notion of similarity between these tokens, and this we have also already encountered (in Figure 2–3 for instance); and (3) spatial arrangement. This last idea has two parts. The one that we have encountered already has to do with density measures of various kinds, and these can be made by counting things in neighborhoods; this gives us image properties 3 and 4 above. But image properties 5 and 6 require a new idea, a new representational primitive on which we can base the analysis of the local configurations of tokens. The information that needs to be made explicit here is the distance between and relative orientation of two similar tokens. To do this, I propose a primitive called the *virtual line,* which is constructed between neighboring similar tokens and has the properties of orientation and length. It also indicates somewhat the way in which the two tokens it joins are similar, so that virtual lines joining two pairs of dissimilar tokens are treated as dissimilar (in the sense of the third physical assumption).

Perceptually, virtual lines are not meant to correspond to subjective contours, although they may be their precursors. Subjective contours, in this theory, are a later construct. They are made in the 2½-D sketch, part of whose business it is to make explicit discontinuities in the distance of visible surfaces from the viewer. Virtual lines, on the other hand, are concerned with representing the organization of images, not surfaces. They are what enables us to see the flow in the Glass patterns (see Figure 2–3) or to see the different rivalrous organizations of Figure 2–5.

The notion of a virtual line is very attractive from a computational point of view, and Stevens (1978) undertook his study of Glass patterns to try to acquire some evidence for the psychophysical existence of such lines and also to explore the idea of tokens in the images—the supposed entities that virtual lines were thought to connect.

Stevens' study was extremely interesting, for in the space of one short experimental investigation he was able to make seven fascinating points, several of them quite unexpected:

1. The local orientation organization in a Glass pattern can be recovered by a purely local algorithm, illustrated in Figure 2–26. The basic idea is to connect neighboring points with virtual lines and then to search locally among these virtual lines for the predominant orientation. By splitting patterns into several portions, each having a different transformation (see Figure 2–27), Stevens showed that perception of the global gestalt, contrary to Glass' (1969) suggestion, is not necessary for recovery of the local orientation.

2. If our perceptual analysis depends, like Stevens' algorithm, on the analysis of the distribution of orientations of virtual lines joining together dots in the pattern, the virtual lines are created between only nearby dots. The reasons for this are twofold; first and more obvious, the predominant local orientation changes as one moves globally over the pattern; second and not quite so obvious, the more virtual lines one creates from each dot, the more random the orientation distribution becomes locally and the finer must be the buckets in the histogram of the local orientation distribution that is being used to discover the predominant local orientation. If

---

Figure 2–26.   (opposite) (a) Stevens' algorithm for recovering the local orientation organization in a Glass pattern has three fundamental steps. Place tokens that are defined in the image are the input to the algorithm, which is applied in parallel to each token. Since, in the case of the Glass dot patterns, each dot contributes a place token, the first step is to construct a virtual line from a given dot to each neighboring dot (within some neighborhood centered on the dot). A virtual line represents the position, separation, and orientation between a pair of neighboring dots. To favor relatively nearer neighbors, relatively short virtual lines are emphasized by means of a simple weighting function. The second step is to make a histogram of the orientations of the virtual lines that were constructed for each of the neighbors. For example, the neighbor *D* would contribute orientations *AD, DF, DG,* and *DH* to the histogram. The final step (after smoothing the histogram) is to determine the orientation at which the histogram peaks and to select the virtual line (*AB*) closest to that orientation as the solution. (b) The results (on the right) of applying the algorithm to the patterns on the left. (Reprinted by permission from K. A. Stevens, "Computation of locally parallel structure," *Biol. Cybernetics 29* (1978), 19–28, figs. 4, 5.)

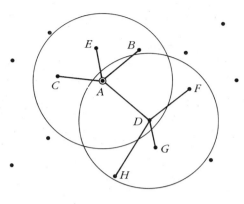

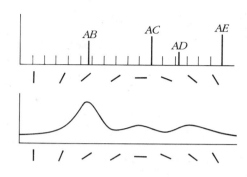

(a)

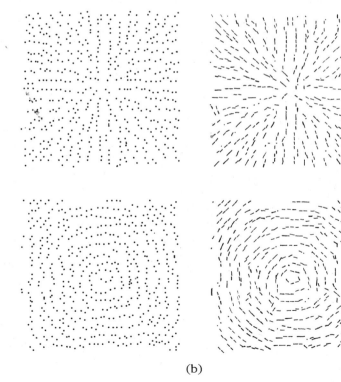

(b)

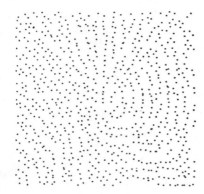

Figure 2–27. The algorithm used by our visual systems for detecting the local orientation structure is also a local one, as one can see from this pattern. Different portions of this pattern have different local orientation structures, and this can easily be discerned. (Reprinted by permission from K. A. Stevens, "Computation of locally parallel structure," *Biol. Cybernetics 29* (1978), 19–28.)

the orientation is analyzed to an accuracy of 10°–15°, then not more than about four virtual lines can be made, on the average, from each dot. Stevens also established that more than one virtual line is made. In a personal communication, he showed that only two have to be made.

3. The phenomenon scales linearly over a range of densities covering two orders of magnitude.

4. The idea that virtual lines join abstract tokens which can be defined in several ways is supported by examples like Figure 2–28, in which one of the sets of dots is replaced by small lines having randomly chosen orientations.

5. The tokens do, however, have to be reasonably similar in order for the analysis to succeed—in our terms, in order for the virtual lines to be inserted (Figure 2–3; Glass and Switkes, 1976). Stevens' own example of this, which I described in Section 2.1, consisted of three superimposed dot patterns, two dim and one bright. We see only the organization inherent in the dim dots. This is evidence both for the idea of tokens and for the notion of similarity. It proves that even at this early stage (Glass patterns can be seen in under 80 ms even with random-dot presentations immediately before and after), the analysis of the image is being carried out in quite abstract terms.

6. Interestingly, if the short lines at the random orientations shown in Figure 2–28 are replaced by short lines having a common orientation, as in Figure 2–29, rivalry appears between the overall orientations due to the short lines and due to the structure of the Glass pattern—in our terms, between the orientations of the real and the virtual lines. This bears upon how more global analysis of the image is implemented and controlled.

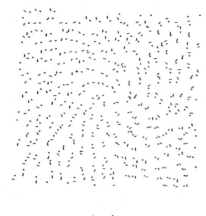

Figure 2–28. As we saw in Figure 2–3, the tokens in the two patterns do not have to be identical in order for their spatial organization to be apparent. They do, however, have to be similar. (Reprinted by permission from K. A. Stevens, "Computation of locally parallel structure," *Biol. Cybernetics 29,* 1978, 19–28).

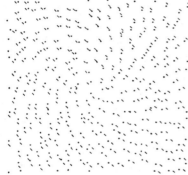

Figure 2–29. Here the superimposed pattern consists of small lines all having the same orientation. Interestingly, we perceive a kind of rivalry between this orientation and the orientation due to the spatial organization of the pattern. (Reprinted by permission from K. A. Stevens, "Computation of locally parallel structure," *Biol. Cybernetics 29,* 1978, 19–28.)

7. Finally, Stevens showed that there is little or no hysteresis in our perception of these patterns. The point at which the organization seems to disappear as the dot patterns are separated is very nearly the point at which the organization reappears as the patterns are brought together again. We were surprised by this. The reason we looked for it was Fender and Julesz's (1967) demonstration of a strong hysteresis effect in stereopsis. This had led Poggio and me to formulate a cooperative algorithm for the stereo matching problem, and the idea of cooperative processes as a way of writing an algorithm directly from constraints was an exciting one that was just emerging then (see also Zucker, 1976). The Glass pattern problem looked very well suited to a cooperative approach based on the constraints of the uniqueness and continuity of local orientation. Stevens' finding, however, showed that our perceptual systems probably do not employ a cooperative algorithm for this problem. Quite soon afterwards, we also realized that

our cooperative stereo algorithm was not the one used by our own visual systems and that matching was probably achieved by an algorithm involving very little cooperativity. Thus the opinion gradually formed that our visual systems do not use cooperative or purely iterative algorithms if it is possible to avoid them. I shall discuss some possible reasons for this later on.

Stevens' study left us somewhat more confident both about the questions we were asking and about some of the details of the primal sketch. At about that time Schatz (1977) argued that the raw primal sketch and virtual lines were by themselves sufficient to explain texture discrimination. The argument did not succeed, however, and to see why, we need to turn our attention to the more complicated levels of image representation that we call the full primal sketch.

2.4   LIGHT SOURCES AND TRANSPARENCY

Although the main stream of our account is concerned with spatial aspects of the image and visible surfaces, it is important not to forget that we are sensitive to other useful physical qualities of the visual world as well. One of these has to do with the detection of light sources—the subjective quality of fluorescence.

An important contribution to the visual detection of light sources was made by Ullman (1976b) in an article of characteristic elegance. He discussed six methods that the visual system might possibly use to help it detect light sources and then explored them empirically using achromatic "Mondrian" stimuli of the type introduced by Land and McCann (1971) in their study of lightness. These stimuli, named after the painter Piet Mondrian, consist of an array of rectangular shapes of black, gray, or white (as in Figure 2–30). In Ullman's display, one of these rectangles was sometimes a light source.

Ullman discussed light-source-detection methods based on the highest intensity in a field, high absolute intensity, high intensity compared with the average in the field, high contrast, and some other parameters. He found that none of these factors defined necessary conditions for the perception of a light source, though a contrast ratio of about 30:1 does provide a sufficient condition. High contrast is not, however, necessary; for example, a light source was perceived in a Mondrian where the ratio of intensities in no place exceeded 3:1.

Ullman then proposed a method based on the idea illustrated in Fig-

Figure 2–30. A Mondrian stimulus of the sort introduced by Land and McCann and used by Ullman in his study of fluorescence.

ure 2–31. In this figure, the x-axis represents distance along a surface illuminated from the right and which consists of three regions, A, B, and C. In A, the surface has reflectance r_1, and in B and C it has reflectance r_2 $< r_1$; in C there is also a source present underneath the surface. A camera looks down at the surface and records the intensity I at different points in the image, and the values of I have been plotted in the figure.

The idea behind Ullman's method is this: At the border between A and B, the intensity I changes and so does the intensity gradient ∇I, but they both change by the same amount so that the ratio $\nabla I/I$ remains constant.

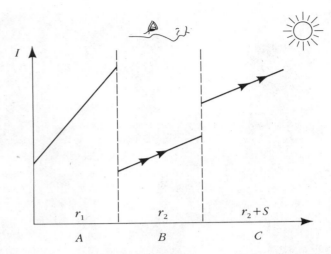

Figure 2–31. The idea behind the visual detection of light sources. Regions *A* and *B* have reflectances r_1 and r_2, and give rise to intensities *I* as shown. The value of *I* and of its gradient ∇I change together between *A* and *B*, so that $\nabla I/I$ remains constant. At *C*, however, a source *S* is added. This changes *I* but not ∇I, as shown. Hence the value of $\nabla I/I$ changes at a source boundary. This fact can be used to detect light sources in Mondrian images.

This is not so at the boundary between *B* and *C*, however, because here all that happens is that the constant-source value *S* is added to *I*. So *I* changes, ∇I does not, and hence $\nabla I/I$ does. So the ratio $\nabla I/I$ changes across a light-source boundary but not across a reflectance boundary.

This idea can be turned into a method for detecting light sources in the simplified Mondrian world, and Ullman satisfied himself that some such algorithm accounted for the perception of light sources in this environment.

Other Light-Source Effects

Forbus (1977) suggested that the operator $\nabla I/I$ could be applied to other illumination effects, including the detection of shadows and the various effects of surface wetness, luster, and glossiness that had so intrigued Beck (1972) and Evans (1974). For example, shadow boundaries behave like light-source boundaries with respect to the measure $\nabla I/I$. In addition, they are often, but not always, somewhat fuzzier than surface or reflectance boundaries, since the intensity change at a shadow is rarely sharp. This can be detected by comparing the slopes of the corresponding zero-crossings from the different-sized $\nabla^2 G$ filters, and a measure of the spatial extent of an intensity change is in fact incorporated into the raw primal sketch as the width parameter associated with an edge.

Glossiness is due to the specular or mirrorlike component of a surface reflectance function, so that one can treat the detection of gloss as essentially the detection of light sources that appear reflected in a surface (see Beck, 1972), and this depends ultimately on the ability to detect light sources. Forbus divided the problem into three categories: (1) the specularity is too small to allow gradient measurements; (2) both intensity and gradient measurements are available, but the specularity is local (as it is for a curved surface or a point source); and (3) the surface is planar and the source is extended. He derived diagnostic criteria for each case.

This topic, like the detection of shadows and light sources themselves, needs further study. The reason is that changes in surface orientation alone can also cause changes in $\nabla I/I$, although the orientation must usually change substantially in order to produce noticeable changes in $\nabla I/I$. This means that $\nabla I/I$ cannot be used as a pure diagnostic for illumination effects without taking changes in surface orientation into account. In preliminary studies we found that although in natural images one can find measurable changes in $\nabla I/I$ that are due to changes in surface orientation alone, most of these changes are small. And if one constructs an artificial image in which $\nabla I/I$ changes by a small amount across a boundary, one does not see it as a change in orientation. In fact, one sees nothing special until the change is quite large, at which point one begins to see one region as a light source.

Transparency

Another interesting phenomenon is transparency, which has attracted considerable popular attention. An example is the *Scientific American* article by Metelli (1974), in which he showed that one has the perception of transparency when a variety of inequalities hold in image intensities.

As one might expect, Metelli's inequalities might be deduced from the physics of the situation. Suppose a surface's reflectance changes from r_1 to r_2 along a boundary and that a sheet is overlaid in the manner shown in Figure 2–32. The effective illumination without the sheet is L_2, and with it (after being attenuated twice) L_1. Plainly, if the intensities in each quadrant are $i_{11}, i_{12}, i_{21},$ and i_{22}, as shown, we have

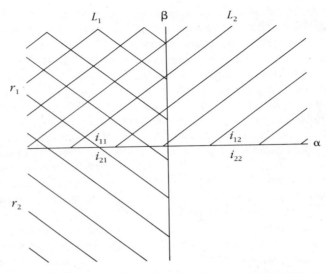

Figure 2–32. Boundary α represents a reflectance boundary, and β a transparency boundary. The quantities r_i represent reflectances; L_i, luminances; and i_{ij} are measured intensity values (for $i, j = 1, 2$).

$$\frac{i_{11}}{i_{21}} = \frac{i_{12}}{i_{22}} = \frac{r_1}{r_2}$$

and

$$\frac{i_{11}}{i_{12}} = \frac{i_{21}}{i_{22}} = \frac{L_1}{L_2}$$

These relations between the intensity values hold at transparency boundaries and at shadow boundaries; they do not hold at general four-way reflectance changes. Unlike shadow boundaries, however, transparency boundaries are almost always sharp (having a "width" of zero), and they do not cause a change in $\nabla I / I$.

Conclusions

Although these studies are incomplete, they suggest that even quite abstract qualities of the physical world, like fluorescence and transparency, can be detected by early autonomous processes. From a representational point of view, this means that one can hope to include these qualities at an early stage, such as in the primal sketch boundaries. Additional primitives will be necessary to represent them, but this poses no great problem. It will be interesting to see what other qualities of the visual world can be detected at the same rather early level of processing.

2.5   GROUPING PROCESSES AND THE FULL PRIMAL SKETCH

Let us now resume our analysis of the spatial organization of images. There are two main goals to the analysis now; (1) to construct tokens that capture the larger scale structure of the surface reflectance function and (2) to detect various types of change in the measured parameters associated with these tokens that could be of help in detecting changes in the orientation and distance from the viewer of the visible surfaces. Roughly speaking, the goals are to make tokens and to find boundaries. Both tasks require selection processes whose function it is to forbid the combination of very dissimilar types of token, and both tasks require grouping and discrimination processes whose function is to combine roughly similar types of tokens into larger tokens or to construct boundaries between sets of tokens that differ in certain ways.

In general terms, then, the approach is to build up descriptive primitives in almost a recursive manner. The raw material from which everything starts is the primitive description obtained from the image that we called the raw primal sketch. One initially selects roughly similar elements from it and groups and clusters them together, forming lines, curves, larger blobs, groups, and small patches to the extent allowed by the inherent structure of the image. By doing this again and again, one builds up tokens or primitives at each scale that capture the spatial structure at that scale. Thus if the image was a close-up view of a cat, the raw primal sketch might yield descriptions mostly at the scale of the cat's hairs. At the next level the markings on its coat may appear—which may also be detected directly by intensity changes—and at a yet higher level there is the parallel-stripe structure of these markings. The whole description would then be organized somewhat as shown in Figure 2–7. At each step the primitives used are qualitatively similar symbols—edges, bars, blobs, and terminations or discontinuities—but they refer to increasingly abstract properties of the image.

Some examples of these primitives appear in Figure 2–7. Other examples are the bloblike groups in the centers of Figures 2–33(a),(b), the small

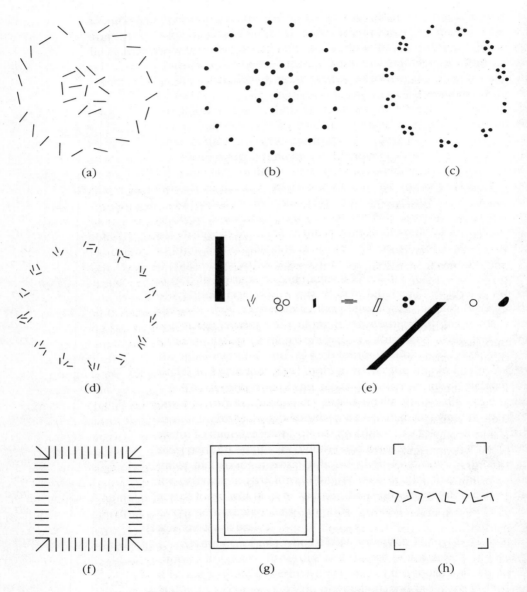

Figure 2–33. The essence of the higher primitives in the primal sketch is their ability to capture a wide range of image items as a group or token and their ability to be arranged into groups and boundaries. These diagrams show some examples of the different ways of defining place tokens and of grouping them. In each one a small line, a group of lines, or a group of dots is being combined and treated as a single unit.

clusters in Figures 2–33(c), (d), the rather heterogeneous collection of items that make up the groups in Figure 2–33(e), the sides of the squares in Figures 2–33(f), (g), and the central line in Figure 2–33(h). Any kind of local cluster or blob or group, the ability to treat it as a single item—these are the fruits of this class of processes, the processes responsible for token formation. The representation of the three-dimensional angles between two lines or the notions of a square or triangle, for example, are not included in the repertoire of the primal sketch, since they concern properties of the real world that form the image, not of the image itself.

Once these primitives have been constructed, they can tell us about the geometry of the visible surfaces—either through the detection of changes in surface reflectance or through the detection of changes that could be due to discontinuities in surface orientation or depth. About the first type of detection, one can say virtually nothing, except to remark that at a change in the surface, the change in the reflectance function is usually so great that almost any measure will detect it. I shall therefore restrict attention here to the second—the detection of boundaries that might be caused by surface discontinuities. There are two rather different ways in which these boundaries can be detected; one is by finding sets of tokens that owe their existence to the physical discontinuity and are therefore organized geometrically along it. An example of this is the lining up of terminations or of discontinuities, as illustrated in Figures 2–25(a), (b). The machinery for finding such things, I think, is also responsible for the circles in Figures 2–33(a) through (d) or the line in Figure 2–33(e).

The second type of clue to surface discontinuity consists of discontinuities in various parameters that describe the spatial organization of an image. In the section before last, we isolated six image properties that are useful to measure, three of them intrinsic to a token—average brightness, size (perhaps length and width), and orientation—and three pertaining to the spatial arrangement of tokens—their local density, distance apart, and the orientation structure, if any, of their spatial arrangement. Changes in any of these will help us to infer the geometry of the visible surfaces, and by our second physical assumption, we shall want to measure such changes at a variety of scales.

Examples of this type of clue appear in Figure 2–34. Figure 2–34(a) shows a boundary that is due to a change in dot density. In Figure 2–34(b) it is due to the change in average size of the squares. In Figure 2–34(c) it is due to a change of 45° in orientation, and in Figure 2–34(d) several of these factors change.

Thus the point of the second type of task is to measure locally (at different scales) the six quantities we defined above and to make explicit, by means of a set of boundary or edge primitives, places where discontin-

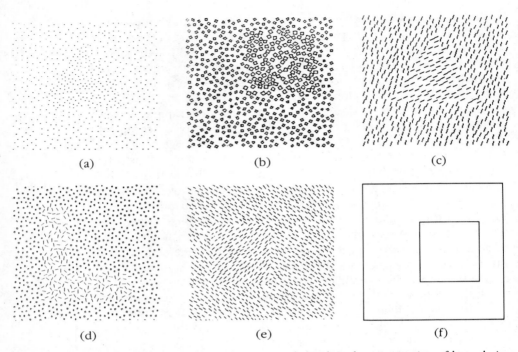

(a)   (b)   (c)

(d)   (e)   (f)

Figure 2–34. Another important aspect of the primal sketch is the construction of boundaries between regions on the basis of cues that could be caused by discontinuities in surface orientation or distance from the viewer. All examples in this figure are due to M. Riley, and they give rise psychophysically to boundaries in the sense defined in the text. The boundaries in (a) to (c) could be of geometric origin, but not in (d). Motion correspondence can be obtained between the boundaries in (e) and (f).

uities occur in these measures. The reason for adding such boundaries to the representation of the image is that they may provide important evidence about the location of surface discontinuities. This point of view has the important consequence that parameter changes likely to have arisen because of discontinuities in the surface ought to be those that give rise to perceptual boundaries, whereas those that probably could not have their origins traced to geometrical causes should be much less likely to produce perceptual boundaries. I call this the *hypothesis of geometrical origin for perceptual texture boundaries.* The principal limitations on its usefulness come from the fact that reflectance functions seldom have a precise geometrical structure. For example, if there is an oriented component to the surface structure, it is usually not very exact. Hence small changes in orientation in an image that may be produced by small changes in surface

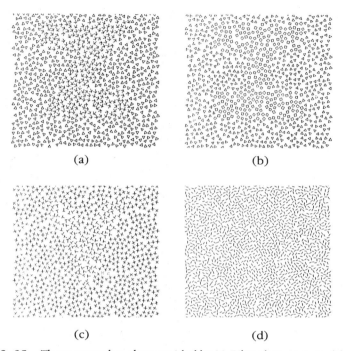

(a)

(b)

(c)

(d)

Figure 2–35. These examples, also provided by M. Riley, show texture differences that could not be of purely geometrical origin. They do not give rise psychophysically to boundaries in the sense defined in the text, even though we are sometimes able to say that one region differs from another in some way. In example (d), the inner region contains lines of just two orientations, whereas the outer region contains lines of all orientations. It is interesting to contrast these examples with those of Figure 2–34.

orientation will not usually produce a clear signal. The same applies to changes in apparent size in an image, although density allows a more sensitive discriminant. Hence, only when an image structure is extremely regular would one expect to find high perceptual acuity for these discriminations. On the whole, we should be pretty bad at them—as indeed we are (see Figure 2–35).

Before summarizing this line of argument, I should perhaps make a final point. Although it is convenient to separate grouping processes into the two categories of token formation and boundary formation, they are not, in fact, quite separate, and the two categories can overlap. In Figure 2–7, for example, some of the dot-density boundaries are boundaries of tokens. The tokens could be constructed either from such boundaries or from the cluster of the cloud of dots there, or, of course, in both ways. In

Figure 2–34(a), the triangle could be made by the linear grouping of nearby dots, by finding a local increase in dot density, or even by a local decrease in average brightness. A single boundary is often defined in many ways, a fact of life that aids its recovery by the visual system but raises difficulties for the experimental psychophysicist.

Main Points in the Argument

The idea, then, is to start with the raw primal sketch and operate on it with processes of selection, grouping, and the discrimination to form tokens, virtual lines, and boundaries at different scales. The approach I have outlined gives the reasons for doing this: It enables us to deduce what types of tokens should be made, what types of selection and grouping should be available, which circumstances should give rise to perceptual boundaries and which should not, and perhaps even how to compare differences in acuity due to different discriminants. For example, when token size is viewed as a discriminant that indicates a change in surface orientation, the resolution of the analysis of token size should be comparable to the resolution of the analysis of token orientation. These arguments provide a physical basis for the suggestion that some types of visual discrimination of texture rest on first-order discriminations acting on the primal sketch (Marr, 1976). We now explore this question in more detail.

The Computational Approach and the
Psychophysics of Texture Discrimination

From a purely psychophysical point of view, it has been difficult to define exactly what is meant by the phrase *texture discrimination*. In his well-known series of articles on the subject, Bela Julesz (for example, see Julesz, 1975) distinguishes between textures that can be immediately distinguished (so-called preattentive perception) and those that cannot be distinguished without close and often prolonged study (so-called scrutiny). He limited his investigations to discriminations of the first kind, those that can be distinguished in under 200 ms—roughly, those that can be distinguished without eye movements.

I should perhaps point out that the approach I have suggested to the problem is somewhat more restrictive, for it also requires that perceptual boundaries be formed at the borders between the textures. Not all of the textures devised by Julesz have this property. None of the examples in

Figure 2–35 do, for instance, whereas all the examples in Figure 2–34 do. Psychophysically, then, our approach requires that the discrimination be made quickly—to be safe, in less than 160 ms—and that a clear psychophysical boundary be present. There are various criteria for this second requirement. One is that, in addition to being able to state that two textures are present in a Julesz display like those in Figure 2–34, one should also be able to give information about the shape of the distinguished region. Schatz (1977), for example, included this condition as one of his experimental criteria.

Another possibility, suggested to me by Shimon Ullman, is to try to obtain apparent motion between texture boundaries that have been generated in different ways in two frames. Frame 1, for example, might consist of Figure 2–34(e), and frame 2, presented after an interstimulus interval of, say, 100 ms, of Figure 2–34(f). If the boundaries appear to move in the obvious way, this is corroborating evidence that they are in fact constructed. If the boundaries obey the same local correspondence rules that are obeyed by intensity boundaries (Ullman, 1979b), this is then very strong evidence that the boundaries are being made explicit. The examples illustrated in Figure 2–34 all pass both the shape and apparent-motion tests.

A third criterion for when a boundary is being constructed perceptually may perhaps be developed from a finding by Kidd, Frisby, and Mayhew (1979). They found, using suitably constructed stereograms, that certain kinds of texture boundary are capable of initiating disjunctive eye movements, which are eye movements that cause the two lines of sight to converge or diverge.

If all these criteria succeed or fail together at the different types of boundary, we shall have a powerful technique for saying when a perceptual boundary is created from a change in visual texture. Similar combined approaches may also help us to determine whether something like the full primal sketch is in fact obtained from the image by telling us what types of tokens are made explicit in preattentive perception.

Finally, it seems to me that psychophysical studies of the relative power of the different discrimination processes can be most convincing if something like Barlow's (1978) absolute measures of efficiency are used. In this study, Barlow asked how sensitively humans could detect targets of greater dot density embedded in backgrounds of random dots. He found that his subjects were able to use about two-thirds of the objective signal-to-noise ratio of the displays, which corresponds to about 50% of the statistical information available. He also suggested an interesting, economical model to explain his results, consisting of "dot-number estimating" elements that are roughly circular and of variable size. They are sufficient in number to cover the central area of vision with neighborhoods 1°–4° in diameter, and with an average mismatch and overlap of 50%. They integrate temporally for about 0.1 s. I hope that studies like this can be extended to other discrimination tasks.

That ends our discussion of how to represent an image. We now turn to the use of these representations in deriving surface information.

CHAPTER 3

From Images
to Surfaces

3.1   MODULAR ORGANIZATION OF
THE HUMAN VISUAL PROCESSOR

Our overall goal is to understand vision completely, that is, to understand how descriptions of the world may efficiently and reliably be obtained from images of it. The human system is a working example of a machine that can make such descriptions, and as we have seen, one of our aims is to understand it thoroughly, at all levels: What kind of information does the human visual system represent, what kind of computations does it perform to obtain this information, and why? How does it represent this information, and how are the computations performed and with what algorithms? Once these questions have been answered, we can finally ask, How are these specific representations and algorithms implemented in neural machinery?

The study of working visual systems can help us in this endeavor, and nowhere is this clearer than in the study of visual processes. At the level of computational theory, the investigator's first question is, What computational problems are being solved, and what information is needed to solve them?

As usual, the point is best made with an example. Because of how our eyes are positioned and controlled, our brains usually receive similar images of a scene from two nearby points at the same horizontal level. If two objects are separated in depth from the viewer, the relative positions of their images will differ in the two eyes. You can see that this is so by holding your thumb at various distances from your eyes against a background. Closing first one eye and then the other will then convince you that objects in the world have somewhat different positions in the images cast upon each of your retinas. The relative difference in position is called *disparity*; it is usually measured in minutes of arc, and the disparity between the images of your thumb and the background in your two eyes increases as you move your thumb nearer to you. One minute of disparity roughly corresponds to a depth difference of 1 in. for an object 5 ft away.

The brain is capable of measuring disparity and using it to create the sensation of depth. For purposes of demonstration, a stereoscope from a souvenir shop will do: When individual views are seen with just one eye at a time, they look flat. However, if you have good stereo vision and look with both eyes, the situation is quite different. The view is no longer flat: The landscape jumps sharply into relief, and your perceptions are clearly and vividly three-dimensional.

How does stereo vision work? Unfortunately, we cannot even begin to ask the right questions from just the evidence described above. The reason is that from the experience of everyday life or even from the small experiment with the stereoscope, it is not at all clear how separate stereoscopic processing is from the more familiar, monocular analysis of each image. If stereo processing were an isolated module, so to speak, then one could tackle it on its own. But it may not be isolated—for example, stereo vision could involve a complicated and gradually increasing interaction between the individual processings of each eye and a comparison of the results between the two eyes. This is not as absurd as it seems. It does not take much imagination to see how such a scheme might work. We could start by finding, for example, the images of an oak tree as seen independently by the left and right eyes. Then we could find the trunk in each image and then, perhaps, the lowest branch on the right hand side of the trunk. Pretty soon we would have correspondences between the small details of the left and right images whose disparity could be measured accurately. And because the match has been obtained in this general-to-specific way, there is never any real problem in deciding what should match what.

This type of approach, incidentally, is typical of the so-called top–down school of thought, which was prevalent in machine vision in the 1960s and early 1970s, and our present approach was developed largely in reaction to it. Our general view is that although some top–down information is

Figure 3–1.   The interpretation of some images involves more complex factors as well as more straightforward visual skills. This image devised by R. C. James may be one example. Such images are not considered here.

sometimes used and necessary (see Figure 3–1 and Marr, 1976, fig. 14), it is of only secondary importance in early visual processing. The evidence for this comes from psychophysics and for some reason was willfully ignored by the computer vision community. The argument suggested by this evidence is a simple one. If, using the human visual processor, we can experimentally isolate a process and show that it can still work well, then it cannot require complex interactions with other parts of vision and can therefore be understood relatively well on its own.

One way of isolating a visual process is to provide images in which, as much as possible, all kinds of information except one have been removed and then to see whether we can make use of just that one kind. Bela Julesz did this for stereopsis by inventing the computer-generated random-dot stereogram, which we met in Figure 1–1. Both the left and right images shown there are computer-generated assemblies of black and white squares that are identical except for a centrally located, square-shaped region shifted horizontally in one image relative to the other. That is, it has a different disparity. The stereo pair contains no information whatever about visible surfaces except for this disparity.

When the pair is viewed stereoscopically and fused, one vividly and unmistakably perceives a square floating in space above the plane of the background. This proves two things: (1) Disparity alone can cause the sensation of depth, and (2) if there is any top–down component to the processing (and, in fact, we think that there probably is a little), it must be of a very limited kind, because neither image contains any recognizable large-scale monocular organization.

This observation—which is qualitative rather than quantitative, not at all technical, and, like many of Julesz's demonstrations, absolutely and strikingly convincing to behold—is fundamental to our approach, for it enables us to begin separating the visual process into pieces that can be understood individually. Computer scientists call the separate pieces of a process its *modules,* and the idea that a large computation can be split up and implemented as a collection of parts that are as nearly independent of one another as the overall task allows, is so important that I was moved to elevate it to a principle, the *principle of modular design.* This principle is important because if a process is not designed in this way, a small change in one place has consequences in many other places. As a result, the process as a whole is extremely difficult to debug or to improve, whether by a human designer or in the course of natural evolution, because a small change to improve one part has to be accompanied by many simultaneous, compensatory changes elsewhere. The principle of modular design does not forbid weak interactions between different modules in a task, but it does insist that the overall organization must, to a first approximation, be modular.

From a theoretical point of view, observations like Bela Julesz's are extremely valuable because they enable us to formulate clear computational questions that we know must have answers because the human visual system can carry out the task in question. It was Julesz's findings that allowed us to formulate our theory of human stereopsis (Marr and Poggio, 1979). The analogous findings of Miles (1931) and of Wallach and O'Connell (1953) allowed Ullman (1979b) to develop his theory of structure from motion. Some other experiments by Julesz (1971, chap. 4), together with Braddick's (1974) identification of a short-range, short-term process in apparent motion, contributed to the formulation of our theory of directional selectivity.

The existence of a modular organization in the human visual processor proves that different types of information can be analyzed in relative isolation. As H. K. Nishihara (1978) put it, information about the geometry and reflectance of visible surfaces is encoded in the image in various ways and can be decoded by processes that are almost independent. When this point was fully appreciated, it led to an explosion of theories about possible decoding processes. This chapter describes the computational theories of those decoding processes that are now quite well understood. These processes are (1) stereopsis, (2) directional selectivity, (3) structure from apparent motion, (4) depth from optical flow, (5) surface orientation from surface contours, (6) surface orientation from surface texture, (7) shape from shading, (8) photometric stereo (the determination of surface orientation and reflectance from scene radiances—the intensity of reflected light—observed by a fixed sensor under varying lighting conditions), and (9) lightness and color as an approximation to reflectance. Of course, other cues are available, like occlusion, but unless I have been able to give a process a reasonably integrated treatment, I have not discussed it here. Not all of the methods described here have biological relevance—photometric stereo certainly has none—but they are all of interest as ways of inferring the geometry and reflectance of visible surfaces from their images.

3.2   PROCESSES, CONSTRAINTS, AND THE AVAILABLE REPRESENTATIONS OF AN IMAGE

Before embarking on a detailed description of the different theories, I should make some remarks about the general nature of these theories and what the reader should look for in them and expect from them.

The first point is to remind the reader that we expect to analyze processes at three levels (remember Figure 1–4)—the levels of computational theory, of algorithm, and of implementation. Of course, the vision problem has not been completely solved yet, so we cannot analyze at all three levels every process within the human visual system. But we can analyze some processes at all three levels, and many of them at one or two—perhaps even most of the processes that discern surfaces from images.

In every case, we start with the first level—the computational theory— because this book is about the computational approach to vision. And at this level the reader should look out for the physical constraints that allow the process to do what it does. The situation is quite like what happened in Chapter 2. There we were dealing with ways of representing the image, and in order to say what would be useful and what would not, we were continually referring to the interaction between the imaging process and the underlying properties of the physical world that gives rise to structure in images. In this chapter, where we deal with processes instead of representations, the situation is entirely analogous but arises in a slightly different way. We have already met an example of this new situation in the theory of how to combine zero-crossings from different-sized filters in order to make the physically meaningful primitives of the raw primal sketch. The critical point was that, in general, there is no reason why the zero-crossings from two channels that do not overlap in the frequency domain should be related. They are related in early vision because intensity changes are caused by markings on a surface, the edges of objects, and so on, and these happen to have the critical property of spatial localization.

This interaction between the imaging process and the underlying properties of the physical world commonly occurs in the study of visual processes, and we shall meet several examples here. Frequently an apparently insoluble problem arises, such as which dots in the left-hand pattern in Figure 1–1 should match which dots in the right-hand pattern. From the image alone one just cannot tell. The critical step in formulating the computational theory of stereopsis is the discovery of additional constraints on the process that are imposed naturally and that limit the result sufficiently to allow a unique solution. Finding such constraints is a true discovery— the knowledge is of permanent value, it can be accumulated and built upon, and it is in a deep sense what makes this field of investigation into a science (Marr, 1977b).

Once we have isolated where the extra information comes from—in what ways, if you like, the information is constrained by the world—we can incorporate it into the design of a process. For combining zero-crossings, for example, this was done by the spatial coincidence *assumption*— that coincident zero-crossings are adequate evidence of a physical edge. Thus, the constraints are used by turning them into an assumption that may or may not be internally verifiable.

This, then, is one aspect of the top-level computational theory of a process, but there is another, almost as important. We saw in Chapter 1 that a process can be viewed as a transformation from one representation to another. Addition, for example, maps a pair of numbers into a number. All the processes that we shall discuss take as their inputs properties of the image and produce as their outputs properties of the surfaces—indicating to us either something about the geometry or the reflectance of the surfaces.

We shall discuss ways of representing the outputs of these processes in the next chapter, but now we are concerned with their inputs. What should serve as the inputs to these processes? We already have four options—the image itself, zero-crossings, the raw primal sketch, and the full primal sketch. Part of the computational theory must indicate which of these four should be used (or if something else entirely is appropriate) and why, and a portion of the investigation of each process will deal with this question.

Ultimately, of course, psychophysics tells us which input representation is used—if the process is in fact incorporated in the human visual system. There is, however, one useful point to bear in mind (Marr, 1974b): Essentially, since the constraints allow the processes to work, and since the constraints are imposed by the real world, by and large the primitives that the processes operate on should correspond to physical items that have identifiable physical properties and occupy a definite location on a surface in the world. Thus one should not try to carry out stereo matching between gray-level intensity arrays, precisely because a pixel corresponds only implicitly and not explicitly to a location on a visible surface.

This point is important. For example, failure to recognize it held Wallach and O'Connell (1953) up for years by their own admission. They could not understand why the shadow of a bent wire should be different from the shadow of a smooth solid object. If a wire is rotated, its shadow moves, and one instantly perceives the wire's three-dimensional shape; if a solid object is rotated, its shadow moves but one cannot perceive its shape. The reason is that the shadow of the wire produces an outline that is effectively in one-to-one correspondence with fixed points on the wire, each having a definite physical location that changes from frame to frame, admittedly, but that always corresponds to the same piece of wire. For the rotating object this is just not true. From moment to moment, the points on the silhouette correspond to quite different points on the object's surface. The image primitives are no longer effectively tied to a constant physical entity. Hence the shape recovery process fails.

On the other hand, the more complex the derivation of a representation from an image, the longer the derivation is liable to take. In real life, time is often of the essence; especially in the analysis of motion, an answer is required as soon as possible—before the image has become out-of-date or before the mover has eaten the viewer. In general, therefore, evolution is prejudiced toward getting things started as soon as possible.

Hence, although processes that operate on the information in an image could use any of a wide variety of input representations in principle, in practice they are likely to use the earliest representations that they possibly can. The range that we have discussed includes the gray-level image, zero-crossings, the raw primal sketch, and the full primal sketch. The earlier ones are not yet "physical," and so a bit unsafe, which might cause us to make mistakes. But for some purposes this possible error is worth the extra speed, for example, in the control of eye movements in response to a sudden change in an image and perhaps also for looming detectors in the theory of directional selectivity (see Section 3.4). Furthermore, just because a boundary is physical does not always make it safe to use. The edges of a uniform cylindrical lamppost give rise to perfectly good edges in the images seen by the left and right eyes, but these edges correspond to different lines on the physical surface. This gives the stereopsis process trouble when, having matched the images, it tries to calculate how far away the lamppost is.

So our rule, then, that the inputs to a process should consist of elements with close physical correlates, is only a general one. It is clearly inappropriate for some things, like shape from shading or photometric stereo, but probably rather important for things like the correspondence process in apparent motion (Ullman, 1978) or the analysis of shape from surface contours or texture. The rule has its attendant dangers, though, and for some processes it is obeyed only marginally—for example, I think that both stereopsis and directional selectivity can use zero-crossings directly. However, the important point is that the rule is sufficiently strong and apparently valid and that violations cannot be allowed to go unnoticed. They have to be defended.

So much, then, for the level of computational theory. The second of the three levels of understanding a process is the level of the algorithm. At this level we formulate a particular procedure for implementing a computational theory. There are two principles that guide the design of algorithms, and they probably ought to be satisfied by any serious candidate for an early visual process in the human visual system. One principle says, roughly, that the algorithm has to be robust: the other, that it must behave smoothly. They are as follows (Marr, 1976):

1. *Principle of graceful degradation.* This principle is designed to ensure that, wherever possible, degrading the data will not prevent the delivery of at least some of the answer. It amounts to a condition on the continuity of the relation between different stages in the processing. For example, it should be required that a rough two-dimensional description of the kind that a vision system might compute out of a drawing enable the system to compute a rough three-dimensional description of what the drawing represents.

2. *Principle of least commitment.* This principle requires not doing something that may later have to be undone, and I believe that it applies to all situations in which performance is fluent. It states that algorithms that are constructed according to a hypothesize-and-test strategy should be avoided because there is probably a better method. My experience has been that if the principle of least commitment has to be disobeyed, one is either doing something wrong or something very difficult.

It would be nice to be able to give general rules about processes at the third level of analysis, the level of neural implementation. Unfortunately, only a few process theories have been developed to the point where specific neural implementations have been proposed, and none of these implementations have been confirmed experimentally in every detail so we are not yet in a position to formulate such rules.

However, one suggestion of a rule can be extracted from our experience with cooperative algorithms for stereopsis and locally parallel organization (Marr and Poggio, 1976; Stevens, 1978). It is only a suggestion, however, and I give it with that caution. It is that, if possible, the nervous system avoids iterative methods—that is, pure iteration in which no new information is introduced at each cycle. Instead, it seems to prefer one-shot methods, like Stevens' (1978) one-shot algorithm for finding the local orientation in Glass patterns. The nervous system also seems to prefer methods that run from the coarse to the fine, doing essentially the same thing at each state but being saved from pure iteration by introducing new information at each cycle. Our stereo algorithm has this form, as we shall see in the next section. And it might be a sound design principle, too, since it effortlessly incorporates the principles of graceful degradation and least commitment.

Yet cooperative methods (a type of nonlinear, iterative algorithm) look very plausible from some points of view. They are very robust, for example, and often have a structure that is readily translatable into the inhibitory and excitatory connections of a plausible neural network. Why, then, are they not used?

One possible explanation may be that cooperative methods take too long and demand too much of the neural hardware to be implemented in any direct way. The problem with iteration is that it demands the circulation of numbers around some kind of loop, which could be carried out by some system of recurrent collaterals or closed loops of neuronal connections. However, unless the numbers involved can be represented quite accurately as they are circulated, errors characteristically tend to build up rather quickly. To use a neuron to represent a quantity with an accuracy of even as low as 1 in 10, it is necessary to use a time interval that is sufficiently long to hold between 1 and 10 spikes in comfort. This means at least 50 ms per iteration for a medium-sized cell, which means 200 ms for four iterations—the minimum time ever required for our cooperative algorithm to solve a stereogram. And this is too slow.

This argument against purely iterative algorithms is not compelling. It is, however, persuasive enough to make me skeptical of them as candidates for processes used by the human visual processor, and it suggests that one should try very hard when designing ways of implementing a process to use algorithms with a more open and flexible structure.

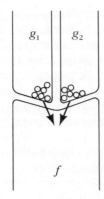

Figure 3–2. The synaptic arrangement considered by Torre and Poggio (1978). Such an arrangement could approximate an AND–NOT gate.

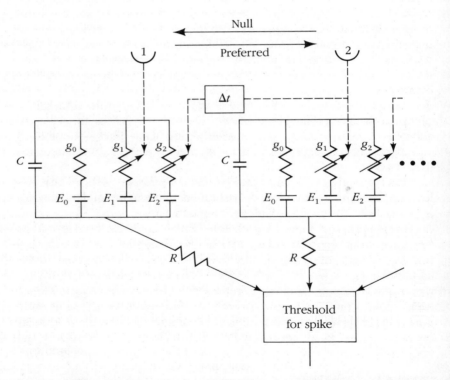

Figure 3–3. The electrical circuit equivalent of the synaptic arrangement shown in Figure 3–2 in the configuration suggested by Torre and Poggio (1978) for implementing directional selectivity. The interaction implemented by the circuit has the form $g_1 - \alpha\, g_1\, g_2$, which approximates a logical AND–NOT gate. A logical AND gate can be implemented by a similar circuit.

One other lesson about neural implementations may perhaps be drawn, this time from the work of Torre and Poggio (1978), who showed how the nonlinear operation AND–NOT could be implemented at the level of synaptic interactions on a dendrite. They showed, using a cable-theoretical analysis, which calculates the time dependent electrical properties of the dendrite from its geometry, that the synaptic arrangement shown in Figure 3–2 has the electrical properties of the circuit shown in Figure 3–3 and the behavior shown in Figure 3–4. It approximately com-

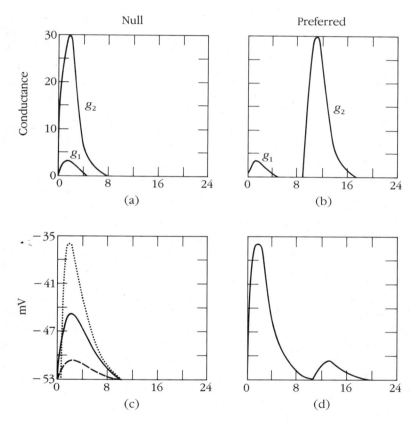

Figure 3–4.   The calculated behavior of the circuit in Figure 3–3. For movement in the null direction, the time course of the inputs g_1 and g_2 is shown in (a), and the output of the circuit is the solid line in (c). The dotted and dashed curves show, respectively, the responses with g_1 and g_2 separately. For motion in the opposite direction, the inputs arrive as shown in (b), and the output of the circuit is shown in (d). Notice how attenuated (c) is relative to (d). In this manner, the output of the system can be made directionally selective. The time courses (horizontal axes) are plotted in units of the membrane time constant.

putes $g_1 - \alpha g_1 g_2$, which behaves like AND–NOT, and they suggested that this might be how the ideas of Hassenstein and Reichardt (1956) and of Barlow and Levick (1965) about directional selectivity in the fly and rabbit retinas are implemented (see Section 3.4). Poggio and Torre (1978) extended this idea, showing that a wide range of primitive, nonlinear operations could be implemented using local synaptic mechanisms.

One message of this work is that neurons might do more than we think. Early models, like those of McCulloch and Pitts (1943), tended to see neurons as basically linear devices that could implement nonlinear functions by means of a threshold, which could perhaps be variable if produced by an inhibitory interneuron. This way of thinking led Barlow and Levick to formulate their model of directional selectivity, and I employed it myself when I was interested in the cerebellar cortex (Marr, 1969). We have already seen, however, that local nonlinearities may be important. For example, the scheme for zero-crossing detection in Figure 2–18 is based on the use of many AND gates. The force of Poggio and Torre's work is that such things as AND gates may not require whole cells for their implementation—they can perhaps be executed much more compactly by local synaptic interactions in small pieces of dendrite.

Enough, then, of generalities; let us turn to the processes themselves. I shall start with stereopsis, since it was the first psychological process to be understood and because it led to much of the general knowledge about early vision already incorporated into my account. I have tried not to be too technical in describing the various processes, my aim being to give the reader a general feel for how they all work and to show some examples of them working. For full details, the reader may consult the original articles.

One final point about the organization of the account. Many of these processes divide naturally into two parts, the first concerned with setting up and making a measurement, so to speak, and the second with using the measurement to recover three-dimensional structure. In stereopsis, for example, the first step is the matching process, which establishes the correspondence between the two eyes so that disparities can be measured; the second is the trigonometry that recovers distance and surface orientation from disparity. The first step is the difficult one; the second is easy. In directional selectivity, the first step is to establish the local direction of movement, and the second is to use this sparse local information to help separate figure from ground. Neither step is particularly difficult. In apparent motion, the first step is to establish a correspondence between successive "frames" so that the displacements between frames can be measured; the second step is to use these measurements to recover three-dimensional structure. Here both steps are difficult.

For this reason I have split several of the sections into two parts. Of course, whether a process is indeed implemented by the human visual processor is sometimes unknown, and, even if it were known, whether it is divided as I have described is still an open psychophysical question. In such cases, I have tried to make clear what the current evidence is and what needs to be done to resolve the open questions.

3.3   STEREOPSIS

We saw earlier that the two eyes form slightly different images of the world. The relative difference in the positions of objects in the two images is called disparity, which is caused by the differences in their distance from the viewer. Our brains are capable of measuring this disparity and of using it to estimate the relative distances of the objects from the viewer. I shall use the term *disparity* to mean the angular discrepancy in position of the image of an object in the two eyes; the term *distance* will refer to the objective physical distance from the viewer to the object, usually measured from one of the two eyes; and the term *depth* I shall reserve for the subjective distance to the object as perceived by the viewer.

I shall divide the account into two parts, the first concerned with measuring disparity, and the second with using it. Both parts are separated into the three levels of Figure 1–4. The articles on which this account is based are by Marr (1974b) and Marr and Poggio (1976), which deal with the computational theory; by Marr and Poggio (1979), which deals with the algorithm thought to be used by the human visual system; and by Grimson and Marr (1979) and Grimson (1981), which describe Eric Grimson's computer implementation of the algorithm. Between 1977 and 1979, the additional work done on zero-crossings (Marr, Poggio, and Ullman, 1979; Marr and Hildreth, 1980) allowed certain simplifications in the implementation of the algorithm; most notably, we found from mathematical arguments that we could use circularly symmetric instead of oriented receptive fields for the initial convolutions. This particular detail was arrived at independently on psychophysical grounds by Mayhew and Frisby (1978a).

Measuring Stereo Disparity

Computational theory

Three steps are involved in measuring stereo disparity: (1) A particular location on a surface in the scene must be selected from one image; (2)

that same location must be identified in the other image; and (3) the disparity between the two corresponding image points must be measured.

If one could identify a location beyond doubt in the two images, for example, by illuminating it with a spot of light, the first two steps could be avoided and the problem would be easy. In practice, we cannot go around carefully shining a spot of light on a surface and noting where its image falls in the two eyes, so we must somehow find a way of identifying a location by the more passive means of sensing the environment.

The reason why the task of identifying corresponding locations in the two images is difficult is because of what is called the false target problem. This occurs in what may be its extreme form in Julesz's random-dot stereograms (see Figure 1–1), and the nature of the problem is illustrated in Figure 3–5. The question is, Which dot corresponds to which? The left eye here sees four dots, and the right eye sees four, but which corresponds to which? A priori, all of the 16 possible matches are plausible candidates but when we observe such a stereo pair, we make the correspondences shown by the filled circles and not any of the correspondences shown by the open circles, which are called false targets.

Although this obviously makes some kind of sense, it is nevertheless surprising. How do we know which matches are correct and which should be ignored? What is more, there is another solution to this particular correspondence problem that seems just as valid. Look at the figure for a moment and try to see what it is. The other answer is the four central vertical matches, in which R_1 is paired with L_4, R_2 with L_3, R_3 with L_2, and R_4 with L_1. But we never see this match perceptually, which would appear as a set of squares in a receding line. Why not? Why only the other one, in which the squares line up, all about the same distance away?

From reading Chapter 2, the reader will immediately suggest using higher-level descriptions of the image—for example, matching first the two rows of dots as units and then, going on to match the individual squares and finally the edges of each square. And I think that something like this happens, but the first point to be clear about is that such a suggestion on its own is only a mechanism. The real question to ask is *Why* might something like that work? For the plain fact is that if we look just at the pair of images in Figure 3–5, there is no reason whatever why L_1 should not match R_3, L_2 match R_1, and even L_3 match R_1.

What we need is some additional information to help us decide which matchings are correct by constraining them in some way, and to do this we have to examine the basis in the physical world for making a correspondence between the two images.

The constraints that we need are the following, and they look deceptively simple; (1) A given point on a physical surface has a unique position

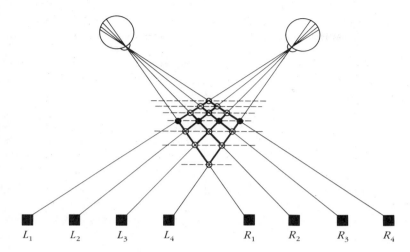

Figure 3–5. Ambiguity in the correspondence between the two retinal projections. In this figure, each of the four points in one eye's view could match any of the four projections in the other eye's view. Of the 16 possible matchings, only 4 are correct (filled circles); the remaining 12 are false targets (open circles). Without further constraints based on global consideration, such ambiguities cannot be resolved. The targets (filled squares) are assumed to correspond to matchable descriptive elements obtained from the left and right images. (Reprinted by permission from D. Marr and T. Poggio, "Cooperative computation of stereo disparity," *Science 194,* October 15, 1976, 283–287. Copyright 1976 by the American Association for the Advancement of Science.)

in space at any one time; and (2) matter is cohesive, it is separated into objects, and the surfaces of objects are generally smooth in the sense that the surface variation due to roughness cracks, or other sharp differences that can be attributed to changes in distance from the viewer, are small compared with the overall distance from the viewer.

These observations are properties of physical surfaces, and they constrain the behavior of the surface position. Hence, if we want to use these observations to help us establish a correspondence between two images of a surface, we must ensure that the items to which we apply them are in one-to-one correspondence with well-defined locations on a physical surface. To do this, we must use image predicates that correspond to surface markings, shadows, discontinuities in surface orientation, and so forth.

These physical considerations were precisely the motivation for the primal sketch, as we saw in Chapter 2, and that is why the primal sketch can be used, because the descriptive items in it—line and edge segments, blobs, terminations and discontinuities, and tokens obtained from these by grouping—usually correspond to items that have a physical existence on a surface. And it is perhaps worth pointing out here that since the grouping processes have to be rather catholic in what they are prepared to group together, the larger and more abstract tokens tend to be less reliable than the very early and primitive things in the raw primal sketch. This is particularly relevant to stereopsis for another reason: Large-scale tokens are quite large, perhaps several degrees, whereas useful disparities tend to be rather small, on the order of minutes. To make accurate measurements, therefore the smaller, more primitive descriptors are preferred. On the other hand, clear statistical effects are likely to be quite a reliable indication of a physical change even at quite high levels so that high-level boundaries of the kind I called texture discrimination boundaries are probably more useful for stereopsis than aggregates at the same high level. We shall meet what I think are some consequences of this later on.

We can therefore rewrite the physical constraints as matching constraints, which restrict the allowable ways of matching two primitive symbolic descriptions, one from each eye. For the matching constraints to be valid, the elements in the matched descriptions must correspond to well-defined locations on the physical surface being imaged. We can think of these elements as carrying only position information, like the black dots in a random-dot stereogram, although for a full image, rules will exist that specify which matches between descriptive elements are possible and which are not. These rules will again be deducible from the physical situation; if the two descriptive elements could have arisen from the same physical marking, then they can match. If they could not have, then they cannot be matched. This is our first matching constraint, which I shall call the *compatibility* constraint.

The second and third matching constraints come from the two physical constraints. The uniqueness constraint means that, except in rare cases, each descriptive item can match only one item from the other image. The exceptions can arise as a result of the imaging process when two markings lie along the line of sight from one eye but are separately visible from the other. The third constraint, *continuity*, means that disparity varies smoothly almost everywhere. This constraint follows because the second physical constraint implies that the distance to the visible surface varies continuously except at object boundaries, which occupy only a small fraction of the area of an image.

These three restrictions, then, are our constraints. We now turn them to our purposes by making what I shall call the *fundamental assumption of stereopsis: If a correspondence is established between physically meaningful primitives extracted from the left and right images of a scene that*

*contains a sufficient amount of detail, and if the correspondence satisfies
the three matching constraints, then that correspondence is physically correct.* It follows immediately from this assumption that the correspondence
must be unique.

But this is all very well, the skeptical reader will say. The matching
constraints look perfectly reasonable and even quite powerful. But to turn
them into a fundamental assumption which asserts that they are not only
necessary consequences of the physical world but also actually *sufficient*
to determine uniquely the correct correspondence—now that is an altogether different matter.

To say this is absolutely correct and hits fairly and squarely upon a
philosophical point that constitutes one of the foundations of the approach.
For to isolate this fundamental assumption and to establish that it is valid
is precisely what I mean by the computational theory of a process. Establishing the sufficiency of this assumption here is more difficult than establishing the sufficiency of the spatial coincidence assumption that we met
in Chapter 2, because that is a rather simple assumption which follows
quite directly from the structure of the physical world.

However, we can establish validity for a wide range of situations. I
shall try to show here in more general terms how the argument runs,
because the underlying methodological point is so important. We shall
meet it at the heart of the theory of every process.

As formulated, the fundamental assumption of stereopsis contains
phrases like "scene that contains a sufficient amount of detail" and "physically meaningful primitives," which are too imprecise for mathematical
demonstrations. So I will replace the phrase "physically meaningful primitives" by employing the special case of a physical surface that is white with
black dots on it, and the first phrase by specifying the condition that the
density—call it v—of the dots be sufficiently high; specifically, we shall
need v to be at least 2% or so for our demonstration to work. By these
somewhat devious means, analogous to spraying the world with black paint
spots, I have converted the real-world situation into images that bear an
uncanny resemblance to one of Julesz's random-dot stereograms. The
matching conditions now obtain between the two binary images, and when
translated, they read as the following three rules:

Rule 1: *Compatibility.* Black dots can match only black dots.

Rule 2: *Uniqueness.* Almost always, a black dot from one image can
match no more than one black dot from the other image.

Rule 3: *Continuity.* The disparity of the matches varies smoothly
almost everywhere over the image.

Our task now is to prove that these rules force a unique correspondence between the two images, and we can do this in the following way. First, note that because the two eyes lie horizontally, we need consider only all the possible matches along horizontal lines; therefore, we can reduce the problem to the simple one-dimensional case illustrated in Figure 3–6(a). L_x shows all possible positions for dots on the left retina, and R_x for dots on the right retina. The continuous vertical and horizontal lines represent the lines of sight from the left and right eyes, respectively; the dotted diagonal lines, marking traversals at the same rate across the left and right images, therefore represent planes of constant disparity.

Our proof is now easy, at least in conception. Rule 1 tells us to consider only black dots. Rule 3 tells us that, on the whole, the correct matches cluster along or close to these diagonal lines, and Rule 2 tells us that, at each point, only the matches along one of these planes should be chosen. The density of dots in each image is v, so on the correct plane the density of possible matches is v. On the incorrect planes it is only v^2. Hence, provided the disparity changes slowly enough so that the area A spent on each disparity plane is big enough for Av to be significantly different from Av^2, the three rules will yield a unique solution. Hence, since the solution is unique (following the Av matches), it is physically correct, since the physically correct situation will yield one solution. That is the gist of the argument. Of course, this version is somewhat baldly stated, and various subtleties have to be attended to.

The arguments I have given have established two things. First, the fundamental assumption of stereopsis is valid, and this is why the constraints that it incorporates were derived from arguments based on the structure of the physical world. And second, the fundamental assumption provides a sufficient basis for defining the matching process, since a matching that satisfies it is guaranteed to be correct. Furthermore, there will always be such a match in normal physical situations. This completes the computational theory of stereopsis.

Algorithms for stereo matching

A cooperative algorithm

In order to drive home the point that more than one algorithm can be designed to implement a given process, I shall give two algorithms for the stereo matching process. The first one (Marr and Poggio, 1976) follows naturally from the thinking of the last section, and it can be understood most easily from the diagrams in Figure 3–6.

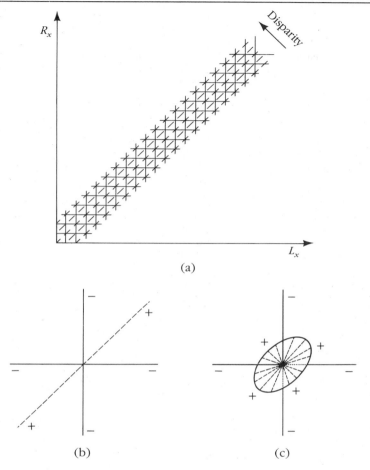

(a)

(b)   (c)

Figure 3–6. In (a), L_x and R_x represent the positions of descriptive elements in the left and right images. The continuous vertical and horizontal lines represent lines of sight from the left and the right eye. The intersections of these lines correspond to possible disparity values. The dotted diagonal lines are lines of constant disparity.

   In the cooperative algorithm described in the text, a cell is placed at each node; then solid lines represent inhibitory interactions, and dotted lines excitatory. The local structure at each node of the network in (a) is given in (b). This algorithm may be extended to two-dimensional images, in which case each node in the corresponding network has the local structure shown in (c). The oval in this figure represents a two-dimensional disc rising out of the plane of the page. (Reprinted by permission from D. Marr and T. Poggio, "Cooperative computation of stereo disparity," *Science 194,* October 15, 1976, 283–287. Copyright 1976 by the American Association for the Advancement of Science.)

As we saw above, Rules 2 and 3 determine the solution to the matching problem. Rule 2 says in effect that only one match is allowed along any of the small vertical or horizontal lines in Figure 3–6(a). Rule 3 says that the correct matches tend to lie along the dotted diagonals.

What we do now is to make a parallel, interconnected network of processors that implements these two rules directly. At each intersection, or node, in Figure 3–6(a) we place a little processor. The idea is that if the node represents a correct match between a pair of black dots, then it should eventually have the value 1. If it represents an incorrect match—a false target, as we called it earlier—then the processor should have the value 0.

We implement the rules by interconnections between the processors. As we saw, Rule 2 tells us that only one match is allowed along each horizontal or vertical line. So, we make all the processors at the nodes along each vertical or horizontal line inhibit each other—the idea being that, in the resulting competition along each line, only one processor will survive to be 1, all the others will be 0, and so Rule 2 will be satisfied. Rule 3 says that correct matches tend to lie along the dotted lines, so we insert excitatory connections between processors in these directions. This gives each local processor the structure shown in Figure 3–6(b). Each such processor sends inhibitory connections to processors along the horizontal and vertical lines shown there, which correspond to the lines of sight from the two eyes, and excitatory connections along the diagonal line, which is the line of constant disparity. We can even extend the algorithm to two-dimensional images, in which case the inhibitory connections remain the same but the excitatory ones cover a small two-dimensional neighborhood of constant disparity. This situation is diagrammed in Figure 3–6(c).

The idea now is to load the network of processors by taking the two images and putting a 1 wherever two black dots could match—false targets and all—and a 0 at all other places. Then we let the network run. Each processor adds up the 1's in its excitatory neighborhood, adds up the 1's in its inhibitory neighborhoods, and subtracts the resulting figures (after multiplying one of the sums with a suitable weighting factor). If the result exceeds a certain threshold, the processor takes the value 1; if it does not, the processor is set to 0. Formally, this algorithm can be represented by the iterative relation

$$C_{x,y;d}^{t+1} = \sigma \left\{ \sum_{x',y';d' \in S(x,y;d)} C_{x',y';d'}^{t} - \varepsilon \sum_{x',y',d' \in O(x,y;d)} C_{x',y';d'}^{t} + C_{x,y;d}^{0} \right\}$$

where $C_{x,y;d}^{t}$ denotes the state of the cell corresponding to position (x,y), disparity d, and time t in the network of Figure 3–6(a); $S(x,y,d)$ is the local excitatory neighborhood, and $O(x,y,d)$ the inhibitory neighborhood. The Greek letter ε is an inhibition constant, and σ is a threshold function. The initial state C^0 contains all possible matches, including false targets, within the prescribed disparity range; here it is added at each iteration. (It does not have to be, but the algorithm converges faster if it is.) Notice how Rules 2 and 3 are implemented through the geometry of the inhibitory and excitatory neighborhoods O and S.

This algorithm successfully solves random-dot stereograms, and an example is shown in Figure 3–7 of how the network gradually organizes itself into the correct solution. The stereograms themselves are labeled Left and Right, the initial state of the network as 0, and the state after n iterations is marked as such. To understand how the figures represent states of the network, imagine looking at the network from above—that is, from the direction of the top of Figure 3–6. The different disparity layers in the network lie in parallel planes, so that the viewer is looking down through them. In each plane, some nodes are on and some are off. Each of the seven layers in the network has been assigned a different gray level, so that a node that is switched on in the top layer (corresponding to a disparity of $+3$ pixels) contributes a dark point to the image, and one that is switched on in the lowest layer (disparity of -3) contributes a light point. Initially (iteration 0) the network is disorganized, but in the final state the order has stabilized (iteration 14), and the inverted wedding-cake structure has been found. The dot density of this stereogram is 50%.

The algorithm defined by the iterative relation above with the parameter values used for the example of Figure 3–7 is capable of solving random-dot stereograms with dot densities from 50% down to less than 10%. For this and smaller densities, the algorithm converges increasingly slowly. If a simple homeostatic mechanism is allowed to control the threshold σ as a function of the average activity (number of on cells) at each iteration, the algorithm can solve stereograms whose density is very low. In the second example, Figure 3–8, the density is 5% and the central square has a disparity of -2 pixels relative to the background. The algorithm fills in those areas where no dots are present, but it takes several more iterations to arrive near the solution than in cases where the density is 50%. When we look at a sparse stereogram, we perceive its shapes as being cleaner than the shapes found by the algorithm. This seems to be due to subjective contours that arise between dots that lie on shape boundaries.

We can see intuitively how the algorithm works from these examples. It never seems to have any trouble with stereograms, but this alone is not sufficient evidence for placing confidence in it. We did, however, manage to make it intellectually respectable; in a mathematical analysis of the algo-

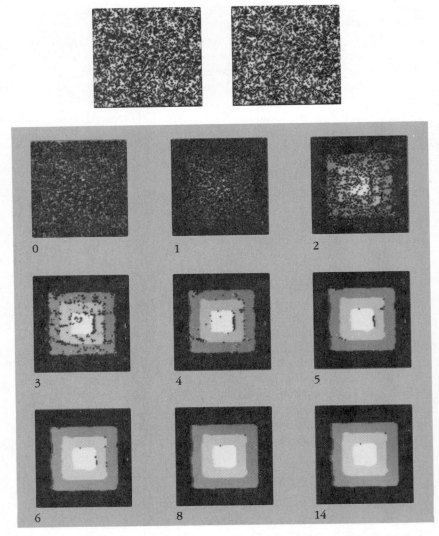

Figure 3–7. The decoding of a random-dot stereogram pair by the cooperative algorithm described in the text. The stereogram appears at the top, and the initial state of the network, which includes all possible matches within the prescribed disparity range, is labeled 0. The algorithm runs through a number of iterations, as shown, and gradually the structure is revealed. The different shades of gray represent different disparity values.

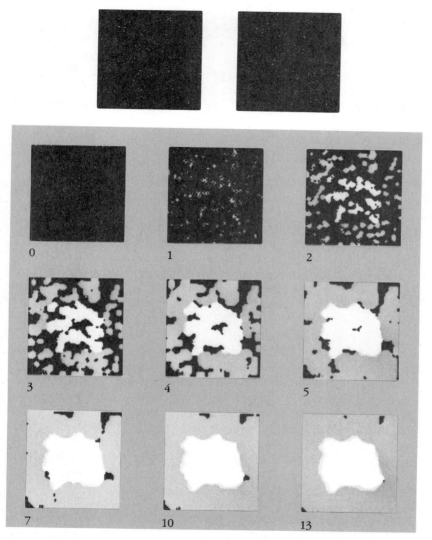

Figure 3–8.   The algorithm used in Figure 3–7 can also decode and fill in very sparse stereograms. This one has a density of 5%.

rithm (Marr, Palm, and Poggio, 1978), we demonstrated that states obeying Rules 2 and 3 were stable states of the algorithm, and we showed that the algorithm converges for a wide range of parameter values.

This is an example of a cooperative algorithm, so-called because of the way in which local operations appear to cooperate in forming global order in a well-regulated manner. Cooperative phenomena are well-known in physics; for example, the Ising model of ferromagnetism, superconductivity, and phase transitions in general. Cooperative algorithms have many characteristics in common with these phenomena.

Cooperative algorithms and the stereo matching problem

Until 1977, almost all of the stereo algorithms put forward as models for human stereopsis were based on Julesz's proposal that stereo matching is a cooperative process (Julesz, 1971, pp. 203ff.; Julesz and Chang, 1976; Nelson, 1975; Dev, 1975; Hirai and Fukushima, 1976; Sugie and Suwa, 1977; Marr and Poggio, 1976). The two exceptions were Julesz's (1963) AUTOMAP program, which used an approach based on cluster-seeking, and Sperling's (1970) model, which is based on gray-level correlations but does make an interesting point of the connection between stereopsis and vergence movements.

There is a rather fascinating moral that one can draw from these attempts: Apart from our own, which was based on the computational approach, not one of these algorithms was accompanied by an analysis of the underlying computational theory of the stereo matching problem. As a direct consequence, not one of them computed the right thing—at least one of the constraints in the fundamental assumption of stereopsis was either missing or incorrectly implemented. Sperling's model was based on gray-level correlation—which, as we have seen, is incorrect—and because this model was not implemented, he failed to specify the area and disposition of the neighborhoods over which the correlation is taken. It is in trying to do this that one comes up against the problems.

Dev's algorithm deserves credit for being one of the first precise attempts to embody Julesz's ideas (Dev, 1975, eqs. 1 and 2). The algorithm realizes Rule 3 but employs an incorrect version of Rule 2. Instead of two lines of inhibition, one down each line of sight, she has one that bisects the angle between the lines of sight. This algorithm, illustrated in Figure 3–9, should be contrasted with the geometry of Figure 3–6. Physically, the connections in Figure 3–9 correspond to something like the rule that any direction out from the viewer meets only one surface. This is not true in general; for example, when one looks into a shallow lake, one sees two surfaces, the lake surface and its bottom. The correct version, shown in

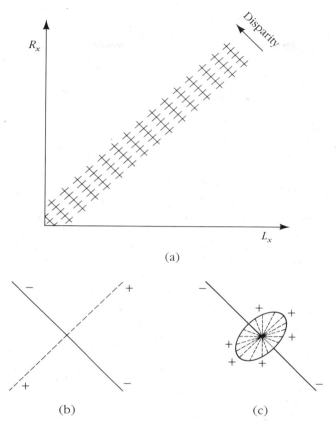

(a)

(b)   (c)

Figure 3–9.   Several of the cooperative stereo algorithms that have been proposed include just one set of inhibitory connections between detectors of different disparities at the same retinal position. If we represent these connections in the same way as in Figure 3–6, it becomes obvious that they implement slightly different constraints. Instead of forbidding double matches down each line of sight, as was the case in Figure 3–6, these connections forbid double matches along the radial out from the viewer. It is incorrect to formulate the stereo correspondence process in this way.

Figure 3–6, says that any particular visible marking will lie either on the lake's surface or on the bottom (or perhaps on a fish swimming by), but only on one of these.

Sugie and Suwa's (1977) algorithm implements only a part of Rule 3 and the same, incorrect version of Rule 2. Nelson (1975) gave no precise algorithm, nor did he implement any form of his ideas, but he also seems to mean an algorithm that implements the wrong form of Rule 2. Hirai and Fukushima (1976) correctly implemented Rule 2 (p. 48, function [1]) but did not implement Rule 3, preferring instead a network that favored solutions with lower parallax.

Julesz's (1963) AUTOMAP fails to implement Rule 2 but implements Rule 3 implicitly in the way it detects clusters. Julesz's dipole model is more interesting. It is defined as a mechanical analogy in which the left and right stereo images are each represented by a network of compass needles (magnetic dipoles), one for each image marking to be matched. The needles are oriented so that they can point to nearby locations in the opposite image's network when the two networks are overlayed. The endpoints of neighboring needles on each side are coupled together by springs and the polarity of each needle (north or south) is chosen according to the intensity of the image (black or white) at that location. The idea is that when the left and right networks are overlayed in rough registration, the magnetic attraction between similarly arranged groups of needles on either side will cause the network to settle into a stable state with the needles pointing towards their correct matches on the other side. While the relation between the polarity of the magnets and the retinal intensity values is unclear except for random-dot stereograms, the dipole model implicitly implements uniqueness, Rule 2, because a given dipole can have only one orientation at a time. Spring coupling between the tips of adjacent dipoles implements the continuity of Rule 3. This model therefore comes the closest to meeting our requirements, but it has the interesting feature that, unlike the other cooperative models, it does not represent explicitly all possible nodes in the diagram of Figure 3–6(a). That is, there is really only one processor for each vertical or horizontal line in that diagram, the different nodes along them being represented by different angular positions of a single dipole. It would be interesting to see whether such a model could be made to work.

The reason for elaborating upon this point is simply to help my overall argument that intellectual precision of approach is of crucial importance in studying the computational abilities of the visual system. Unless the computational theory of a process is correctly formulated, the algorithm will almost certainly be wrong.

Finally, none of these algorithms has been shown to work on natural images. Gray-level correlation works some of the time, but it makes mistakes that a human operator has to correct. The other proposals make no specific suggestions about what their input representations should be, although Marr and Poggio (1976) suggested that the primal sketch is suitable.

Biological evidence

All of these algorithms are designed to select correct matches in a situation where false targets occur in profusion. Consequently, apart from early versions of Julesz's dipole model perhaps, they do not critically rely upon eye movements, since in principle they have the ability to interpret a random-dot stereogram without them. However, eye movements seem to be important for human stereo vision. Without them, in fact, one can see very little depth—the range over which one can fuse two images (called Panum's fusional area) is small, about 6′–18′ of arc (Fender and Julesz, 1967; Julesz and Chang, 1976)—and almost no structure can be perceived (Richards, 1977), except for small disparities (Mayhew and Frisby, 1979). For complex stereograms such as Julesz's spiral (1971, fig. 4.5–4), eye movements are probably essential (Frisby and Clatworthy, 1975; Saye and Frisby, 1975). In fact, in view of Fender and Julesz's early findings, it is quite surprising that so little psychophysical attention has been given to eye movements until very recently.

There are several other psychophysical phenonema that would be difficult to explain in terms of the type of algorithms we have been discussing. Some subjects, for example, can tolerate a 15% expansion of one image (Julesz, 1971, fig. 2.8–8). If one severely defocuses one of the pair in a stereogram, fusion is easy to obtain (Julesz, 1971, fig. 3.10–3). This is only the most striking demonstration of a phenomenon that can be shown in several other ways. In fact, one can simultaneously experience both binocular rivalry and fusion of different spectral components in a stereogram, as the reader may experience in Figure 3–10 (Kaufman, 1964; Julesz, 1971, sec. 3.9 and 3.10; Julesz and Miller, 1975; Mayhew and Frisby, 1976). Such findings raise the interesting possibility that disparity information is conveyed at some stage by independent stereopsis channels that are tuned to different frequencies and are roughly one and a half octaves wide—very reminiscent, in fact, of the different-sized $\nabla^2 G$ operators that we met in Chapter 2.

Other interesting findings are the physiological, clinical, and psychophysical evidence about Richards' two-pools hypothesis (Richards, 1970, 1971; Richards and Regan, 1973; Poggio and Fischer, 1978; Clarke, Donaldson, and Whitteridge, 1976). Richards' basic finding was that stereo blindness manifests itself as a blindness to all convergent disparities, all divergent disparities, or both—and some kind of stereo incapacity, incidentally, is extraordinarily common, having an incidence of about 30%. In other words, stereo detectors seem to be organized into two pools, one dealing with convergent and the other with divergent disparities, with perhaps a third pool dealing with zero disparity. The neurophysiologists report some-

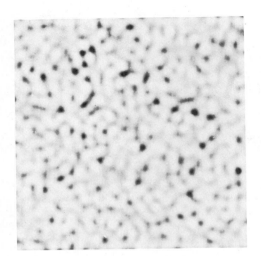

Figure 3–10. The high-frequency spectral components of this stereogram are rivalrous, yet the low-frequency components are not and can be fused. This suggests that independent spatial-frequency-tuned channels are involved in stereopsis. (Reprinted, by permission, from B. Julesz and J. E. Miller, "Independent spatial-frequency-tuned channels in binocular fusion and rivalry," *Perception 4,* 1975, 125–143, fig. 6.)

thing similar—roughly three classes of disparity-tuned neurons, one class broadly tuned to convergent (the so-called near neurons), and another broadly turned to divergent (far neurons), and a third sharply tuned to near-zero disparities. This goes against what one would expect of a neural implementation of the algorithms I discussed above, since, apart from the dipole model, all require many "disparity-detecting" neurons, whose peak sensitivities cover a range of disparity values that is much wider than the tuning curves of the individual neurons.

Finally, a remark about the motivation for the cooperative algorithm approach. As I have mentioned, these ideas were all inspired by Fender and Julesz's (1967) exhibition of hysteresis in stereopsis. In their experiment, they stabilized the images against eye movements and showed that once fusion was achieved, the two images could be "pulled" apart by up to about 2° of disparity before fusion "broke." However, once fusion had broken, the images had to be brought back to the 6′–14′ range before they would refuse. Hysteresis is one property of cooperative algorithms, and so is filling-in, which also seems to occur in stereopsis—as the reader has already seen, sparse stereograms like Figure 3–8 give the appearance of a smooth, solid surface, not of a few dots hanging isolated in space. Hence everybody, including Julesz and ourselves, searched for a cooperative algorithm.

But not very sensibly. After all, the critical point of the Fender and Julesz experiment was that the hysteresis occurred over 2° of disparity, whereas matching only occurred under 20′. It therefore seems unlikely that the hysteresis is a consequence of the matching process, and much more likely that it is due to a cortical memory that stores the results of the matching process but is distinct from it. Fender and Julesz even suggested such a thing. Of course, this does not forbid the presence of cooperativity in the matching process, and the so-called pulling effect, described later by Julesz and Chang (1976), is probably evidence for its existence; however, the lesson is that we should probably deemphasize our ideas about cooperative processes and look instead for a rather different approach to the problem of stereopsis.

A second algorithm

The basic problem to be overcome in binocular fusion is the elimination or avoidance of false targets, and its difficulty is determined by two factors: the abundance of matchable features in an image and the disparity range over which matches are sought. If a feature occurs only rarely in an image, the search for a match can cover quite a large disparity range before false targets are encountered, but if the feature is a common one or the criteria for a match are loose, false targets can occur within quite small disparities.

For a given disparity range, then, if we want to simplify the matching problem, we have to decrease the incidence of matchable feature pairs; that is, we have to make features rare. There are two ways to do this. One way is to make them quite complex or specific, so that even if their density in the image is high, there would be so many different kinds that there would seldom be a compatible pair. The other way is to reduce drastically the density of all features in the image, for example, by decreasing the spatial resolution at which it is examined.

We know from Julesz's work on random-dot stereograms that the prospects for the first approach are rather slim. We know that the matching is carried out locally, yet all the edges are exactly vertical or horizontal and all have the same contrast, so even forcing very specific criteria onto them would not help us much. Furthermore, doing so would severely impair performance on real images, for which the orientations and contrasts of two corresponding edges can differ by surprising amounts. The reader can see for himself that stereograms with different contrasts can be fused by

Figure 3–11.   The left and right images have different contrasts, yet fusion is still possible.

looking at Figure 3–11. The contrasts must, however, have the same sign. The criteria for orientation are also quite lax.

However, the other possibility is more promising. Indeed, the existence of independent spatial-frequency-tuned channels in binocular fusion now acquires a new and special interest, because it suggests that several copies of the image, obtained by successively finer filtering, are used during fusion, providing increasing and, at the limit, very fine disparity resolution at the cost of decreasing disparity range.

A notable feature of a system organized along these lines would be its reliance on eye movements for building up a comprehensive and accurate disparity map from two viewpoints. The reason for this is that the most precise disparity values are obtained from the high-resolution channels, and eye movements are therefore essential so that each part of a scene can ultimately be brought into the small disparity range within which high-resolution channels operate. The importance of vergence eye movements is also attractive in view of the extreme precision with which they may be controlled (Riggs and Niehl, 1960; Rashbass and Westheimer, 1961a).

These observations suggest the following scheme for solving the fusion problem: (1) Each image is analyzed through channels of varying coarseness and matching takes place between corresponding channels from the two eyes for disparity values of the order of the channel resolution; (2) coarse channels control vergence movements, thus causing fine channels to come into correspondence.

This scheme contains no hysteresis and therefore does not account for the observations of Fender and Julesz (1967). According to our emerging theory of intermediate visual information processing, however, a key

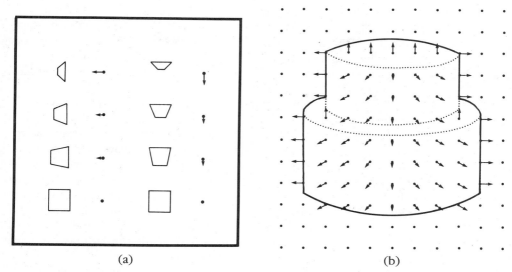

Figure 3–12. Illustration of the 2½-dimensional sketch. In (a), the perspective views of small squares placed at various orientations to the viewer are shown. The dots with arrows symbolically represent the orientations of such surfaces. In (b), this symbolic representation is used to show the surface orientations of two cylindrical surfaces in front of a background orthogonal to the viewer. The full 2½-dimensional sketch would include rough distances to the surfaces as well as their orientations; contours where surface orientations change sharply, which are shown dotted; and contours where depth is discontinuous (subjective contours), which are shown with full lines. See Chapter 4 for more details. (D. Marr and H. K. Nishihara, 1978.)

goal of early visual processing is the construction of something like an orientation-and-depth map of the visible surfaces around a viewer (see Chapter 4). In this map, information is combined from a number of different and probably independent processes that interpret disparity, motion, shading, texture, and contour information. These ideas are illustrated by the representation shown in Figure 3–12, which Marr and Nishihara (1978) called the 2½-D sketch.

Suppose now that the hysteresis that Fender and Julesz observed was not due to a cooperative process during matching but was in fact the result of using a memory buffer, like the 2½-D sketch, for storing the depth map of the image as it is discovered. Then the matching process itself need not be cooperative (even if it still could be): it would not even be necessary for the whole image ever to be matched simultaneously, provided that a depth map of the viewed surface was built and maintained in this intermediate memory.

Our scheme can now be completed by adding to it the following two steps: (3) When a correspondence is achieved, it is held and written down in the 2½-D sketch; (4) there is a reverse relation between the memory and the channels, acting through the control of eye movements, that allows one to fuse any piece of surface easily once its depth map has been established in the memory.

The idea of matching coarse, widely separated features first, and then with the information so obtained, repeating the matching process at successively finer scales of resolution sounds promising, but what features should we match at these different resolutions? We have seen enough of early visual processing to suggest various possibilities. Are they zero-crossings, the raw primal sketch, the full primal sketch, or some combination of them all? Poggio and I proposed that the input representation for the stereo matching process consists of the raw zero-crossings, labeled by the sign of their contrast change and their rough orientation in the image, and of terminations—local discontinuities—also labeled by contrast and perhaps very rough orientation.

The matching process. The choice of input representation leads to the matching algorithm illustrated in Figures 3–13 and 3–14. These figures show Eric Grimson's computer implementation of the algorithm running on a pair of random-dot stereograms, which represent one of the most difficult kinds of input for the algorithm.

The left and right images, forming a random-dot stereogram with density 50%, appear at the top of Figure 3–13. The first step in the algorithm is to apply a large $\nabla^2 G$ filter to each image and obtain the zero-crossings, just as we did in Chapter 2. Although in theory the elements to be matched between images include both zero-crossings and terminations, it is only the zero-crossings that cause difficulties with false targets. Thus Figure 3–14 shows only the zero-crossings and in fact horizontal segments are ignored, since they cannot be easily matched.

In addition to their locations, the zero-crossings have been given a sign and a rough orientation. The sign corresponds to the sign of the contrast change from left to right across the zero-crossing, and it is indicated by the shade of the zero-crossing in the figure. Two zero-crossings are matchable if they have the same sign and their local orientations are within 30° of each other. Matching itself is carried out point by point along the zero-crossings.

The convolution values and signed zero-crossings for three sizes of the $\nabla^2 G$ filter appear in Figure 3–14. The reader can see that far more zero-crossings are obtained from the smallest channel than from the largest, which means that the disparity range considered can be greater for the larger channels without any increase in the incidence of false targets.

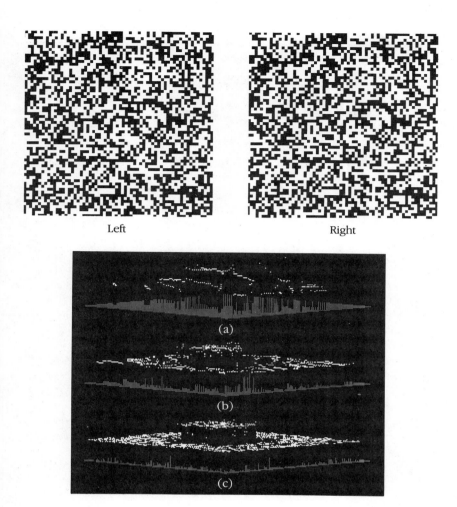

Left   Right

(a)

(b)

(c)

Figure 3–13.   The solution of a 50% random-dot pattern. The left and right images
are shown at the top. The three lower figures indicate an orthographic view of the
disparity maps obtained by matching the zero-crossing descriptions of Figure 3–14.
A point in the image with coordinates (x,y) and an assigned disparity value of d is
portrayed in this three-dimensional system as the point (x,y,d). Here the heights of
the bright points above the plane indicate their disparity values.

In general terms, then, the overall structure of the algorithm is clear
from Figures 3–13 and 3–14. First, the coarse images are matched; the
results of this are illustrated in Figure 3–13(a), which shows an orthographic view of the resulting disparity map. This rough result is used as
the starting point for the same matching process applied to the medium-
sized channel. The decrease in the allowed disparity range is offset by the
knowledge, obtained from the large channel, of its approximate value. This

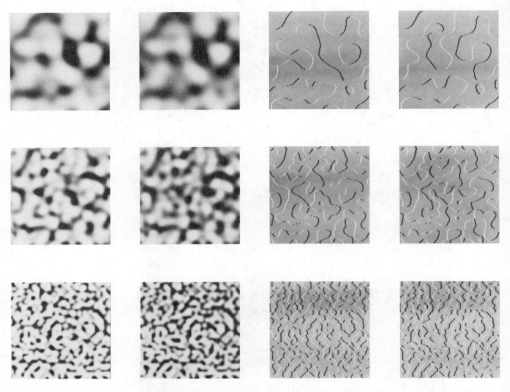

Figure 3–14. The convolutions and zero-crossings involved in solving the stereogram of Figure 3–13. The two left columns indicate the convolutions of the left and right images with masks of size $w_{2-D} = 35$, 17, and 9, respectively, from top to bottom. The two right columns indicate the zero-crossings obtained from the convolutions in the left two columns. Notice how much more detail the smaller masks reveal.

gives the disparity map shown in Figure 3–13(b). Second, the smallest channel is considered, yielding the accurate disparities made possible by its small disparity range, and the results appear in Figure 3–13(c). In this example, the central square has a disparity of 12 pixels, and each black square is 4 × 4 pixels. In the final disparity map, less than 0.1% of the points are incorrectly matched, and these all occur at the borders of the square.

More properties of zero-crossings. In this algorithm, the false target problem is solved essentially by evasion, but exactly how it is solved is

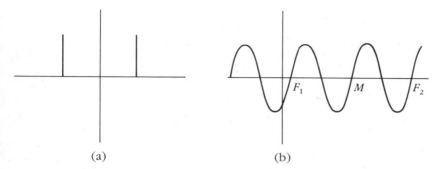

(a)   (b)

Figure 3–15.   The positive (or negative) zero-crossings of a pure sine wave are guaranteed to be λ apart, where λ is the wavelength. See discussion in text.

interesting and, from the point of view of psychophysics, quite important. I shall not give the proofs here, but the general argument can be conveyed without much technical detail.

The central idea is illustrated in Figure 3–15. Suppose, for sake of argument, that the intensity variation in the image was purely sinusoidal, consisting solely of a vertically oriented sinusoidal grating. Such a signal has the Fourier transform shown in Figure 3–15(a) and passes unscathed through $\nabla^2 G$, giving the same curve shown in one-dimensional cross-section in Figure 3–15(b). Now the problem is to match the zero-crossings between the two filtered images, so let us suppose that we have fixed on a particular positive-going zero-crossing from the left image whose true match is the one marked M in Figure 3–15(b). Then F_1 and F_2 are false targets. But since they also have to be positive-going zero-crossings, they must be at least a distance λ away, where λ is the wavelength of the sinusoid. Hence, provided that we restrict our search for possible matches to a disparity range of at most λ, we are guaranteed to find only one possible match, and provided that we know by some other means roughly where to carry out the search, we can be sure that the one match we find will be correct.

That is the basic idea, but the real world is not restricted to pure sine wave gratings. A sine wave is, however, only the extreme case of a band-pass function, in which the bandwidth is zero. The same qualitative argument holds for wider bandwidths, and this can be seen roughly from Figures 2–19 and 3–16. For example, consider the case of an ideal one-octave band-pass filter of the type whose Fourier transform appears in Figure 2–19(b). A portion of a typical signal from such a filter is illustrated in Figure 2–19(c). The average value of this signal is zero, so the signal

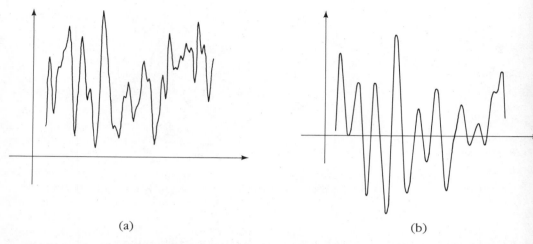

(a)   (b)

Figure 3–16.   The signal in (a) varies randomly in the range 0 to 100. After being passed through the filter $\nabla^2 G$, it has the appearance shown in (b), with more or less regularly occurring zero-crossings. A similar example is given in Figure 2–19 for a pure one-octave band-pass filter. For general band-pass signals, like those passed by $\nabla^2 G$ or a pure one-octave filter, the zero-crossings cannot on average occur too closely together or too far apart. The intervals between zero-crossings are governed by the statistical rules illustrated in Figure 3–17.

crosses zero quite frequently, like the sine wave. However, because it is a band-pass signal, its zero-crossings cannot occur too far apart. On average, they occur at the frequency corresponding to the middle of the filter's range.

The important point for us is that zero-crossings cannot on average occur too close together, and this is true for any band-pass filter. The filter $\nabla^2 G$, however, is also roughly a band-pass filter—the reader may care to look once again at its one-dimensional Fourier transform, shown in Figure 2–9(c). The results of passing a random one-dimensional signal (Figure 3–16a) through $\nabla^2 G$ are shown in Figure 3–16(b), and the reader can see that it has the same qualitative features as Figure 3–15, its average value is zero, and the zero-crossings lie neither very close to nor very distant from their neighbors.

The general lines of the argument are now quite straightforward, and they are the same as those of the argument for the sine wave. Since $\nabla^2 G$ is roughly a band-pass filter, its zero-crossings are usually separated by some minimum distance. Provided we know approximately where to look for a match, and provided we do not search over too large a range, we shall find a unique candidate for the match and it will be correct.

This shows us a promising approach to the matching problem, but it also raises another rather exciting possibility. From the point of view of psychophysics, $\nabla^2 G$ is monocular, but matching is binocular. That is, the parameters of the $\nabla^2 G$ filters—their widths w_{1-D}, for instance—are obtained by purely monocular measurements. The disparity range for matching, usually called Panum's fusional area and which I shall denote by ∇, is essentially a binocular phenomenon. If our theory is true, it will predict a clear and unexpected relationship between these a priori unrelated quantities, which are measured in completely different ways. This will therefore provide an excellent way of testing the theory.

It is therefore important to derive the precise quantitative relationship that we expect should hold between w_{1-D} and ∇. In order to do this, we need a quantitative model for the channels used in early processing and some way of estimating the probable distances between zero-crossings. The idea of using zero-crossings, it should perhaps be said, came from early work on the primal sketch (Marr, 1976) in which many of the cells early in the visual pathway were thought of not as feature detectors but as differential operators. Hubel and Wiesel's (1962) definition of a cortical simple cell as linear led us to think, for example, of a bar-shaped receptive field as an oriented second-derivative operator from which one subsequently found zero-crossings. Only later did we come to realize that the simple cells themselves are probably the zero-crossing detectors, as in Figure 2–18 (see also Section 3.4). This slight confusion does not matter from a mathematical point of view, because under only very weak assumptions the two points of view are equivalent (see Marr and Hildreth, 1980, app. A). With respect to their implementation and consequently to psychophysics, the two things are rather different. I shall return to this point later on.

For our analysis, then, we need a quantitative hold on the distances between zero-crossings for the filters that the visual system actually uses. At the time the present stereo theory based on matching at different scales of resolution was formulated, we did not know that $\nabla^2 G$ was the optimal filter to use, but we knew something just as good, because Hugh Wilson at Chicago had just formulated his four-mechanism model for the structure of the channels. He described their structure using DOG's—differences of Guassians—which are almost indistinguishable from $\nabla^2 G$, as we saw in Figure 2–16.

We were also very lucky with the mathematics of the problem because obtaining estimates of the probable distances between the zero-crossings of band-pass signals turns out to be very difficult. Various mathematicians had already worked on it, starting with Rice in 1945 and more recently M. Longuet-Higgins (1962) and Leadbetter (1969). The problem itself is interesting, because it relates to a number of physical phenomena, some quite important and some less so. The important ones include the effects of Brownian noise due to the random motion of electrons in electrical circuits—and some amplifiers, for example, switch as the voltage crosses zero—and the analysis of the distribution of wave heights in the sea, which is of particular interest now that people are trying to tap this source of energy. On a more frivolous note, the same type of mathematics is involved in the study of twinkles, which are the places in the sea that happen to reflect the sun back into your eyes, causing the surface to glitter and, well, twinkle.

We can therefore analyze the spatial distribution of zero-crossings, at least for one-dimensional band-pass signals. The results are illustrated in Figure 3–17 for two cases: first, the example shown in Figure 2–19 of a pure one-octave band-pass filter (left column), and second, the case, illustrated in Figure 3–16, of a $\nabla^2 G$ filter which closely approximates the filters that Wilson concluded are present in the early stages of the human visual system (right column).

The legend explains the details, but the important graphs are the two in Figure 3–17(c). They show the probability given a zero-crossing at the origin, of encountering another zero-crossing of the same sign at distance ξ away. The units in which ξ are plotted, in the biologically interesting case of the right-hand column, are such that w_{1-D} has the value 2.8. Two values of this probability are worth remembering: at distance w_{1-D} it is about 5%,

---

Figure 3–17. (opposite)   Interval distributions for zero-crossings. A "white" Gaussian random process is passed through a filter with the frequency characteristic (transfer function) shown in (a). The approximate interval distribution for the first (P_0) and second (P_1) zero-crossings of the resulting zero-mean Gaussian process is shown in (b). Given a positive zero-crossing at the origin, the probability of having another within a distance ξ is approximated by the integral of P_1 and shown in (c). In the left column, these quantities are given for an ideal band-pass filter one octave wide and with center frequency $\omega = 2\pi/\lambda$; in the right column, these quantities are given for the case of the receptive field described by Wilson and Giese (1977). The ratio of space constants of excitation and inhibition is 1:1.5. The width w of the central excitatory portion of the receptive field is 2.8 in the units in which ξ is plotted. For the case portrayed in the left column, a probability level of $\int P_1 = 0.001$ occurs at $\xi = 2.3$ and a probability level of 0.5 occurs at $\xi = 6.1$. The corresponding figures for the case illustrated in the right column are $\xi = 1.5$ and $\xi = 5.4$. If the space-constant ratio is 1:1.75, the values of $\int P_1$ change by not more than 5%. (Reprinted by permission from D. Marr and T. Poggio, "A computational theory of human stereo vision," *Proc. R. Soc. Lond. B 204,* 301–328.)

Ideal one-octave band-pass filter

Wilson–Giese receptive field

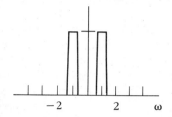
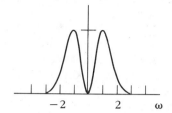

(a)

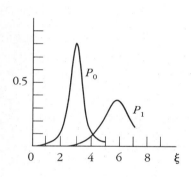
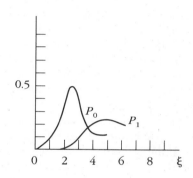

(b)

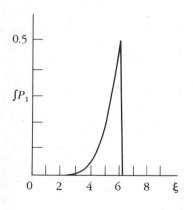
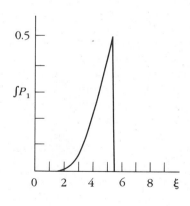

(c)

and at distance $2w_{1-D}$ it is about 50% and increasing rapidly. Moderate changes in the shape of the underlying filter do not change these numbers by very much.

The matching algorithm. Given this background, we can now formulate the matching algorithm and prove that it will work. Let us first examine a simple case, in which false targets are essentially avoided. It is best explained by looking at Figure 3–18(a). Here we have a zero-crossing from the left image, marked L, which matches another of the same sign in the right image, which is displaced by a disparity of amount d. The correct match is labeled R, and a possible false target F, shown dotted, is shown lurking nearby. Provided that we consider only the disparity range $w/2$, however, we are safe, because even if R is right at one end of the range— if $d = w/2$, for example—our statistical analysis assures us that, with a probability of 95%, it will be the only zero-crossing of its type within a disparity range that extends over w. Even if we ignore all cases in which two candidates are present, we shall still succeed over 95% of the time.

This assumes, of course, that R *is* the correct match, that is, that the correct match lies in the range of $w/2$ that the procedure examines. However, we can tell when the correct match does not lie in this range, because if the visible surface has disparity in this range, almost all zero-crossings from the left image will find matches in the right image, and all of them will find at least one candidate match. If the surface has a disparity lying outside this range, then the probability that a zero-crossing from the left image will find a candidate match within range from the right image is, for all intents and purposes, simply the probability that a zero-crossing of the appropriate sign falls by chance within the particular spatial interval $w/2$ in the right image. This probability is about 40%. Hence, if the surface lies outside the disparity range, only 40% of the matches will be achieved versus nearly 100% if the surface falls within this range. It is therefore easy to tell when the matching process is succeeding. And notice, incidentally, that we rely on the third constraint, continuity, of our fundamental assumption, since it is assumed that we can look over a neighborhood in the image that is large enough to enable us to measure the difference empirically between a situation with a 40% probability of matches and one of, say, 95%. Such a neighborhood does not have to be very large, but it has to exist, and this is why we need the continuity assumption.

Now that this simple algorithm has given us the basic idea, we can improve on it, and by doing so increase the allowed disparity range from $w/2$ to w. Figure 3–18(b) shows our zero-crossing L in the left image, but this time its match R in the right image has a disparity d that can be as much as w. The first point to note is that if d is positive, then by the same

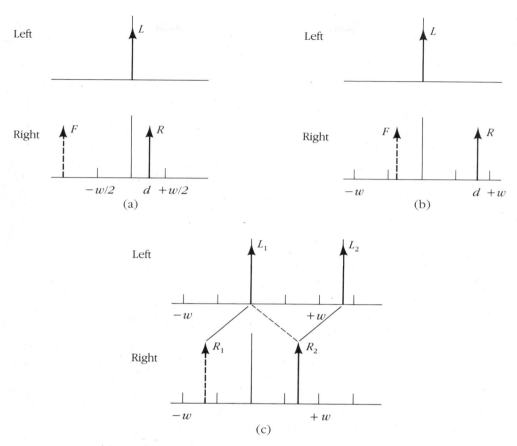

Figure 3–18.   The matching process driven from the left image. A zero-crossing L in the left image matches one R displaced by disparity d in the right image. The probability of a false target within w of R is small, so provided that $d < w/2$ (a), almost no false targets will arise in the disparity range $w/2$. This gives the first possible algorithm. Alternatively, all matches within the range w may be considered (b). Here false targets, designated F, can arise in about 50% of the cases, but the correct solution is also present. If the correct match is convergent, the false target will with high probability be divergent. Therefore, in the second algorithm, unique matches from either image are accepted as correct, and the remainder as ambiguous and subject to the pulling effect, illustrated in (c). Here L_1 could match R_1 or R_2, but L_2 can match only R_2. Because of this and because the two matches have the same disparity, L_1 is assigned to R_1. (Reprinted by permission from D. Marr and T. Poggio, "A computational theory of human stereo vision," *Proc. R. Soc. Lond. B 204*, 301–328.)

arguments as before, R is at least 95% certain to be the only candidate in the disparity range 0 to w. Second, we know from our statistics that the likelihood of a false target in the $2w$ disparity range from $d = -w$ to $d = +w$ is at most 50%, even when the correct match lies at one extreme of this range. Putting these two facts together, we see that at least 50% of the time the match will be unambiguous and correct and that remaining cases will be ambiguous, consisting mainly of two alternatives, one convergent (in the range $(0,w)$) and one divergent (in the range $(-w,0)$), one of which will be correct. In the ambiguous cases, selection of the correct alternative can be based simply on the sign of neighboring matches (note the use of continuity here). Notice, incidentally, that if a match is near zero disparity, it is likely ($p > 0.9$) to be the only candidate, again according to the statistics of the situation. Hence the notion of three disparity ranges— one convergent, one divergent, and one around zero—follows naturally from this matching technique.

Once again, if the surface lies within the disparity range, nearly 100% of the zero-crossings will find matches; if it does not, the figure in this instance is 70% instead of 40%, but this is still different enough from 100% to enable us to tell when matching is succeeding.

We cannot improve much on the range w without resorting to more powerful techniques for removing false targets, because the probability of false targets occurring increases quite sharply above the range $2w$. The percentage of unambiguous matches, for example, is already down to 20% at $1.5w$.

Uniqueness, cooperativity, and the pulling effect

Eric Grimson (1981) made the important point that matching can be carried out from either image or from both images. In Figure 3–18(c), for example, if matching is initiated from the left image, the match for L_1 is ambiguous, but for L_2 it is unique. From the right image, matching is unique for R_1 but ambiguous for R_2. Together the two unique matches provide the correct solution.

That the two unique matches should be correct rather than contradictory is a consequence of the uniqueness property embedded in the fundamental assumption of stereopsis. As a result, the algorithm can be designed to accept unambiguous matchings by starting from either image. However, this design does have some fascinating consequences, for it means that the uniqueness assumption is no longer internally verifiable by the algorithm, whereas the continuity assumption is.

This fact is determined in the following way. We have already seen that the algorithm needs to check the proportion of local candidates that are matched in order to tell whether the surface lies within the disparity range under examination. If the proportion is near 100%, everything is satisfactory. If it is not (in which case it is probably 70%), the solution is rejected. It is extremely difficult to fool this test, and since it relies on continuity for its validity, it amounts to an internal check that continuity is being locally satisfied by the visible surfaces.

Not so uniqueness. If the algorithm accepts unique solutions from either image, this allows it to fuse patterns such as the Panum's limiting case example (Figure 3–19) not only for rare occurrences across an image, as in Figure 3–19(a), but also for frequent ones. Oliver Braddick investigated this point by constructing stereograms like Figure 3–19(b), in which each dot from the right image matches two from the left. Matching initiated from the left image is unique, so one accepts it, and the resulting percept is of two planes, one behind the other. The visual system is not particular about which eye it operates from, and one can mix the doublets up so that some of them are in the right image, and some are in the left. It makes no difference.

Physically, of course, this situation is effectively impossible to produce with two real surfaces, which is perhaps why we have not evolved an internal check for uniqueness. The general point here is an interesting one, though; some assumptions can be and are checked internally, like continuity here; some could be but are not, like uniqueness; and some cannot be even in principle. We shall meet some examples of this later on, but it may be worth mentioning here that the Ames room illusion may be one. Without stereopsis or motion cues, the assumptions of right angles cannot be tested internally.

Finally, there are situations in which matching is ambiguous from both eyes. In this case, the ambiguity can be resolved by consulting the signs of the neighboring matches and choosing the matches with the same sign. There is, however, an important distinction between the two most obvious ways of doing this. Either we consult the signs of the neighboring matches that were unambiguous from the start, or we consult the signs of the neighboring matches that have so far been assigned. The second scheme introduces cooperativity, the first does not.

To see this, imagine a stereogram cleverly constructed so that every match is ambiguous except for an unambiguous region located, for example, at the border. With the first scheme, none of the matches in the interior region of the stereogram will ever be disambiguated, because there are never any unambiguous matches to start from. With the second scheme, however, the disambiguation will gradually propagate from the borders, where the matches are determined, into the interior, where matches will eventually be chosen whose signs are those of the matches at the border.

Julesz and Chang (1976) did just this experiment, and an example of the type of stereogram they used appears in Figure 3–20. It transpired that

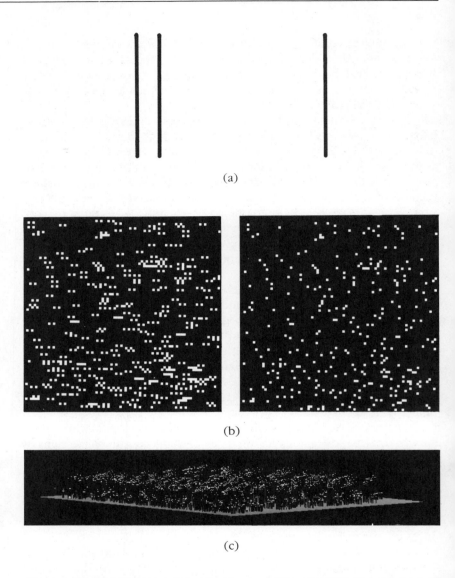

(a)

(b)

(c)

Figure 3–19. (a) Panum's original limiting case. When fused, the impression is of two lines separated in depth. In (b), each dot in the right image is paired with two in the left image. When fused, the viewer sees two planes. The doubling does not have to be restricted to one image. (c) The results of running the stereo algorithm on (b), disparity being displayed according to the same conventions as were used for Figure 3–13. Two planes are found.

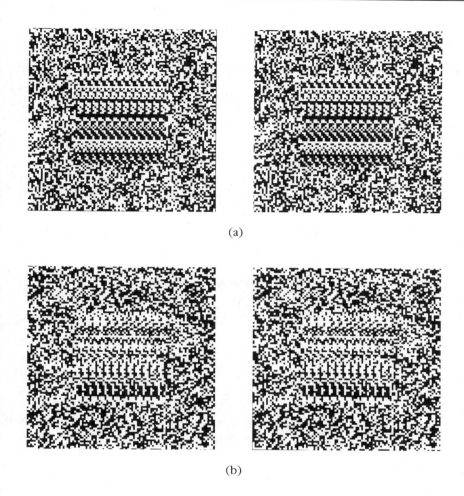

(a)

(b)

Figure 3–20.   (a) There are many possible ways of matching the center of this stereogram, but usually only the matches having the smallest disparity are perceived. However, the particular match found can be biased by inserting unambiguously matchable dots at a particular disparity. In (b), 6% of the dots in the upper half of the square have unambiguous matches at a two-dot crossed disparity— shifted in the nasal direction, whereas the lower half is biased with a two-dot uncrossed disparity. Even a bias inserted into the border will pull fusion to one of the possible solutions in the center. This is evidence for some cooperativity in the human stereo matching algorithm. (Reprinted by permission from B. Julesz and J. J. Chang, "Interaction between pools of binocular disparity detectors tuned to different disparities", *Biol. Cybernetics 22,* 1976, 107–120, figs. 1, 2.)

information from the border could pull the matching going on in the interior one way or the other. This suggests that our visual systems use the second of the two alternatives outlined above.

Panum's fusional area

By using the second of the above schemes, matching may be assigned correctly for a disparity range of w. The precision of the disparity values thus obtained should be quite high and a roughly constant proportion of w (which can be estimated from stereoacuity results to be about $w/20$). For Wilson's foveal channels, this means 3′ disparity with a resolution of 10″ for the smallest and perhaps up to 20′ for the largest with a resolution of 1′. At 4° eccentricity, the range is 5.3′ to about 34′.

Under these assumptions, the predicted values apparently correspond quite well to available measures of the fusional limits without eye movements. Mitchell (1966) used small, flashed line targets and found, in keeping with earlier studies, that the maximum amount of convergent or divergent disparity without diplopia is about 10′–14′ in the fovea and about 30′ at 5° eccentricity. The extent of the so-called Panum's fusional area is therefore twice this.

Under stabilized image conditions, Fender and Julesz (1967) found that fusion occurred between line targets (13′ by 1° high) at a maximum disparity of 40′. This value probably represents the whole extent of Panum's fusional area. Using the same technique on a random-dot stereogram, Fender and Julez arrived at a figure of 14′ (6′ displacement and 8′ disparity within the stereogram). Since the dot size was only 2′, we expect more energy in the high-frequency channels than in the low, which would tend to reduce the fusional area. Julesz and Chang (1976), using a 6′ dot size over a visual angle of 5°, routinely achieved fusion up to 18′ disparity. Taking all factors into account, these figures seem to be consistent with our expectations.

A critical prediction of the theory is that the maximum fusible disparity should scale with the spatial frequency of the stimulus, since the lower spatial frequencies will be detected by only the larger channels. There are already hints that this might be so (Felton, Richards, and Smith, 1972).

Impressions of depth from larger disparities

We have assumed that Panum's area corresponds to pure stereoscopic fusion. One still gains some impression of depth outside this disparity range, however, although this impression does not accurately reflect the disparity that is present. There are two interesting cases to examine.

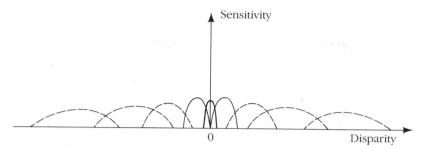

Figure 3–21. In addition to the three narrow disparity pools (solid lines) required by the matching algorithm, sets of outlying disparity detectors may exist, as illustrated here by the dotted lines. Their function would be to estimate whether the fusion plane lies convergent or divergent, so that vergence eye movements can be initiated in the appropriate direction.

The first is diplopia, in which one sees double but still senses depth. The stereo matching algorithms I described above are designed to work when the images are complex. When they are very sparse, there is no real trouble with matching them, because there are no false targets to be avoided. If, for example, there are no possible matches at all in the range w, detectors operating outside this range, possibly sensitive to any match over a broad interval may be consulted. The idea would be that if some indication of the sign of the disparity was available, this would be enough to initiate vergence eye movements in the correct direction so as to bring the images into the fusible range.

There is another way in which such detectors could be used. As we saw in the subsection on the computational theory of stereopsis, if the image contains matchable features with a density of v, the density of matches at the correct disparity is v, whereas at incorrect disparities it is only v^2. If there is a range of disparity detectors and we want only to extract the sign of the disparity where the correct matches lie, we could conceive of a scheme in which the total number of convergent matches—false targets and all—is summed and compared with the corresponding number of divergent matches. We can think of various ways of doing this. For example, adding up over the whole convergent and divergent disparity ranges simultaneously would be the simplest, but just conceivably the range of summation might be gradually extended until a significant difference is obtained. In any case, in a biologically plausible implementation of the kind illustrated in Figure 3–21, we would expect the number of detectors to decrease as the disparity increases. This would, for statistical reasons, produce a psychophysical interdependence between the disparity in an unfused stereogram and the area needed to detect the disparity's sign.

Interestingly, Tyler and Julesz (1980) have reported that such a relationship holds for dynamic random-dot stereograms. These are stereograms that change in their patterns but not necessarily their disparity at rates around 30 frames per second. The sign of disparity can be detected, but not, for example, the shape of the disparate pattern at up to several degrees of disparity. Their finding, that detection ability depends on the square root of the area, $\sqrt{A}$, could be explained by the kind of scheme I suggested, in which the density of disparity detectors falls off with the inverse of disparity, $1/d$. This produces a $\sqrt{A}$ dependence (Marr and Poggio, 1980). Of course, there are other possible explanations of these findings, based on things like motion cues or possible nonlinear, temporal summation at the receptor level between successive frames.

Finally, we shall return to what I still regard as something of a puzzle about stereopsis; namely, Why should one use zero-crossings as the input representation for the matching process? Why not wait and use the raw and full primal sketches, using a scheme that has the same general characteristics but which replaces the low-spatial-frequency zero-crossings by the rough, large-scale primitives in the primal sketch, and the high-spatial-frequency zero-crossings by the raw primal sketch. The findings of Julesz and Miller (1975), for example, about the independent fusion of different spatial frequencies seem the best evidence for the pure zero-crossings approach, but they can probably be explained by this other scheme. The reason is that since, in Julesz and Miller's patterns, like the one reproduced in Figure 3–10, information from different regions of the spatial-frequency spectrum does not come from a common source, the spatial coincidence assumption will be violated, and so independent descriptions for each will appear in the primal sketch.

In addition to this, we have the evidence of Kidd, Frisby, and Mayhew (1979), which I described in Chapter 2, that some texture boundaries can drive vergence movements in stereopsis. This is definite evidence that some of the later primal sketch descriptions are used for stereo vision.

On the other hand, however, the same group found that, in some sense, stereo fusion can preempt and therefore probably precede texture vision discriminations (Frisby and Mayhew, 1979, figs. 1b, c, and d). Figure 3–22 shows some examples. When viewed monocularly, the differently textured regions are clearly visible. When viewed binocularly, however, they disappear. This is slight but not incontrovertible evidence for the zero-crossings approach.

My own view is that some combination of the two is in fact used, although it is based mainly on the zero-crossings approach. The decisive advantages of zero-crossings are probably speed, since they are the first

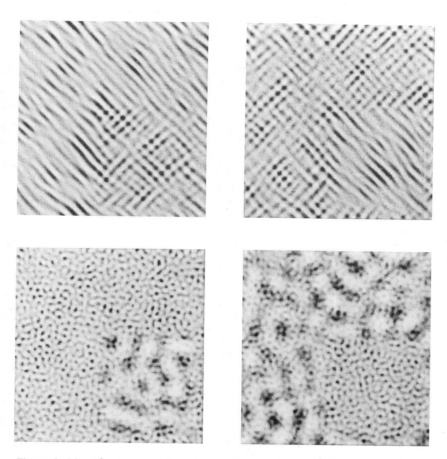

Figure 3–22.   The texture differences, which are clearly visible monocularly, disappear when the two images are fused stereoscopically. (Reprinted by permission from J. P. Frisby and J. E. W. Mayhew, "Does visual texture discrimination precede binocular fusion?" *Perception 8,* 1979, 153–156, figs. 1, 2.) Figure 3–22 continues on next page.

things to be obtained, and precision, since they can be very accurately localized. The theoretical reservations one has about them—that they are only approximately and not strictly tied to physical changes—are not very strong points because zero-crossings are pretty physical (much more so than gray levels, for example). In fact, we know that they are sufficiently physical, because the computer implementation of the zero-crossings theory works well on natural images (Grimson and Marr, 1979; Grimson, 1981).

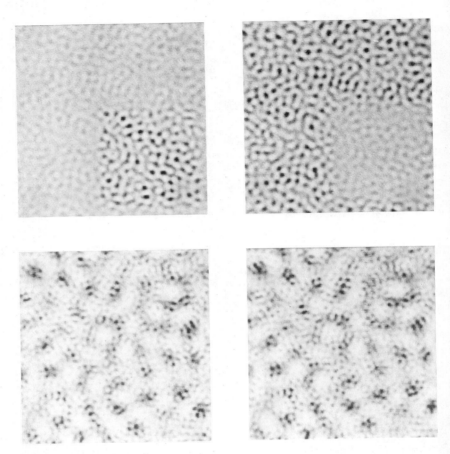

Figure 3–22 (continued).

Have we solved the right problem?

The basic issue that faces the designer of a stereo matching algorithm is, What are the difficult problems and what are the easy ones? A neurophysiologist could, with some justification, object that the matter of stereo fusion is not very difficult at all, and that the really remarkable thing about our stereoscopic vision is its precision, which can be as great as $2''$ of arc for a 75% success rate, that is, roughly one-twelfth the diameter of a foveal cone (Berry, 1948). The false target problem, he might argue, is not difficult if we match special features that occur only rarely.

I disagree with these arguments for the following reasons. In stereo matching, the critical questions are, of course, How rare is rare, and how is rarity related to the disparity range that is consulted? The psychophysical evidence is that the features that can be matched are low level and not very specific for contrast or for orientation. Thus, random-dot stereograms must create false targets, yet we can fuse them. The theory of our second algorithm is in fact largely devoted to precisely the question of how rare is rare, and it is specifically tied to the suggestion that the input representation for stereo fusion is the roughly oriented, signed zero-crossings.

Stereoscopic acuity, on the other hand, although quite remarkable, is an engineering and not a theoretical problem. It occurs at the third of our three levels, the level of implementation mechanisms, because the only question that it raises is, How accurately are the zero-crossings localized? That they can be located to 2″ is remarkable but easy to incorporate in a computer program, for example. We simply have to calculate quite precisely the positions at which the $\nabla^2 G$ convolution passes through zero. No issue of principle is raised here. That neural hardware can do this calculation is remarkable, and it probably means that very many small cells are at some stage used to find and locate these positions, but this calculation is not a theoretical problem in the same way that stereo fusion is. I shall return to the problem of acuity in the neural implementation subsection.

Vergence movements and the 2½-D sketch

According to the second stereo matching theory, once zero-crossing matches have been obtained between $\nabla^2 G$ filtered images using masks of a given size, they are represented in a temporary buffer. These matches also control vergence movements of the two eyes, thus allowing information from large masks to bring small masks into their range of correspondence. The control of vergence could be direct, deriving from the matching neurons themselves, or it could be indirect, routed through the memory buffer or (most likely) through both paths.

The reasons for postulating the existence of a memory are of two kinds, those arising from general considerations about early visual processing and those concerning the specific problem of stereopsis. A memory like the 2½-D sketch (see Figure 3–12) is computationally desirable on general grounds, because it provides a representation in which information obtained from several early visual processes can be combined (see Chapter 4). The reason associated specifically with stereopsis is the computational simplicity of the matching process, which requires a buffer in which to preserve its results as disjunctive eye movements change the plane of fixation and as objects move in the visual field. In this way, the 2½-D sketch becomes the place where global stereopsis is actually achieved, combining the matches provided independently by the different channels, making the resulting disparity map available to other visual processes, and forming the representational basis for the subjective impression that we obtain from stereograms of visible geometrical surfaces.

I shall discuss the 2½-D sketch in detail in the next chapter; here I shall make a few brief remarks about the control of eye movements during stereo vision.

Disjunctive eye movements, which change the plane of fixation of the two eyes, are independent of conjunctive eye movements (Rashbass and Westheimer, 1961b), are smooth rather than saccadic, have a reaction time of about 160 ms, and follow a rather simple control strategy. The (asymptotic) velocity of eye vergence depends linearly on the amplitude of the disparity, the constant of proportionality being about 8°/s per degree of disparity (Rashbass and Westheimer, 1961a). Vergence movements are accurate to within about 2′ (Riggs and Niehl, 1960), and voluntary binocular saccades preserve vergence nearly exactly (Williams and Fender, 1977). Furthermore, Westheimer and Mitchell (1969) found that tachistoscopic presentation of disparate images led to the initiation of an appropriate vergence movement but not to its completion. These data strongly suggest that vergence movements are not ballistic but rather are continuously controlled.

The hypothesis is that vergence movements are controlled by matches obtained through the various channels by means of the mechanisms described earlier that can give a rough sense of depth and by means of some higher types of boundary acting either directly or indirectly through the 2½-D sketch. This hypothesis is consistent with the observed strategy and precision of vergence control, and it also accounts for the finding that perception times depend to some extent on the distribution of disparities in a scene (Frisby and Clatworthy, 1975; Saye and Frisby, 1975). A stereogram of a spiral staircase ascending toward the viewer does not produce the long perception times associated with a two-planar stereogram of similar disparity range. This is to be expected within the framework of the theory, because scenes like a spiral staircase, in which disparity changes smoothly, allow vergence movements to scan a large disparity range under the continuous control of the outputs of even the smallest masks. On the other hand, two-planar stereograms with the same disparity range require a large vergence shift but provide no accurate information for its continuous control.

The long perception times for such stereograms may therefore be explained in terms of a random search strategy by the vergence control system. In other words, vergence movement control is a simple, continuous, closed-loop process that is usually inaccessible from higher levels. The stereograms in Figure 3–23 will enable the reader to see for himself that this is at any rate subjectively true.

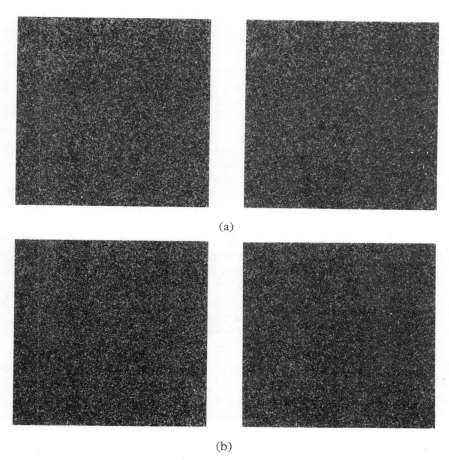

(a)

(b)

Figure 3–23.   These two stereograms have about the same disparity range, but in (a) disparity varies continuously while (b) consists of just two disparity planes. It takes longer to see this second one, presumably because the vergence control system has less information about how to cover the disparity range.

Interestingly, there is some evidence that an observer can learn to make an efficient series of vergence movements (Frisby and Clatworthy, 1975). However, this learning effect seems to be confined to the type of information used by the closed-loop vergence control system. A priori, verbal or high-level cues about the stereogram are ineffective, as, incidentally, they seem to be at all levels of processing up to and including the 2½-D sketch.

Neural implementation of stereo fusion

A complete neural implementation of the second stereo matching algorithm just described has not yet been formulated. One reason is that such a formulation was not worth the considerable work involved until we were reasonably certain from implementation studies and psychophysics that the algorithm works and is roughly correct. However, the first steps have been taken in the analysis of possible neural mechanisms underlying the computation of $\nabla^2 G$ and of the detection of zero-crossings (Marr and Hildreth, 1980; Marr and Ullman, 1979).

The problem of the binocular combination is still an open question—the first of many that we shall be able to formulate. We can, however, allow ourselves some preliminary remarks on the topic. First, disparity sensitivity should not arise before zero-crossing detection. Hence, if the simple cells of area 17 (the striate cortex), which are the first cortical cells in the visual pathway, are disparity sensitive, as seems likely in the cat (Barlow, Blakemore, and Pettigrew, 1967), then they must also detect zero-crossings.

This can occur in several ways, and Figure 3–24 shows two examples. In the first, two independent zero-crossing detections are proposed to take place in the dendrites, each much as shown in Figure 2–18 and relying on local synaptic mechanisms of the Poggio and Torre (1978) type.

Figure 3–24. (opposite)   Two possible neural implementations of disparity detectors. In the first, the cell (a) detects zero-crossings of a given sign independently in two dendrites, one driven by each eye. It then combines the result through an AND gate, which has the effect of making the cell fire whenever an appropriate zero-crossing simultaneously appears anywhere in the cell's left-eye and right-eye receptive fields. This is illustrated in (b). However, such a scheme can provide only a rather rough type of disparity detection; it has the disadvantage, for example, that the range of disparities to which it is sensitive varies with the position of the zero-crossing in the left-eye receptive field. Circles represent excitatory inputs; squares, inhibitory inputs. Open synapses (circles and squares) represent on-center inputs; filled ones, off-center inputs. L and R denote left-eye and right-eye inputs.

The second scheme does not suffer from this disadvantage, since it accurately signals the sign of the disparity, but it operates on only a small disparity range. A zero-crossing is detected in the left image by the AND dendrite of the cell in (c), and the sign of the disparity is assessed by examining the sign, at the zero-crossing, of the difference—computed by a linear process—in the values of the $\nabla^2 G$ convolution for the left and right eyes. This yields a detector of disparity sign that is independent of the position of the left-eye zero-crossing, at least for a small range (d). As illustrated in (e) and (f), if the difference at the zero-crossing is positive, the disparity has one sign; if it is negative, the disparity has the opposite sign.

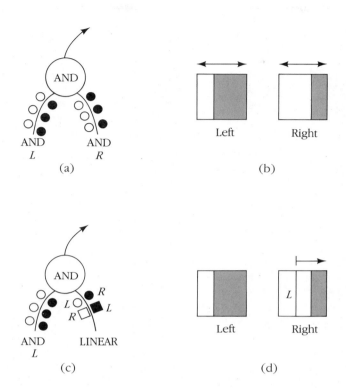

(a)

(b)

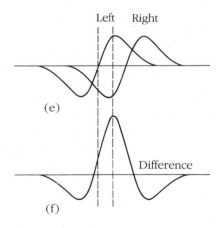

(c)

(d)

(e)

(f)

Such a mechanism does have disadvantages. First, it is not very sensitively tuned to disparity, because the zero-crossings in each eye are located with an accuracy of not much better than w_{1-D}. Second, the range of disparities to which the mechanism responds depends upon the exact position of the zero-crossing in the left eye, because the range of positions in the right eye is also fixed by the geometry of the connections there.

The second model in Figure 3–24 shows another possibility. The cell is left-eye dominant, being driven by a zero-crossing from the left eye. However, it is gated by the difference between the left- and right-eye convolutions at the zero-crossing. If this difference is negative, the disparity will usually be of one sign, and if it is positive, the disparity will usually be of the other sign, as explained in Figure 3–24. For an edge that goes from light to dark as one moves from left to right in the visual field, a negative difference corresponds to divergent (near) disparities. This mechanism removes some of the imprecision associated with the first mechanism, since it measures quite directly whether the right image's zero-crossing (of fixed sign) is to the left or to the right of the left image's. It has its disadvantages, however, since for too closely occurring zero-crossings or for very different contrasts in the two eyes, it can be unreliable.

Unfortunately, the technical problems associated with the neurophysiology of stereopsis are considerable, and rather few quantitative data are currently available—certainly too few to enable us to rule out either or both of the mechanisms of Figure 3–24. Since Barlow, Blakemore, and Pettigrew's (1967) original paper, relatively few examples of disparity tuning curves have been published. Recently, however, Poggio and Fischer (1978) and von der Heydt and others (1978) have published properly controlled disparity curves for the monkey and cat, respectively. On the whole, these studies favor the idea that disparity detectors are organized into three pools—convergent, near zero, and divergent—and recently Clarke, Donaldson, and Whitteridge (1976) have found that, in the sheep, these detectors are organized into columns, as Hubel and Wiesel (1970) suggested they might be in area 18 of the macaque. However, the size of the disparities involved are surprisingly large—7° in the sheep and up to a degree or even several in the monkey. The precise role of these detectors in stereopsis is therefore not yet clear.

Curiously enough, even the owl, which diverged from the monkey probably before stereopsis evolved, appears to use an algorithm similar to the monkey's. Pettigrew and Konishi (1976) have found that although the anatomical organization of the owl's wulst is quite different from that of the monkey's visual pathway, the physiological responses of the cells are very similar. The owl, however, is unable to move its eyes very much, so at first it might be thought to be deprived of the ability to make the vergence movements that are so essential for this approach to stereopsis. Nature, though, has found a way—the owl's horopter is sloped, passing through its feet at the bottom of the visual field and extending to infinity roughly straight ahead. The owl can therefore attain the effect of vergence eye movements, together with the simultaneous impression of a profound and grave wisdom, by the gentle but deliberate nodding of its head.

Finally, there is the problem of stereo acuity, which, like all human hyperacuity abilities, requires an underlying mechanism that is able to localize small, isolated features in an image to within about 5" of arc for an average subject (Westheimer and McKee, 1977). Crick, Marr, and Poggio (1980) discussed the neurophysiological implications of these findings and suggested that one possible solution might be based on the high-resolution spatial reconstruction of the $\nabla^2 G$-filtered image as it enters the visual cortex from the optic radiations. Barlow (1979) made the suggestion first, and we amended it slightly, saying that the reconstruction need not be completely accurate. It will suffice to reconstruct accurately only those parts of the signal lying around the zero-crossings.

The natural candidate for performing the reconstruction is the granule cell population of layer IVCβ in area 17. Worst-case estimates suggest that, for each type (on center and off center) and each eye, there is easily one granule cell for every 5" of arc for the smallest channel. David Hubel furthermore reports that these cells are all center–surround, so far indistinguishable from geniculate fibers, and that their spatial arrangement is very precisely retinotopic—nearby cells correspond to nearby points on the retina. These are all properties that we would expect of cells engaged in reconstruction. It would therefore be of great interest to know whether their responses differ physiologically in any way from those of the lateral geniculate fibers, for example in their spatial or particularly in their temporal characteristics.

Computing Distance and
Surface Orientation from Disparity

Computational theory

Distance from the viewer to the surface

Suppose a point P lies at distance l from the viewer's left eye L and at angle ω to his forward line of sight, as illustrated in Figure 3–25. Let the distance between the viewer's eyes be δ_T; then, because the line of sight to P does not lie directly ahead, the effective distance between the two eyes is only $\alpha = \delta_T \cos \omega$. Writing $\beta = \delta_T \sin \omega$, we see from the figure that ϕ, the angle between the lines of sight from the two eyes, is given by

$$\tan \phi = \frac{\alpha}{(l + \beta)} = \frac{\delta}{l}$$

For small values of ϕ, we can write

$$\phi \cong \frac{\delta}{l} = \frac{\alpha}{l + \beta}$$

Now take two points P and P' along the same line of sight from the left eye, with P at distance l and P' at distance l' as in Figures 3–25(a) and (b). It follows that the disparity $\triangle\phi$ between P and P' is $\phi' - \phi$. Hence, if we let

$$q = \frac{l' + \beta}{l + \beta}$$

then

$$\triangle\phi \cong \left(\frac{1}{q} - 1\right)\frac{\alpha}{l + \beta} = (1 - q)\frac{\delta}{l}$$

We can rewrite this as

$$(1 - q) \cong \left(\frac{l}{\delta}\right)\triangle\phi$$

In other words, the fractional change in distance for a given disparity depends upon the distance away. This fact can be important for depth-judging experiments and, as we shall shortly see, for the perception of surface orientation, because it shows that if the human visual system does its job properly, the proportional change in perceived depth obtained for a given disparity should depend on l, that is, on what the observer happens to think the current true depth is.

Surface orientation from disparity change

The trigonometry of the recovery of surface orientation is rather tedious. However, the resulting formulas are interesting, so I shall discuss them here. We need to consider two cases, one in which the surface slopes

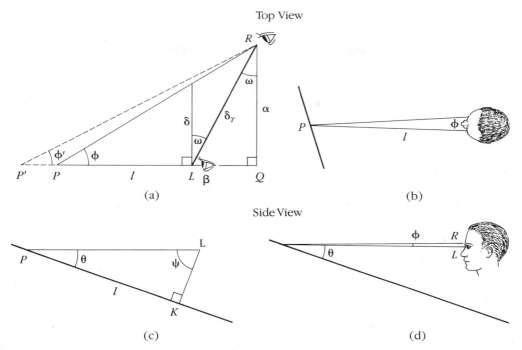

Figure 3–25. The trigonometry of recovering depth from disparity. (a) shows a top view of the geometry of the two eyes looking at a point P distance l from the left eye, as illustrated in (b). The line of sight is not necessarily perpendicular to the line joining the two eyes L and R, and the difference is described by the angle ω as illustrated. The true interocular distance is δ_T, and the effective interocular distance for this line of sight is $\delta_T \cos \omega$. The angle between the lines of sight from the two eyes is ϕ, and it is the differences in the values of ϕ for different points P' that are normally called disparities. The lengths $\alpha = \delta_T \cos \omega$ and $\beta = \delta_T \sin \omega$ are useful geometrical quantities.

(c) shows a side view of the same situation, illustrated in (d). The point P is shown lying on a plane that slopes vertically, and its slope at P is described by the angle θ. Only the left eye L is shown in this diagram, and again the distance l refers to the distance from the left eye. In order to recover surface orientation, it is necessary to recover the angle θ.

in the horizontal direction, as in Figures 3–25(a) and (b), and one in which it slopes away in the vertical direction, as in Figures 3–25(c) and (d). These situations differ because our eyes are positioned horizontally, not vertically. In both cases, we need the formulas that relate surface orientation, which I denote by θ, to the rate of change of disparity ϕ with visual angle ψ, which I write $\partial\phi/\partial\psi$. The formulas are as follows:

For surfaces changing in depth in the vertical direction:

$$\frac{\partial \phi}{\partial \psi_V} = \frac{-\alpha l \cot \theta}{\alpha^2 + (\beta + l)^2}$$

For surfaces changing in depth in the horizontal direction (the formula is perforce more complicated):

$$\frac{\partial \phi}{\partial \psi_H} = \frac{\alpha^2 + \beta(\beta + l) - \alpha l \cot \theta}{\alpha^2 + (\beta + l)^2}$$

There are two points to be noted about these formulas. First, like estimates of fractional depth, they depend on the viewing distance l, roughly as $1/l$. Hence, if the brain is doing its task, a given rate of change of disparity should be perceived as an increasingly steep surface as its distance away is increased. The reader can see this by looking at the stereogram in Figure 3–26 from different distances. Disparity and viewing angle change together, so $\partial \phi / \partial \psi$ is constant for all viewing distances. Hence, the surface should appear to steepen as one moves the stereogram further away, and it does. This also shows, incidentally, that the brain has a pretty good idea of where the stereogram actually is and uses this information.

Second, when the horizontal rate of change of disparity $\partial \phi / \partial \psi_H$ reaches 1, the line of sight from the other eye must fall directly along or in front of the actual physical surface. The viewer then sees a discontinuity

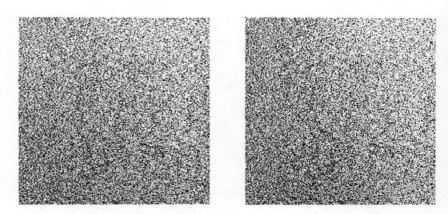

Figure 3–26. Notice that if the viewing distance from this stereogram changes, the perceived surface orientations change. This is to be expected if the visual system is calculating its trigonometry correctly. (Bela Julesz, 1971, p. 156, fig. 5.4–2)

in depth from the second eye. This can be checked by putting $\theta = -\phi$ into the horizontal disparity-change formula; then $\partial\phi/\partial\psi_H = 1$. In this situation, all the change in viewing angle from the first eye is a change in disparity, so $\partial\phi/\partial\psi_H$ remains equal to 1 until the other eye starts seeing the surface again. This fact can be used to help us to find discontinuities in viewing distance from stereopsis.

Algorithm and implementation

Nothing is known about how these formulas are implemented, although the example of Figure 3–26 suggests that approximations to them are and that the approximations may be quite accurate. It is perhaps worth emphasizing that the effects I pointed out, of a dependence of perceived depth and surface orientation on viewing distance and direction, are wholly to be expected and are not some strange psychophysical phenomena that need complex explanations.

3.4   DIRECTIONAL SELECTIVITY

Introduction to Visual Motion

Motion pervades the visual world, a circumstance that has not failed to influence substantially the processes of evolution. The study of visual motion is the study of how information about only the organization of movement in an image can be used to make inferences about the structure and movement of the outside world. Again there are two basic parts to the problem: How are the raw measurements of the changes produced by motion made, and is this information used? Neither is at all easy to solve, and perhaps because the first is so difficult, the second is to some extent a study of the minimum information necessary from the first part in order for subsequent computations to deliver any sort of useful results.

   The psychophysical study of visual motion is old. Most people would probably trace its origins to members of the Gestalt movement (Werthei-mer, 1923; Koffka, 1935), who, like their followers Gibson and Julesz (Gibson et al., 1959; Julesz, 1971, ch. 4), were interested in the effects of motion on the separation of figure and ground and on eye movements. Miles (1931) and Wallach and O'Connell (1953) introduced the problem of determining three-dimensional structure from motion, a problem dealt with at length in the recent and remarkable book by Shimon Ullman (1979b). Gibson (1966) was interested in the problem of optical flow, a problem that has only recently received the mathematical attention it deserves (Longuet-Higgins and Prazdny, 1980).

Table 3–1. Determinants of apparent motion found with two perceptual criteria.

Criterion of segregation in random-dot display	Criterion of smooth apparent motion for isolated element
Spatial displacement must be 15' arc or less (Braddick, 1974).	Spatial displacement may be many degrees (e.g., Neuhaus, 1930; Zeeman and Roelofs, 1953).
ISI must be less than 80–100 ms (with 100 ms stimulus exposure) (Braddick, 1973).	ISI may be at least 300 ms (e.g., Neuhaus, 1930).
Segregation abolished by bright uniform field in ISI (Braddick, 1973).	Motion perceived whether ISI is bright or dark.
Successive stimuli must be delivered to the same eye or to both eyes together (Braddick, 1974), as must bright field for effective masking (Braddick, 1973).	Successive stimuli may be delivered to the same or different eyes (Shipley, Kenney, and King, 1945).
Pattern defined by chromatic but not luminance contrast is inadequate (Ramachandran and Gregory, 1978).	Stimuli may be defined by chromatic contrast alone (Ramachandran and Gregory, 1978).

Note: ISI = interstimulus interval.

The first important psychophysical finding I wish to emphasize, however, is quite recent, and it bears upon the question of how many different motion modules or processes there are, what they do, and how rich the information is that they run on. Following Julesz's (1971, ch. 4) example, Braddick (1973, 1974) used random dots and lines to explore the psychophysical properties of apparent motion. For example, he found a number of strange differences between what happens over short times and small displacements and what happens over long times and large displacements. He concluded that there were two different processes characterized by different perceptual criteria, and which have the properties listed in Table 3–1 (from Braddick, 1979).

These properties were found in experiments of the following kind. Two patterns are used as displays, each composed of random dots or lines. Outside a central rectangle, the two patterns are uncorrelated, as illustrated in Figure 3–27. Inside the central rectangle, the dots are displaced in one pattern relative to the other, in the manner of Figure 3–28. The two patterns

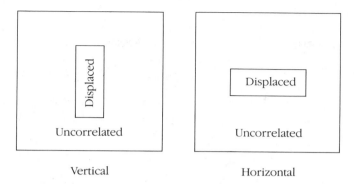

Figure 3–27.   The discrimination task for Braddick's short-range phenomena. A vertical or horizontal rectangle has to be discriminated against an uncorrelated background.

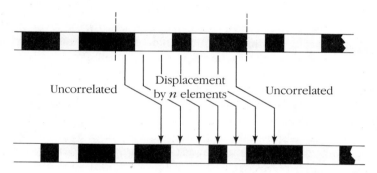

Figure 3–28.   The rectangles in Figure 3–27 are created in a pair of successively presented random-dot displays by displacing a rectangular region by a few elements. The rest of the display is uncorrelated between frames.

are alternated at some rate with an interstimulus interval (ISI) during which other masking fields are sometimes shown. The question is, For what rates and displacements does the subject perceive the rectangular region well enough to say whether it is horizontal or vertical?

The second kind of experiment was like those extensively used by Ullman, in which one or a few lines are presented in frame 1, followed by the ISI, followed by a second few lines, as illustrated in Figure 3–29. Here the question is, Does the subject smoothly perceive one line mapping to another line or lines, and if so, how does the mapping go? Ullman's (1978) experiments have warned us to be wary of smoothness, but the actual mapping itself is a reliable and useful phenomenon.

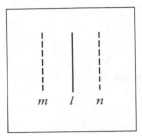

Figure 3–29. The second type of display, extensively used by Ullman, also consists of two frames, but they are much simpler than those in Figures 3–27 and 3–28. The first might consist of the line l shown here, and the second of two lines m and n. The observer is asked, Does l go to m, n, or both?

What Braddick found was that if one does various things to the two types of display—things like changing the displacement or the ISI or flashing a bright uniform field during the ISI—perceptions of the displays are very different. Conditions that easily disrupt the first experimental task do not disrupt the second. For example, to discern the rectangle successfully, the angular displacement must be small (less than 15'), the ISI short (less than 80 ms), and no masking field may intervene. Not so the second task; the angular displacement may be many degrees, the ISI may be 300 ms or more, and the masking field may be bright or dark. These and other differences are summarized in Table 3–1.

What could be the significance of these distinctions? Perhaps the key to the puzzle is that in the analysis of motion—more so, perhaps, than any other aspect of vision—time is of the essence. This is not only because moving things can be harmful, but also because, like yesterday's weather forecast, old descriptions of the state of a moving body soon become useless. On the other hand, the detail of the analysis that can be performed depends upon the richness of the information on which the analysis is based, and this in turn is bound to depend upon the length of time that is available to collect the information. In an instantaneous view, for example, everything is static, so no information about motion is available. After a 60 ms wait, information derived from observed changes may enable a much more thorough analysis, and in a third look in yet another 60 ms perhaps everything about the motion can be recovered, provided that computation is powerful enough.

Perhaps one of the most primitive types of motion analysis is the type concerned with noticing that something has changed, where the change is in the visual field, and perhaps something about the direction of movement involved, though this is arguably a more complex matter. Such analysis we have already met in our earlier discussion of the visual system of the housefly. Another case where similar mechanisms are thought to operate is in the directionally selective cells of the rabbit's retina (Barlow and

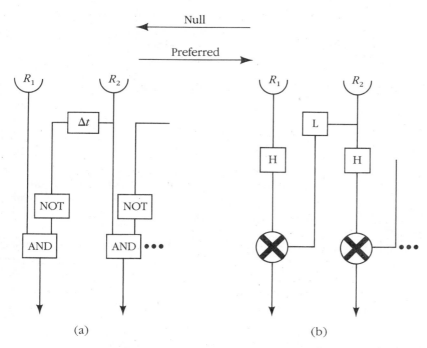

Figure 3–30.   (a) Barlow and Levick's (1965) model for directional selectivity connects two detectors to an AND–NOT gate, one via a delay. Thus the network does not respond to stimuli moving with roughly the right speed in the null direction. (b) Hassenstein and Reichardt's (1956) model operates on the same principle except that the delay is replaced by a temporal low-pass filter (L). H = high-pass filter.

Levick, 1965), the frog's retina (Barlow, 1953; Maturana et al., 1960), the pigeon's retina (Maturana and Frenk, 1963), and perhaps the mammalian retinal W cells.

These mechanisms all have various things in common. They probably all operate at the earliest possible stage—that is, directly on the gray-level image intensity values—and their underlying mechanism is something equivalent to combining a time delay (or temporal low-pass filter) and an AND–NOT gate*. The basic idea is illustrated in Figure 3–30(a). Two receptors are connected to an AND–NOT gate, one directly and one through a delay. If a bright spot moves first across the right-hand receptor R_2, then

---

*A logical device that gives an output only when its first input is on and its second input is off.

across the other one, R_1, signals from the two will arrive at the gate roughly simultaneously, causing it to remain silent. This is called the null direction. A white spot moving the other way will cause the gate to fire.

If the intensity detectors are replaced by a center–surround operator, this difficulty goes away—we get a directionally selective bug detector or edge detector—but it still has characteristic problems. First, if a stimulus is moved very slowly in the null direction or is stopped and restarted halfway between the two receptors, the gate will give a response. Second, and again relating to the delay, the range of spatial frequencies over which the device operates reliably depends on how fast the pattern moves. To the device, a thick sinusoidal grating moving fast looks like a thin one moving slowly. Our own visual systems exhibit similar properties (for example, Kelly, 1979). To maintain reliability, we must make sure that the mechanism looks only at the appropriate portion of the range of spatiotemporal possibilities.

The reason that detectors of the type shown in Figure 3–30 fail to be reliable is a deep one. Fundamentally, they are reading a receptor in one place at one time and another in a nearby place a little later; if anything happens at one and then at the other the correct interval later, the detector implicitly assumes that the two changes are due to the same physical cause. This, in fact, is our first real introduction to the *correspondence problem of apparent motion.* The unreliability of these detectors arises for the same basic reasons that make a fast, clockwise-turning wagon wheel in a Western movie seem to be turning slowly counter-clockwise. The implicit assumption, that the nearest spoke in the next frame is the same one as in the last frame, is wrong because the wheel is turning too fast relative to the movie frame rate.

Such schemes, as I indicated, are still useful for saying where in a visual field a relative movement has occurred and for giving some information about its direction, if one is careful. However, if we also wish to analyze the shape of a moving patch, it seems more sensible to try to combine the analysis of movement with the analysis of contours (Marr and Ullman, 1979). This view, incidentally, is diametrically opposed to current physiological and psychophysical thinking, according to which the sustained and transient channels in human early vision are separated into two parallel systems, one concerned with the analysis of form or pattern, and the other with movement (Tolhurst, 1973; Kulikowski and Tolhurst, 1973; Ikeda and Wright, 1972, 1975; Movshon, Thompson, and Tolhurst, 1978). For eye movement control, of course, there is no need to combine them, but to see the shape of moving patches, it would seem sensible to do so.

We have now discussed the two types of information that can be gleaned from motion—(1) noticing a movement and finding its position in the visual field and (2) determining its two-dimensional shape. As we might have expected, neither requires very sophisticated measurements, and in principle they can both be carried out very quickly given reasonably accurate measurements. What, then, about determining three-dimensional structure? This is clearly more valuable, but intuitively we would have thought that more information from the images would be necessary.

In fact, more information is required, and the basic improvement needed is a good solution to the correspondence problem, rather than the half-baked guess at it which suffices for the simpler tasks. To recover three-dimensional structure, we need to be able to say that point A in the image at time t_1 corresponds to point B in the image at time t_2 for the equivalent of three frames in Ullman's (1979a) style of analysis, or, almost equivalently, we need the exact instantaneous positions and velocities in the image for the simpler task of analyzing the optical flow induced by the observer's movement through a rigid environment. Whether either or both of these theoretical possibilities are incorporated into the human visual system is a matter for psychophysics. As we shall see, the evidence for Ullman's scheme is strong; that for a Gibson-style analysis of optical flow is somewhat weaker, but the theory is nevertheless interesting.

This and the next section in this chapter deal with the different parts of the motion analysis problem. In this section we look first at directional selectivity from the point of view of using it to separate figure from ground and recovering the two-dimensional shape of the figure. We shall then explore Ullman's theory of the interpretation of three-dimensional shape from visual motion in Section 3.5, and shall briefly discuss the problem of optical flow.

Computational theory

The theory of directional selectivity is the theory of how to use partial information about motion—specifically, only its direction defined to within 180°—in order to discern the two-dimensional shapes of regions in the visual field based on their relative movement.

The background to this problem from a computational point of view comes from asking, How much of this information can one gain from motion without solving the full correspondence problem—that is, without being provided with the full instantaneous position and velocity field for the whole image? The motivation for studying what direction alone can tell us comes from something that we call the *aperture problem,* illustrated in Figure 3–31. If a straight edge is moving across the image in direction *b,* as indicated by the arrow in Figure 3–31, this fact cannot be discerned by local measurements alone. As the figure shows, the only motion that can

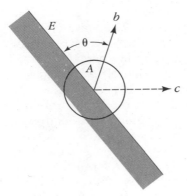

Figure 3–31. The aperture problem. If the motion of an oriented element is detected by a unit that is small compared with the size of the moving element, the only information that can be extracted is the component of the motion perpendicular to the local orientation of the element. For example, looking at the moving edge E through a small aperture A, it is impossible to determine whether the actual motion is in the direction of b or of c.

be detected directly through a small aperture placed over the edge is motion at right angles to that edge—just one bit of information, indicating whether it is moving forward or backward. Of course, if there is only a point or blob or a termination of some recognizable kind, more information can be recovered. And if one somehow knows θ, the angle between the edge and the direction of motion b, then the speed s can be recovered by measuring the component $s \sin \theta$ perpendicular to the edge. But the very simple case in which just the sign is available has at least a theoretical interest.

Various experiments suggest that this simple case is also of interest for understanding one of the visual system's ways of analyzing motion. The experimental situation is like that used by Braddick (1973, 1974), and the stimuli are shown in Figure 3–32. These experiments fall into the first of his two classes, being concerned with short-range, short-term phenomena.

In Figure 3–32(a), the individual dot speeds in the central square are all constant at twice the dot speeds in the surround, but the directions of movement are all random. The central square proves invisible, so we cannot use only speed of movement to separate the patches. Julesz (1971, ch. 4) described a similar effect. In Figure 3–32(b), the surround moves randomly, while the center dots all move in the same direction but with different speeds, spanning a factor of 4. The square can be seen clearly, and where the neighboring speeds are very different, the dots appear to have some relative movement as well.

The remarks about the aperture problem tell us what we want to measure and why we want to measure it. These psychophysical experiments suggest that the visual system uses information about direction alone to help carve up the visual field. We therefore explored algorithms for

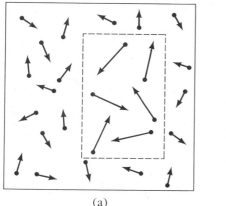
(a)

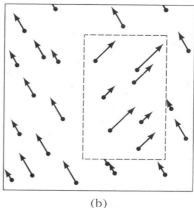
(b)

Figure 3–32. Two experiments showing that Braddick's (1979) short-range system uses only limited information to decompose the image. In (a), the speeds in the central rectangle and in the surround are different and uniform, but the directions of motion are random. Discrimination is not possible. In (b), the directions in the central rectangle are the same but the speeds differ. Discrimination is easy.

quickly detecting the sign of movement direction at the level of local edge segments or their precursors. The earliest stage at which this could be carried out is at the level of zero-crossing segments, and as we shall later see, the physiological data support this possibility.

An algorithm

To construct a directionally selective zero-crossing detector, we must somehow determine the direction of movement of an oriented zero-crossing segment of the type defined in Chapter 2. There we saw that a zero-crossing segment is defined as a locally oriented segment of the zero values of the convolution $\nabla^2 G * I$. A cross section of this convolution appears in Figure 3–33 for the image intensity profile illustrated there.

There are several ways of building a directionally selective unit from this, one of which is to use two zero-crossing detectors as the inputs to a device like Barlow and Levick's (1965). As we have already seen, however, such devices suffer from the stop–restart false response in the null direction, and directionally selective cortical simple cells are known not to do this (Goodwin, Henry, and Bishop, 1975). Marr and Ullman (1979) therefore suggested the following algorithm:

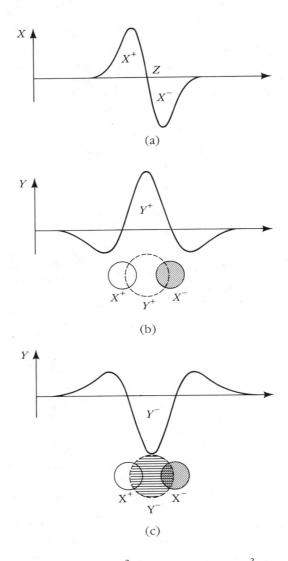

Figure 3–33. The value of $X = \nabla^2 G * I$ and of $Y = \partial/\partial t (\nabla^2 G * I)$ in the vicinity of an isolated intensity edge. (a) The X signal as a function of distance. The zero-crossing Z in the signal corresponds to the position of the edge. (b) The spatial distribution of the Y signal when the edge is moving to the right, and (c) when it is moving to the left. Motion of the zero-crossing to the right can be detected by the simultaneous activity of $X^+Y^+X^-$ in the arrangement shown in (b). Motion of the zero-crossing to the left can be detected by the $X^+Y^-X^-$ unit in (c).

Step 1. Measure the time derivative $\partial/\partial t(\nabla^2 G * I)$.

Step 2. If this is positive at Z, the zero-crossing is moving to the right; if it is negative, it is moving to the left. If the edge has opposite contrast, the directions are reversed.

The truth of these statements can be seen from Figures 3–33(b) and (c), which plots $\partial/\partial t(\nabla^2 G * I)$, the time derivative of Figure 3–33(a), for the two cases of movement to the right and to the left, respectively. The sign of the time derivative is constant over the whole width w_{1-D} between the peaks of the original convolution $\nabla^2 G * I$, so the algorithm is robust.

This scheme has several positive features. (1) It requires only local measurements. (2) No time delay is involved beyond that required to compute the derivative. (3) The method can be made extremely sensitive. The lower limit to the displacement that can be detected is set by the unit's sensitivity, and the upper limit, which depends on the temporal filter, is high if the time constants are small. Hence, a single unit can be made sensitive to a wide range of speeds, and since the only really important part of the measurement of $\partial/\partial t(\nabla^2 G * I)$ is its sign, this can be exploited by making the measuring unit extremely sensitive. It does not matter if it saturates early. (4) Finally, within this range and for a sufficiently isolated edge, the unit will be completely reliable.

The critical difference between the Barlow and Levick type of scheme and this one is that this system does not have to wait until the zero-crossing has passed from the first detector to the second. It can therefore respond instantaneously, and it is sensitive to very small displacements. In addition, unlike systems based on a pair of detectors, it does not have to "guess" that the zero-crossing exciting the left-hand detector now is the same one that excited the right-hand detector a short time ago; and so, at the price of delivering less information, it avoids the difficulties inherent in the full correspondence problem.

Neural implementation

I would not, of course, have suggested this scheme without an idea of how it might be implemented. We have already seen that the detection of zero-crossing segments (Figure 2–18) rests on the idea that the lateral geniculate X cells carry the positive and negative parts of $\nabla^2 G$ via on-center and off-center cells, respectively. Finding a zero-crossing is simply a matter of connecting the on- and off-center X cells via a logical AND gate.

But how to measure the time derivative?—here is an interesting and fascinating point. The psychophysical studies of the transient channels and the neurophysiological recordings of the Y cells, to which the transient channels are thought to correspond, essentially demonstrate that these channels measure this time derivative, $\partial/\partial t(\nabla^2 G * I)$! Interestingly, so far as we are aware, the behavior of these channels has never been formulated as a time derivative, presumably because no one ever thought that such a thing might be a useful function so early in the visual pathway.

Let us look at the evidence a little more closely. Ideally, to obtain a time derivative, we subtract from the current value of a signal its value an infinitesimal time ago. In practice, these measurements must be taken over finite intervals of time. Hence, the impulse response of the device in the time domain should be composed of a positive phase followed by a phase of a similar shape but opposite sign. In the frequency domain, the power spectrum should be roughly linear in frequency over the range in which the device is to operate.

A temporal filter composed of about a 60-ms positive phase followed by a negative phase was explicitly suggested by Watson and Nachmias (1977) and further supported by Tolhurst (1975), Breitmeyer and Ganz (1977), and Legge (1978). The negative phase may be somewhat longer than the positive one, or it may be followed by damped oscillation of small amplitude (see Breitmeyer and Ganz, fig. 3) without significantly affecting the results.

In the frequency domain, the temporal modulation transfer function (MTF) measured by Wilson (1979) for the transient U channel can be accurately described up to range $\omega = 10$ Hz by $F(w) = 16\omega - \omega^2$. This is consistent with an operator that approximates the first derivative of its input, provided that the input signal has no significant power above 8 Hz. Since the U channel attenuates spatial frequencies above 3 cycles/deg, the channel will signal the derivative for edges and bars that drift across the retina with a velocity of up to about 3 deg/s. Figure 3–34 shows how closely

---

Figure 3–34. (opposite)   The computed response of the transient U channel to a light edge, a thin bar, and a wide bar all moving at 3 deg/s. (a) The output of the spatial filter $(\nabla^2 G * I)$ when the U channel parameters from Wilson and Bergen (1979) are used. The *y*-axis represents the normalized response, and the *x*-axis represents distance, the entire range being 3°. The *x*-axis in (b), (c), and (d) represents time; the entire range is 1 s. (b) The theoretically predicted output of the temporal filter if the transient channel carries $\partial/\partial t$ $(\nabla^2 G * I)$. (c) The output of the temporal filter if Wilson's contrast-sensitivity curve is used and the filter is antisymmetric. (d) Comparison of (b) and (c). The thin bar is 2′ wide, and the thick bar is 40′ wide. In all cases the agreement between the curves derived from the time derivative hypothesis and the curves derived from the empirical observations is satisfactory. Hence for isolated bars and edges, the psychophysical evidence is consistent with the idea that the transient channels approximate the function $\partial/\partial t$ $(\nabla^2 G * I)$.

Edge   Thin bar   Wide bar

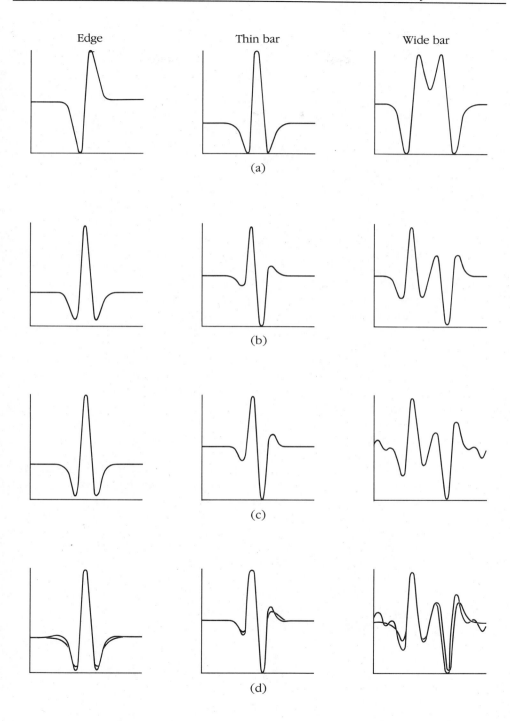

(a)

(b)

(c)

(d)

the measured characteristics of the transient channels match the expected behavior of the time derivative $\partial/\partial t(\nabla^2 G * I)$ for an isolated edge and a thin and a wide bar.

Turning to the neurophysiology, Rodieck and Stone (1965) described retinal ganglian cells whose response to a moving spot was "directly correlated with the gradient of the receptive field as defined by flashing lights" (p. 842). Of course, no physical device can take a perfect time derivative over the entire temporal frequency range. However, the published response curves of retinal and geniculate Y cells to bars and edges moving at moderate velocities closely agree with the predictions based on the time derivative operation $\partial/\partial t(\nabla^2 G * I)$. Figure 3–35 compares the predicted responses of on- and off-center Y cells with their observed responses to various stimuli. All the stimuli were light (that is, light edges and light bars); the thin bars were about ½° wide, and the thick bars 5°. The traces are taken from Dreher and Sanderson (1973). The predicted traces show pure values of $\partial/\partial t(\nabla^2 G * I)$, and as in Figure 2–17, the thickness of the thin and thick bars was, respectively, $0.5w$ and $2.5w$. The observed responses closely agree with the predicted ones, even in cases where both are elaborate (as with the wide bar).

The idea, that the X cells signal $\nabla^2 G$ and the Y cells its time derivative, which enables the construction of directionally selective, oriented zero-crossing segment detectors, offers a precise explanation for part of the function of the retina, and poses a fascinating challenge to the retinal anatomists and neurophysiologists—namely, How are these signals measured? Convolving with $\nabla^2 G$ is easy to imagine, but measuring $\partial/\partial t(\nabla^2 G * $ I$)$ or even just determining its sign is quite a complicated task and requires both spatial and temporal comparisons: The center must be compared with the surround, and the result at a given time compared with the result a short time earlier, which means there must be a 60-ms memory there. In the retina, some of these components may be distorted, especially since comparing the values at two different times requires a delay. Hochstein and Shapley's (1976a) findings suggest, for example, that the Y-cell surround receives a delayed contribution from nearby units about the size of the centers of local X-cell receptive fields, and that this delayed input may be a major source of the observed nonlinearity. The nonlinear effects are induced primarily by gratings (Enroth-Cugell and Robson, 1966; Hochstein and Shapley, 1976a, 1976b). For isolated edges and bars moving at moderate velocities, however, the Y cells approximate $\partial/\partial t(\nabla^2 G * I)$ quite well, as we saw in Figure 3–35.

Provided that the Y channels deliver $\partial/\partial t(\nabla^2 G * I)$ and that positive and negative values are separated into different channels, the zero-crossing segment detector of Figure 2–18, reproduced in Figure 3–36, requires only

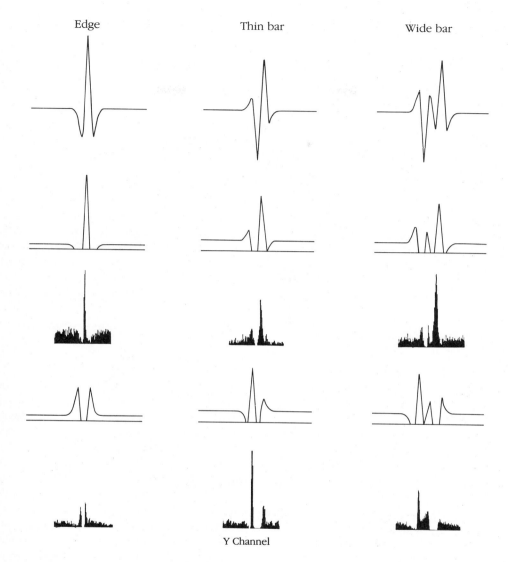

Edge   Thin bar   Wide bar

Y Channel

Figure 3–35. Comparison of the predicted responses of on- and off-center Y cells to electro-physiological recordings. The first row shows the response of $\partial/\partial t\ (\nabla^2 G * I)$ for an isolated edge, a thin bar (bar width $= 0.5w_{1\text{-}D}$, where $w_{1\text{-}D}$ is the width projected onto one dimension of the central excitatory region of the receptive field), and a wide bar (bar width $= 2.5w_{1\text{-}D}$). The predicted traces are calculated by superimposing the positive (in the second row) or the negative (in the fourth row) parts of $\partial/\partial t(\nabla^2 G * I)$ on a small resting or background discharge. The positive and negative parts correspond either to the same stimulus moving in opposite directions, or stimuli of opposite contrast—for example, a dark edge versus a light edge—moving in the same direction. The observed responses (third and fifth rows) closely agree with the predicted ones, even in cases where both are elaborate (such as for the wide bar).

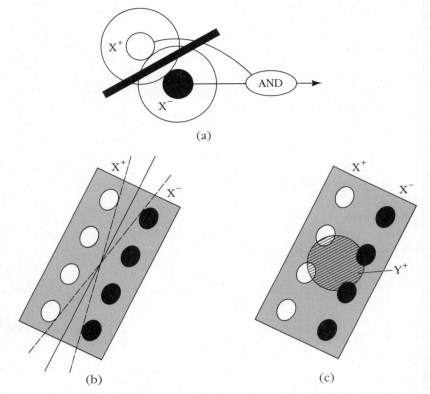

Figure 3–36. The detection of a moving zero-crossing. (a) X⁻ and X⁺ subunits are combined through a logical AND operation. Such a unit would signal the presence of a zero-crossing of a particular sign running between the two subunits. A row of similar units connected through a logical AND would detect the presence of an oriented zero-crossing within the orientation bounds given roughly by the dotted lines in (b). In (c), a Y unit is added to the detector in (b). If the unit is Y⁺, it would respond when the zero-crossing segment is moving in the direction from the X⁺ to the X⁻. If the unit is Y⁻, it would respond to motion in the opposite direction.

the addition of one Y-cell input, again via an AND gate, in order to make it directionally selective.

The basic unit is shown in Figure 3–36(c), which is Marr and Ullman's (1979) XYX model for the simplest type of cortical simple cell. Its receptive field has three components, sustained on-center X inputs, sustained off-center X inputs, and a Y input. The X units need to be all the same size and arranged in two parallel columns not more than $w_{2\text{-}D}/\sqrt{2}$ apart (where $w_{2\text{-}D}$ is the diameter of the central excitatory regions of the X-cell receptive fields). The Y-cell input can in principle be satisfied by a single input whose receptive field is positioned centrally or a little toward one side (toward the positive column for on Y units and the negative column for off Y units).

The ideal scheme requires a strict logical AND operation between the outputs of the subunits. In practice, this could be implemented by a strong multiplicative interaction between the columns and the Y input, and a weaker nonlinearity down the columns. Such a unit would respond optimally to a moving zero-crossing segment that extended along the entire length of the columns, but it would also respond to shorter stimuli and even to moving spots of light. More complicated receptive fields (for example, moving bars or slits) can be built up from these units. A critical empirical characteristic of such a unit would be that if its Y-cell input is abolished, the cell either fails to fire at all or, if it does fire, it loses its directional selectivity. It is not yet known whether this is true of directionally selective units. Otherwise, the model's properties are in overall agreement with the available facts (Hubel and Wiesel, 1962, 1968; Schiller, Finlay, and Volman, 1976a, 1976b [called S_1 cells there]). The paper by Marr and Ullman (1979) contains a fuller account of the properties of and predictions from this model.

Using Directional Selectivity to Separate Independently Moving Surfaces

Computational theory

The movement of an object against its background can be used to delineate the object's boundaries, and the human visual system is very efficient at exploiting this fact. If the complete velocity field is given (that is, speed and direction at each point of the image), object boundaries will be indicated by discontinuities in this field, since the motion of rigid objects is locally continuous in space and time. The continuity is preserved by the imaging process and gives rise to what I earlier called the principle of continuous flow, according to which the velocity field of motion within the image of a rigid object varies continuously everywhere except at self-occluding boundaries. Since the motions of unconnected objects are generally unrelated, the velocity field will often be discontinuous at object boundaries. Conversely, as we saw in Chapter 2, lines of discontinuity are reliable evidence of an object boundary.

Unfortunately, the complete velocity field is not directly available from measurements of small oriented elements. Because of the aperture problem, only the sign of the direction of movement is available locally. This means that an additional stage is necessary for detecting discontinuities in the velocity field. In this section, we ask how and to what extent the limited raw information (the sign of the direction only) may be used to detect these discontinuities.

The sign of the local direction of motion determines neither the movement's speed nor its true direction, but it does place constraints on what the true direction can be (see Figure 3–37). The constraint is that the true direction of motion must lie within the 180° range on the allowed side of

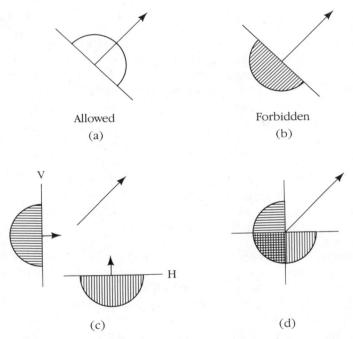

Allowed   Forbidden
(a)   (b)

(c)   (d)

Figure 3–37. The combination of local constraints from directionally selective units to determine the direction of motion. The constraint placed by a single such unit is that the direction of motion must lie within a range of 180° on the allowed side (b). (c) The forbidden zones for two oriented elements (V = vertical; H = horizontal) moving along the direction indicated by the arrow. The forbidden zone horizontal) moving along the direction indicated by the arrow. The forbidden zone of their common motion is the union of their individual forbidden zones, as indicated in (d). The direction of motion is now constrained to lie within the intersection of their allowed zones, that is, the first quadrant.

the local oriented element (Figure 3–37a), or, alternatively, it is forbidden to lie on the other side (Figure 3–37b). The constraint thus depends on the orientation of the local element. Hence, if the visible surface is textured and gives rise locally to many orientations, the true direction of movement may be rather tightly constrained.

   Constraints can be combined as illustrated in Figures 3–37(c) and (d) for the simple case of two local elements. The true direction of motion is diagonal here. The vertically oriented directionally selective unit V sees motion to the right, and the horizontally oriented unit H sees motion upward. If these two units share a common motion, we can combine the constraints they place on the direction of that motion by taking the union of their forbidden zones (Figure 3–37d). The result is that the direction of motion is now constrained to lie in the first quadrant, as illustrated. Additional units can further constrain the true direction of motion by expanding the forbidden zone.

   The diagram also shows how the motion of two groups of elements may be incompatible. If the allowed zone for one group of elements is completely covered by the forbidden zone of another, their motions clearly cannot be compatible. Notice in this connection that only the direction of movement, not its speed, is used here. A system that segments a scene in this way will be relatively insensitive to variations in speed.

   The final observation that we need in order to use this scheme is that objects are localized in space. If the objects are also opaque, their images will have an interior within which the forbidden zones in diagrams like Figure 3–37(d) are consistent, provided that those forbidden zones draw their elements from small neighborhoods. Exceptions can occur, for example the center of a rotating disc, but only rarely. Hence, the method will be reliable. It is not, of course, exhaustive—if two surfaces are relatively stationary, this method will fail to separate them.

Algorithm and implementation

The diagrams of Figure 3–37 contain essentially all the information we need to know here, for the algorithm must consist of searching for neighborhoods with locally compatible directions of motion. Figures 3–38 to 3–40 show some results from a computer implementation of such an algorithm, written by John Batali. The first example, Figure 3–38, shows the detection of a moving pattern embedded in a pair of random-dot images. A central square in Figure 3–38(a) is displaced to the right in Figure 3–38(b), while the background moves in the opposite direction. Figure 3–38(c) depicts the zero-crossing contours of Figure 3–38(a) filtered through $\nabla^2 G$. Figure 3–38(d) represents the values of the transient channel if the two frames shown in Figures 3–38(a) and (b) are presented

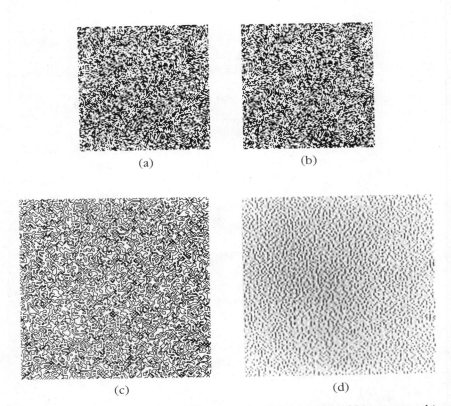

(a)   (b)

(c)   (d)

Figure 3–38.   Separating a moving figure from its background by using combinations of directionally selective units. A central square in (a) is displaced in (b) to the right. The background in the two pictures moves the opposite way. (c) The zero-crossing contours of (a) filtered through $\nabla^2 G$. (d) The convolution of the difference between (a) and (b) with $\nabla^2 G$. If (a) and (b) are presented in rapid succession, the function shown in (d) approximates the value of $\partial/\partial t(\nabla^2 G * I)$. The images are 400 × 400 pixels, the inner square is 200 × 200, each dot is 4 × 4, and the motions are 1 pixel. (Courtesy John Batali.)

in rapid succession. Figure 3–40(a) shows the results of applying the XYX-motion-detection operation to the zero-crossings of Figure 3–38(c). The direction of movement has been coded, as indicated by the star in the figure. As can be seen, black represents motion to the right, and white represents motion to the left. The central square is clearly delineated by discontinuities in the direction of motion.

The same analysis was also applied to the natural images shown in Figure 3–39, which are two successive frames taken from a 16-mm film of

Figure 3–39.   Two successive frames from a 16-mm movie of a basketball game. The same analysis was applied as to the random-dot patterns in Figure 3–38. (Courtesy BBC.)

a basketball game. The results appear in Figure 3–40(b). For example, the left arm of player 7 moved downward and to the left, and the rightmost player moved to the right. Because of the extreme sensitivity of the method, small registration errors, more or less unavoidable because of the way the two images are digitized, sometimes give rise to spurious motion of the background.

Psychophysically, the XYX-motion-detection scheme fits well into the first of Braddick's two categories. For example, the phenomenon should occur only over short ranges (around $w/\sqrt{2}$ or $15'$ at $5°$ eccentricity) and short ISI's (not more than the total time course of the temporal component of the transient channel, about 120 ms), according to Wilson's channel data. If speed and not direction were the only available discriminant, separation should be impossible, which we have found psychophysically (Figure 3–32).

In addition, the amount of information that can be obtained from directional selectivity depends on the direction of movement and on the orientation of the moved elements. Hence, the same velocity field may be seen as coherent or incoherent, depending on the orientations of the moved elements. The reason is that two nearby velocity vectors will produce the same directional sign on an element oriented roughly perpendicular to them but different signs on an element whose orientation bisects them. We also found this to be true psychophysically. Moreover, if the formation of coherent groups proceeds roughly in the manner of Figure 3–37, one might expect to see clusters of locally coherent motions in even purely random display sequences—and, in fact, one does. Such a mechanism also produces Anstis' (1970) reversed phi phenomenon, whereby simultaneous movement and contrast reversal can give rise to the illusion of movement in the opposite direction (see Marr and Ullman, 1979).

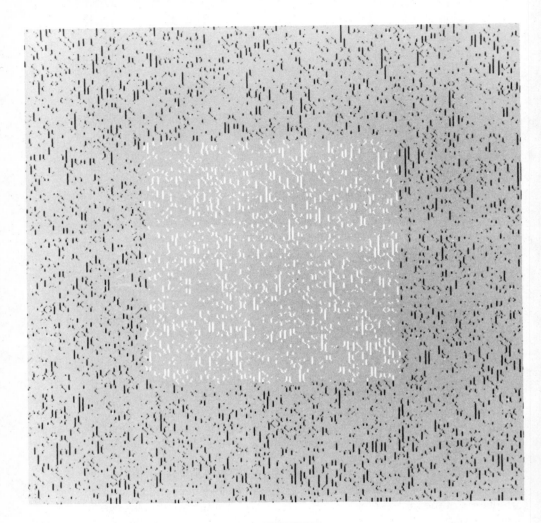

(a)

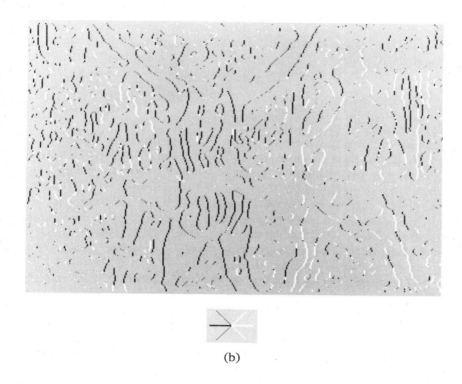

(b)

Figure 3–40.   Motion assigned to zero-crossings from the images of Figures 3–38 and 3–39. The direction of motion was assigned according to the rules described in the text and the result is displayed here using shades of gray. The keys below (a) and (b) indicate the shade of gray assigned to each direction. In (a), the central square clearly moves right, while the surround moves left. In the zero-crossings from the basketball game (b), the left arm of player 7 moves to the left and down, while the player to his right moves to the right. (Courtesy John Batali.)

Finally, the use of color but not luminance boundaries or the inter-position of a white field during the ISI could disrupt the mechanism, as Braddick requires, by interfering with the retinal machinery for measuring the time derivatives traveling up the Y channels.

Looming

There is another way in which the outputs of directionally selective units might prove useful, because combining directionally selective units from the two eyes yields a different kind of information (Marr and Ullman, 1979). Suppose that a particular zero-crossing has been identified and assigned incompatible motions in the two images. Then the zero-crossing is moving in depth either toward the viewer if it is moving away from the nose on both retinas, or away if the motion is toward the nose. If motion is to the right on both retinas, the object will pass safely to the viewer's left, and vice versa (Regan, Beverley, and Cynader, 1979).

For this type of analysis, it is not necessary to combine constraints in the manner of Figure 3–37; the raw output of the directionally selective units can be used. The difficulty in this case lies in ensuring that both left and right detectors are looking at the same zero-crossing; establishing this match is the essence of the stereo matching problem. However, if inaccur-acies from time to time are tolerable, a fast looming detector can be designed that does not have to wait for the results of stereo matching. For example, a simple looming detector can be constructed by comparing the signs of motion at corresponding retinal points. Such points will often but not always correspond to nearby points on the same moving object.

Such a scheme might rely at some point on a cell that has binocular receptive fields close by in the visual field, but not truly disparity sensitive, and whose preferred motions in the two eyes are opposite. There is some evidence for the existence of such cells (Regan, Beverley, and Cynader, 1979).

3.5   APPARENT MOTION

In the last section, we saw how very limited information about the motion in the visual field could be used at quite a primitive stage in the processing to provide certain rather rough information about how to decompose the scene into different surfaces. We also saw that this task can be done rather fast. With a little more time and care, however, visual motion can be made to yield a much richer harvest of information. Although the experiments of Miles (1931) and of Wallach and O'Connell (1953) preceded it, Ullman's (1979b) counterrotating cylinders demonstration (illustrated later in Figure 3–52) is the most telling demonstration so far contrived of what our visual systems can obtain from visual motion.

The demonstration consists of a sequence of frames, each of which is a projection of a set of dots on two concentric, counterrotating cylinders. Only the dots appear in each frame, and their positions change from frame to frame. As in the case of random-dot stereograms, each individual frame has no visible structure. However, when the frames are shown as a movie sequence, a vivid impression of the two counterrotating cylinders is obtained.

From this demonstration, it is clear that our visual system has remarkable powers to recover the shapes of unknown structures simply from the way their appearances change in the image. In his recent book on the subject, Shimon Ullman (1979b) has gone far toward constructing a complete theory of how this may be done, and he includes supporting psychophysical evidence. This section consists of a summary of Ullman's work, together with one or two general points that I wish to raise about it in the context of vision in general.

Why Apparent Motion?

Movement is an inherently continuous process that usually produces smooth changes in an image. Indeed, one might think that this is a rather important intrinsic property of movement with regard to its perceptual analysis, since its very continuity should help in the task of following pieces of an object around in an image to find out how they are moving. Why, then, is this section based on the study of apparent motion, whose essence is a discrete, discontinuous presentation of a fairly rapid succession of frames? Surely something is lost in the transformation from the continuous to the discrete. The theories I shall describe in fact apply to both kinds of motion, continuous and framed (or apparent). But that is not quite a satisfactory answer, and it is worth a little discussion to see that for the type of situation of interest here, one probably can think in terms of framed stimuli.

The first point is that we are no longer dealing with almost instantaneous phenomena, as we were in the last section. We are out of the realm of detection tasks here. Instead of finding out something simple but possibly important within 50 ms, we can afford to take quite a long time—say, ¼ to ½ s, which is large by perceptual standards—to allow the image to change by a reasonable amount. For we want not just to detect the change but also to measure its extent and use this information. So our fundamental approach is to contrast the positions of items at one time in the image with their positions at a sufficiently later time for the differences to be measured reliably, and we shall use the differences to make calculations about the underlying shapes and movements.

Thus, it is in our interests to delay matters, at least up to a point, but not too much, or the image will have changed beyond recognition—visible portions of the surface may become occluded or may rotate out of view. But at least in principle, it is the changes over a period of time that we need here, and they must be determined quite accurately.

That may be, one can reply. But the fact is that even if we want only to know where things have moved after 100 ms or so, surely it is easiest to find this out by smoothly following them. And isn't this made more difficult by cutting the sequence into distinct frames? Well, up to a point this must be true. On the other hand, if the frame rate is sufficiently fast compared with the time constants in, say, the cones (which are of the order of 20 ms or so), the two situations will be indistinguishable. We all know, too, that we can watch movies perfectly well and that the motion there looks quite normal. Yet they are split into only 24 frames per second, and one cannot discern these facts from perceptual evidence alone. In addition, psychophysical presentations consisting of just two frames separated by as much as 300 ms can give the subjective impression of smooth motion.

So, although the continuous problem may be slightly simpler than the recovery of structure from apparent motion, it is probably not much simpler and we can certainly do the harder problem involving apparent motion. The apparent motion problem is also much easier to formulate and to investigate empirically, and its results can be applied to the continuous case. It therefore seems sensible to solve this problem first and then to take stock of where we stand.

The Two Halves of the Problem

Our goal here, then, is not so much to detect the changes induced by motion but to measure and use them to recover the three-dimensional structures in motion. Broadly speaking, this introduces two kinds of task that, at least superficially, look rather different and somewhat analogous to what we met in stereopsis. The first task is to follow things around as they move in the image and to measure their positions at different times. This is the *correspondence problem,* and at its heart is the question, Which item in the image at time t_1 corresponds to which item at time t_2? The second task is to recover three-dimensional structure from the measurements supplied by the results of the first task, and this is called the *structure-from-motion problem.*

Apparently, these two problems are solved independently by the human visual system, and it is a great stroke of good fortune that they are separate. The critical empirical evidence for this is that none of the measurements on which the correspondence process rests involve three-dimensional angles or distances—they are all two-dimensional measurements made on the image (Ullman, 1978). Thus, there is no deep need for any feedback from the later task to the earlier.

The two tasks may therefore be dealt with independently. We shall first examine the correspondence problem and then alternative approaches to the second task. By now the reader can formulate for himself the critical preliminary question—What are the primitives on which the process operates, or, in our earlier terms, what is the input representation for the process? And since the measurements of changes in position must refer to the changes in position of an identifiable surface location, these primitives need to be as physical as possible. So, the reader will not be surprised to learn that the elements in the primal sketch seem to be used, although various interesting side issues arise in the details.

Then we must formulate the relationships that should hold between the positions of the primitives in adjacent frames (remember, we shall be dealing with apparent motion). In general terms, it is not hard to see that the closer and more similar two items are in successive frames, the more likely it is that they correspond. This simply reflects some kind of a statistical rule of the universe, and it will hold provided that the interframe interval is not too long in relation to the velocities of and distances involved with the visible motions. It turns out that the human visual system incorporates a permanent or "hard-wired" table of similarities by which the similarities and dissimilarities in the various parameters may be compared. For example, in experiments that test the similarity of two lines of the same contrast in successive frames, a change in length by a factor of 3/2 produces the same change in similarity as a change in orientation of 45°.

This similarity Ullman called the affinity measure, and it is based on two-dimensional measurements. However, this does not by itself determine the correspondence process. In order to do so, one has to take account of extra factors. For example, suppose one has two lines A and B in the first frame, and two lines a and b in the second. There are four possible pairings.

(1) $A \rightarrow a$ and $B \rightarrow b$
(2) $A \rightarrow b$ and $B \rightarrow a$
(3) $A \rightarrow a$ and $B \rightarrow a$
(4) $A \rightarrow b$ and B $\rightarrow$ b

This list omits possibilities like $A \rightarrow a$, and B goes nowhere. The question is, How does one decide which of the possible pairings actually occurs?

The obvious answer is, take that solution which maximizes the overall similarity between the frames. This similarity can be measured by means of some standard cost function that gives a similarity value to each pairing in a given solution, the overall similarity being the sum of the values for each pairing. The cost function tells us roughly how many quite poor pairings should be accepted in order to avoid an abysmal pairing or to acquire an excellent one in the overall match.

An approach of this type, which involves finding a solution that achieves an overall or global minimum, is analogous to part of what the Gestalt movement became interested in during the first third of this century, although several different phenomena were probably involved in the experiments that the Gestaltists actually carried out. They had the idea of an attraction among elements that bound them into wholes and governed the interaction between successive frames, but they were unable to see how much such an approach could account for the complexity that they saw in the correspondence process. Their basic difficulty was this: In a display such as Figure 3–41, they saw that $A \rightarrow A'$ and $B \rightarrow B'$; but if A and B' were removed, $B \rightarrow A'$. Hence, they reasoned, movements of wholes are of critical importance, and the phenomenon cannot possibly be explained in a purely local way. In large measure, this type of argument killed the school, because the Gestaltists viewed the problem of the formation of wholes as intractable.

There are two fundamental misconceptions here, and I shall make a point of them in order to draw a moral. The first is the point of basic mathematical ignorance. Certainly examples like Figure 3–41 show that the correspondence process involves more than finding purely local minima; if the problem can be formulated in this way at all, the minimum one wants is a global minimum. But—and here is the first point—there are many systems in which global minima can be found using only local inter-

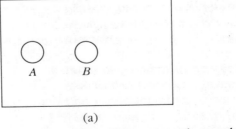
(a)

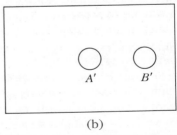
(b)

Figure 3–41. One of the patterns that puzzled the Gestaltists. (a) shows frame 1, and (b) frame 2. Perceptually, A goes to A' and B to B', so that B seems to move. (Courtesy of Shimon Ullman.)

actions, so the Gestaltists' findings did not force the conclusions they drew about the insufficiency of local interactions. In particular, the most obvious way out of the Gestaltists' problem with Figure 3–41 is to say that the total cost of $(A \rightarrow A') + (B \rightarrow B')$ turns out to be less than $(A \rightarrow B') + (B \rightarrow A')$. The idea seems even simpler when we observe that it is linear, and linear systems are extremely well-behaved—basically, they cannot get caught in local minima. In fact, Ullman's correspondence theory is essentially a linear one.

The second misconception was that the Gestaltists lacked the idea of a process. They thought of groupings as being subject to various types of rules—the principles of closure, good continuation, regularity, symmetry, simplicity, and so forth (see Koffka, 1935, p. 110)—which were summarized as the Gestalt law of Prägnanz. This law was to them like a physical law. If they had had the idea of embodying such principles in a number of grouping processes—for example, as constraints on what should or should not be grouped together—they might not have abandoned the other half of their endeavor, the systemization of the formation of wholes.

The moral here is this. We saw in Chapter 1 something of the perils to purely computer vision workers of ignoring the biological evidence about how the human visual system is organized. The basic difficulty is that such an oversight can lead to trying to solve problems which are not really problems at all, but which happen to arise because of the particular limitations of sensors, or hardware, or available computer power. Here we see the opposite, in which mathematical ignorance (which could have been avoided) and a failure to think more in terms of processes (which is more excusable) led to the failure of a school of thought that had actually made a number of valuable insights. The moral is that ignorance in any of these three fields can be damaging. Just as the modern physicist has to know some mathematics, so must the modern psychologist, but the psychologist must also be familiar with computation and have a clear idea of its abilities, its limitations, the fruitful ways in which to think about processes, and, most importantly, what it takes to understand these processes.

This, then, is roughly the current state of affairs concerning the correspondence problem. Ullman formulated it as a linear minimization problem and showed how this can account for much of the psychophysics. We shall explore his ideas in some detail and see some even more recent ideas about their biological implementations based on the higher-level primal sketch primitives. As for the topic as a whole, it is not yet solved at any of our three levels; however, a substantial amount is known about it, and a complete computational theory of it cannot, I think, be too far off.

The second half of the problem, the structure-from-motion theory, is in better shape and has essentially been solved at the level of computational theory (Ullman, 1979a). The form of the theory is by now familiar—it is the same as we saw in Chapter 2 (for the primal sketch) and earlier in this chapter, although chronologically Ullman's was one of the early theories. The critical additional constraint that he used was rigidity, and he made a very precise formulation of its use and showed how the recovery of three-dimensional structure may proceed from the measurements made available by a successful correspondence process. The underlying mathematics consists of a theorem stating essentially that three views of four rigid, noncoplanar points are sufficient for recovering their three-dimensional dispositions and motion. We shall see how this result may be used as the cornerstone of the interpretation of visual motion. Longuet-Higgins and Prazdny (1980) used a similar approach in their study of optical flow.

One final comment is perhaps in order as a conclusion to this brief survey. Although the geometry of three-dimensional space has been studied since the time of Euclid, some relatively simple theorems still appear to be unknown. The four-points, three-views theorem was one, and we shall meet another when we discuss the recovery of shape information from silhouettes (Marr, 1977a). It is difficult to believe that there are not others. These two have recently been formulated because the imaging process occurs in three dimensions, and hence certain types of geometrical relationships, if known and used, can be incorporated into processes for interpreting images. It may be well worth a mathematician's time to look again into the subject of three-dimensional Euclidean geometry.

The Correspondence Problem

Empirical findings

What is the input representation?

On general grounds, we require that the tokens on which the correspondence process operates, which we shall call *correspondence tokens,* be physically meaningful. This eliminates raw gray-levels, and one can directly demonstrate that, in the human visual system, gray-level correlation does not form the basis for the correspondence process. Figure 3–42 shows how. The maximum gray-level correlation between the two frames in Figure 3–42(a) occurs at zero displacement, as can be seen from the correlation graph, Figure 3–42(b). If the sharp edges are matched, however, one would expect edge E in frame 1 to jump to F in frame 2, and this is in fact what happens.

This demonstration establishes that the correspondence takes place above the level of gray-level intensity values, but how far above does one

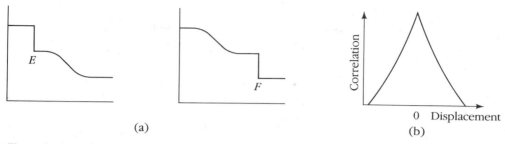

(a)   (b)

Figure 3–42.   Correspondence is not established between gray-level images. If it were, two frames with the intensity profiles shown in (a), when presented in succession, would give no impression of movement, since the maximum value of the correlation between them occurs at zero displacement (b). Instead, edge *E* is seen to move to edge *F,* suggesting that edges, not gray-level images, are the tokens used in the correspondence process. (Reprinted from Shimon Ullman, *The Interpretation of Visual Motion* by permission of The MIT Press, Cambridge, Massachusetts, Figure 1.1, Copyright © 1979 by The Massachusetts Institute of Technology.)

go? Is the correspondence established between relatively small and simple parts of a scene, largely independently of shape and form, or are much more complicated descriptions involved, like the interpretation of the whole of a shape from one frame, before different frames are compared?

Figure 3–43 is one of a series of demonstrations that rules out the second alternative. The figure illustrates two successive frames, one denoted with full lines, and the other with dotted lines. If the whole pattern was analyzed from one frame, with the shape of the wheel extracted and then used to match the elements in the next frame, the observer should perceive the frames as a whole wheel rotating when they are presented in rapid succession. Notice, however, that the inner and outer parts of the wheel have their closest neighbors in one direction, whereas the central ring has its closest neighbors in the other direction. Because of this, if the matching is carried out purely locally, the observer should see the central ring rotating one way and the inner and outer rings rotating the other (as shown with arrows in Figure 3–43). When appropriately timed, this is in fact what happens.

This begins to suggest primal sketch elements, and the next demonstration shows that terminations play a role (as they do in stereopsis). In Figure 3–44(a), a correspondence is established between the ends of the two lines. This breaks down if the distances between the corresponding ends are much greater than that between the line segments, as they are in Figure 3–44(b), in which case a correspondence is established between the short line and only the nearest part of the long one. It has not yet been

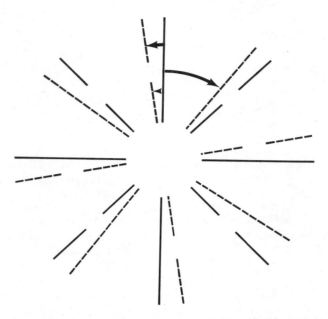

Figure 3–43. Evidence that the correspondence problem for apparent motion involves matching operations that act at a low level. Frame 1 is shown with full lines and frame 2 with dotted lines. Instead of appearing as a single wheel rotating, the wheel splits when appropriately timed, the outer and inner rings rotating one way and the center rotating the other as indicated by the arrows. This suggests that matching is carried out on elemental line segments and is governed primarily by proximity. (Reprinted from Shimon Ullman, *The Interpretation of Visual Motion,* by permission of The MIT Press, Cambridge, Massachusetts, fig. 1.3. Copyright © 1979 by The Massachusetts Institute of Technology.)

firmly established whether discontinuities of the type shown in Figure 3–44(c) are matched, but the question is obviously of interest.

Figure 3–45 adds to the evidence that correspondence is determined by quite low-level tokens and not by the shape or form of the corresponding figures. In Figure 3–45(a), the square *A* goes to the larger square *B*. In Figure 3–45(b), it goes to the larger triangle *B*, not to the smaller square *C*. Thus in these cases the motion of the constituent elements rather than the similarity between the complete forms governs the matching process. Ullman (1979b, p. 27) concluded that (1) differences in the tendency of different figures to fuse is consonant with the motions established between their components, and (2) there are no indications that structural figures are part of the basic elements or that the correspondence process is based on figural similarity.

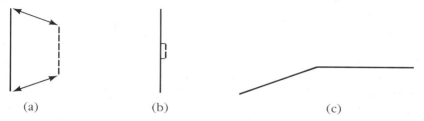

(a)   (b)   (c)

Figure 3–44.   Terminations can also act as correspondence tokens, provided that the two lines in successive frames do not differ too much in length (a). If they are very different (b), correspondence is established between the short line and a segment of the long line. It is not yet known whether orientation discontinuities such as the one shown in (c) can act as correspondence tokens. (Reprinted from Shimon Ullman, *The Interpretation of Visual Motion,* by permission of The MIT Press, Cambridge, Massachusetts, fig. 2.10. Copyright © 1979 by The Massachusetts Institute of Technology.)

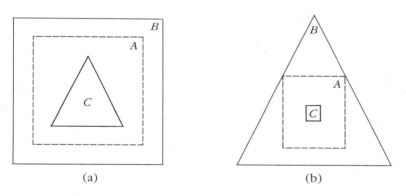

(a)   (b)

Figure 3–45.   In (a), the square *A* goes to the larger square *B*, yet in (b) it goes to the larger triangle *B*, not to the smaller square *C*. This is more evidence that correspondence is governed by the motions of constituent elements, not by complete forms. (Reprinted from Shimon Ullman, *The Interpretation of Visual Motion,* by permission of The MIT Press, Cambridge, Massachusetts, fig. 1.6. Copyright © 1979 by The Massachusetts Institute of Technology.)

As a result of discussions between Shimon Ullman, Michael Riley, and myself, Riley has found that matches can, for example, be established between oriented clouds of dots or between groups of parallel lines, when in neither case do the constituents match. Two illustrations of this phenomenon appear in Figures 3–46(b) and (c). In such cases, the matching rules appear to be governed by parameters like the overall orientation and size of the group. Borders like those in Figure 3–46(a) can also be matched, although there can be no question here of any kind of constituent match.

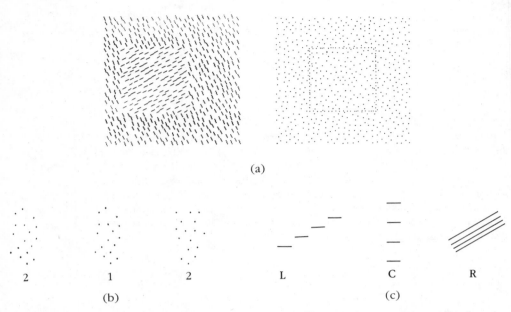

(a)

2   1   2   L   C   R

(b)   (c)

Figure 3–46.   Matching can take place between higher-order borders or tokens even when the constituents do not match. For example, correspondence can be established between the two kinds of border surrounding the squares shown in (a). (b) An experiment in which one cloud of dots is presented in frame 1, and two clouds in frame 2 (as marked), with the property that one of the clouds in the second frame is identical to the cloud in the first frame, whereas the other cloud is not. No preference for the identical cloud is exhibited. In (c), this idea is carried further. The first frame consists of group C, consisting of short horizontal lines. The second frame consists of two groups, L comprised of short horizontal lines and R of long diagonal lines. The observer sees no preference for motion from C to L, which proves that the correspondence in this case is not being carried out between the constituents of the groups but between descriptions of their overall structure.

The ISI's here are around 100 ms, much shorter than the ⅓ s or so required for shape to begin influencing matching.

So Ullman's conclusions may need slight modification so that these more abstract image descriptors from the full primal sketch can be included. However, his main point—that no elaborate form analysis precedes the correspondence process—still stands. And the limitations implied by the word *elaborate* here effectively allow the things that are allowed in the full primal sketch—overall length, size, orientation of a token, and so forth—but exclude the things that are excluded there—for example, any explicit representation of an internal angle in the token, the noticing of right angles, and so forth. It will be interesting to see how far the analogies can be taken between correspondence tokens and primitives in the full primal sketch.

Two-dimensionality of the correspondence process

The local behavior of the correspondence process for small numbers of isolated elements can be studied in experiments like that shown in Figure 3–47(a), in which the first frame (dotted line) contains one element and the second, two (solid lines), and the observer is asked to which line in the second frame the line in the first frame appears to go. Riley recently modified this scheme to the form shown in Figure 3–47(b), which has many copies of the same problem. The extended display has the advantage of being somewhat more sensitive.

Figures 3–47(c), (d), and (e) show stimuli for these experiments; in each case, frame 1 is dotted and frame 2 is not. The examples shown all have approximately the same affinity for the original. Figure 3–47(c) shows how length trades off against distance, Figure 3–47(d) shows how vertical displacement trades off against distance, and Figure 3–47(e) shows how orientation trades off against displacement. The relative weights of the different parameters for a 3 line configuration are tabulated in Figure 3–47(f) (from Ullman, 1979b, table 2.1).

For our brief survey of this problem, the detailed values of the table in Figure 3–47(f) do not matter so much, but the fact that the process uses measurements made on the image and not measurements of objective, three-dimensional quantities is important. This was established by Ullman (1978) in the type of experiment shown in Figure 3–48. In frame 1 of the experiment shown in Figure 3–48(a), for example, all the lines had the same brightness except for C. In frame 2, only L and R were brighter, and motion was induced from C to L or R. In this example, the two-dimensional relations between C and L and between C and R are identical. Their three-dimensional distances apart, however, are very different. In Figure 3–48(b), an experiment along the same lines is shown in which the three-dimensional distances are the same but the two-dimensional ones are very different. Similarly, in Figure 3–48(c) the two-dimensional and three-dimensional angles are different.

From experiments like this, Ullman concluded that three-dimensional measures were irrelevant to the correspondence process; everything he found could be predicted from the two-dimensional configurations. He was also able to make another fascinating point, about the smoothness of apparent motion. When one looks at two frames, the transitions from one

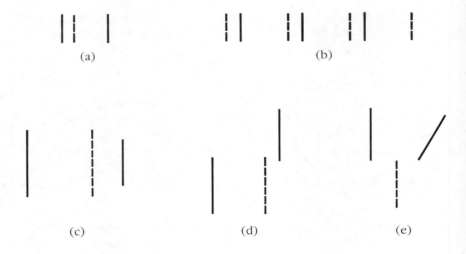

	Relative Weight				
	0	1	2	3	4
Orientation difference (in degrees)	15	30	45	60	75, 90
Distance ratio	1.1	1.2	1.6	2.25	2.7, 3.8
Length ratio	1.04	1.13	1.5	2.1	2.5
$1/\cos \alpha$	1.04	1.15	1.41	2.0	(2.3)

(f)

Figure 3–47. (a) shows a typical two-frame experiment of the kind used to measure affinities, and (b) shows a more sensitive version of the same experiment. In (c)–(e), frame 1 is shown with dotted lines and frame 2 with full lines, and the two stimuli in frame 2 have about the same affinity for the original. (c) How length trades off against distance; (d) how displacement trades off against distance, and (e) how displacement trades off against orientation. The measured affinity values are tabulated in (f). (Reprinted from Shimon Ullman, *The Interpretation of Visual Motion,* by permission of The MIT Press, Cambridge, Massachusetts, figs. 2.5–2.9 and table 2.1. Copyright © 1979 by The Massachusetts Institute of Technology.)

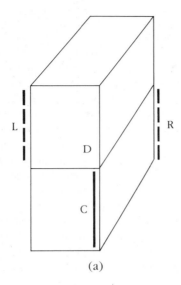

(a)

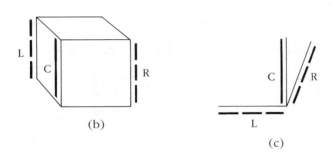

(b)

(c)

Figure 3–48. Only two- and not three-dimensional measures are used by the correspondence process. In (a), a correspondence is established between C (frame 1) and L and R (frame 2), which have identical two-dimensional relationships to C but different three-dimensional ones. They behave identically. In (b), L is preferred over R. (c) tests angles, and again it is two-dimensional angles that determine the correspondence. (Reprinted from Shimon Ullman, *The Interpretation of Visual Motion,* by permission of The MIT Press, Cambridge, Massachusetts, fig. 2.22. Copyright © 1979 by The Massachusetts Institute of Technology. Part reprinted by permission from Shimon Ullman, "Two dimensionality of the correspondence process in apparent motion," *Perception* 7 (1978), 683–693, fig. 1.)

to another sometimes seem to happen smoothly and sometimes not. Studies like those by Corbin (1942) and Attneave and Block (1973) had found that smoothness of motion was determined predominantly and perhaps entirely by perceived three-dimensional distance rather than by objective two-dimensional distance. Yet Kolers (1972, ch. 4 and 5) was only the most recent of a line of researchers who studied correspondence strength using smoothness of motion as a criterion.

Plainly, there was some inconsistency, because the three claims—(1) smoothness of motion depends on perceptual distance, (2) correspondence strength depends on two-dimensional distance, and (3) smoothness of motion reflects correspondence strength—are incompatible. Ullman (1978, experiment 5) resolved the dilemma by constructing a situation like Figure 3–47(a), in which motion one way was smoother but motion the other way was stronger and won. Smoothness and correspondence strength are therefore different phenomena, and the correspondence process relies on two-dimensional measurements only, probably after allowing for the effects of eye movements (Rock and Ebenholtz, 1962).

Ullman's theory of the correspondence process

In more complex displays, an element does not always map to the element with highest affinity, as we have seen in Figure 3–41. Mappings are affected by inter-element interactions as well. In his empirical approach to this, Ullman introduced the notion of *correspondence strength* (CS), which is derived from the local affinities but also incorporates the effects of various kinds of local competition and which determines the final mapping. Figure 3–49 illustrates this idea. First, the affinity between each pairing is measured, and then local interactions take place on these to produce the CS. The interactions weaken the CS when splitting or fusion occurs, for example, and so these conditions are avoided. In a particular numerical example (appendix 4 of his doctoral thesis), Ullman showed that this same simple scheme could account for several examples that were considered challenging to motion perception theories (Kolers, 1972; Attneave, 1974; Ullman, 1979b, sec. 2.4.1).

These points, though, primarily showed that the kind of thinking that had been used when examining the capabilities of local interactions was still often seriously flawed, sometimes in the same way as was the Gestaltists', by a failure to appreciate the complexity of functions that can be computed by local interactions. More interesting was Ullman's attempt to formulate a theory for the correspondence process, which he called the minimal mapping theory. It is, in fact, a maximum likelihood theory.

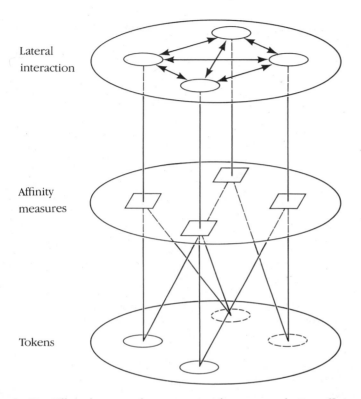

Lateral
interaction

Affinity
measures

Tokens

Figure 3–49.   Ullman's approach to correspondence strength. Raw affinity values
are measured between correspondence tokens, and then local interactions take
place between them to obtain the final correspondence strengths.

There are three main assumptions behind the theory. The idea is to
provide a way of judging the relative merits of pairing tokens between
frames. Since the underlying argument is probabilistic, we need to assume
that different pairing decisions are independent. That is the first assumption. The second is that each token in frame 1 is paired with at least one
token in frame 2, and vice versa. We do not explicitly demand a one-to-
one relationship (that is how splits and fusions are allowed), but since each
pairing costs something, the final answer keeps splits and fusions to a
minimum. Thus, the second assumption is that the set of pairings should
cover both sets of tokens.

The third idea is the interesting one. Of course, the range of true
velocities in the world varies widely—sometimes a viewer moves fast,

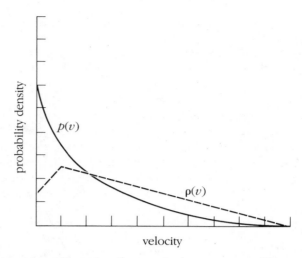

Figure 3–50.   The average distribution of velocities in an image. For almost any reasonable velocity distribution for objects in the world such as $\rho(v)$, after projection into an image, $p(v)$, small velocities will predominate. See discussion in text. (Reprinted from Shimon Ullman, *The Interpretation of Visual Motion,* by permission of The MIT Press, Cambridge, Massachusetts, fig. 3.11. Copyright © 1979 by The Massachusetts Institute of Technology.)

sometimes slowly, sometimes objects move by quickly, sometimes not. But almost whatever is chosen for distribution of velocities in the world, the projections of those velocities in the image will usually be small rather than large, simply because of the imaging process. This point is illustrated in Figure 3–50. The dashed line $\rho(v)$ shows one choice for the probability distribution for true velocities in space. The solid line $p(v)$ shows the corresponding projected velocity distribution. Thus on only very general grounds, mappings that prefer nearest neighbors will be more likely.

The theory is now straightforward: The entropy $q(v)$ of a given velocity v is defined as $-\log p(v)$, where p is its probability. The maximum likelihood solution is the solution that minimizes the total entropy (just as in statistical mechanics), and we can find this simply by letting $q(v)$ be the "cost" of assuming velocity v and then discovering the mapping that minimizes the total cost. This is a linear problem that can be solved by a simple local network, in which one can incorporate additional penalties for deviations from one-to-one mappings if desired. The cost function is the affinity function that we discussed earlier, and the interactions of Figure 3–49 that produce the CS in effect find the minimum total cost, that is, the most likely mapping given the statistics of the universe. This scheme generalizes naturally from the discrete case of successive frames to the continuous case, where the image is represented more as an incoming stream of tokens.

A critique of Ullman's theory

As a first attempt, this theory of the correspondence process is an extremely valuable contribution, and it provides a welcome and refreshing sip of clarity after the confusions and obfuscations of the preceding 50 years. Its importance is that it enables us to formulate a number of empirical questions that would not otherwise arise, and it opens the way for a rational investigation of the phenomenon rather than the confused cataloguing of its phenomenology.

Leaving aside the empirical aspects of the theory for the moment, there are a few points that should be raised, especially in a book whose primary business is the theory of the visual system. The first point is that the independence assumption, necessary for a probabilistic development, is empirically not quite true, at least in its simplest terms. In Figure 3–51(a) we do have independence—the unambiguous match of C_2 to R_2 does not affect the ambiguity of the behavior of C_1. In the situation shown in Figure 3–51(b), however, the behaviors of C_1 and C_2 are related—in fact, as Ullman pointed out, they behave as if they formed the endpoints of line C in Figure 3–51(c). They do not so behave when the induced grouping is different, as it is in Figure 3–51(d).

So it seems that the correspondence process can, to some extent, operate on groups as well as on their constituents. Although the grouping process does not involve explicit descriptions of the internal structure of the groups, and although matching between overall groups does not preclude additional matchings between their constituents, they can perhaps act to constrain those matchings. Specifically, matchings that are compatible with the grosser group matching are allowed, whereas those that are not are disallowed. This type of internal structure in a theory can be accommodated by a probabilistic framework, but it is awkward and indicates that we may not yet have found the most useful approach.

The second point we have already met—that correspondence may be established between groups without correspondences being established between their constituents. Ullman himself noted that this could happen (1979b, sec. 2.4.2), and more recent work with M. Riley has confirmed and extended this finding. Interactions such as these between higher-order units can, of course, be simply added on to the theory in the way that Ullman suggests, but they do not follow from it naturally and are not at all

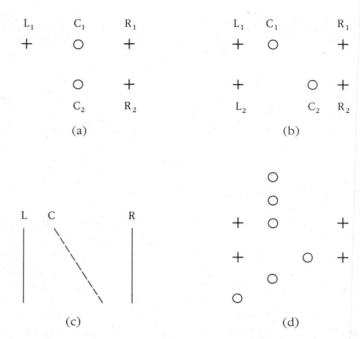

Figure 3–51. In this figure, frame 1 is shown with circles, frame 2 with crosses. In (a), the presence of C_2 does not affect the behavior of C_1. In (b) it does, however, the pair C_1C_2 acting like the line C in (c)—it goes either to L or to R. If the token configuration is disrupted by the presence of another organization as in (d), the central pair is no longer treated like the line C. (Reprinted from Shimon Ullman, *The Interpretation of Visual Motion,* by permission of The MIT Press, Cambridge, Massachusetts, fig. 2.20. Copyright © 1979 by The Massachusetts Institute of Technology.)

predicted by it. In fact, they run almost counter to it, since the whole thrust of the theory is to show how the sometimes confusing and complex behavior of the correspondence process on different patterns can arise from purely local interactions between simple processors that are associated with the constituents of the patterns.

For the third point, we need to adopt a slightly different perspective, that of the theory builder. What, we might ask, does the probabilistic approach buy? And the answer is, essentially, linearity. The practical consequence here is that purely local interactions are guaranteed to yield the global minimization that we seek. This is of great didactic value, because it shows that, as in our first cooperative stereo algorithm, the right global effects can be gained by purely local interactions. At first sight, this is exactly what we should try for, since, on the whole, the tangential connections in the cerebral cortex are known to be quite short (for example, Szentagothai, 1973).

Our experience with stereopsis and locally parallel organization has, however, warned us to beware of these arguments because of the problems associated with iteration. We must be careful here because Ullman's theory is not meant as an algorithm—it is a top-level theory—and there certainly are noniterative ways of implementing it. Nevertheless, the fact that it can be implemented with only local connections is an advantage only if it actually is implemented in this way. Unfortunately, if we take the theory at face value, which suggests an implementation, then I think a major objection must be that the rate of convergence for this type of calculation is slow—slower than, for example, the first stereopsis algorithm. To be sure, the rate depends on the starting point—and the rough grouping with large tokens could help here—but even so one would need, say, 10–70 iterations for reasonable convergence. This argument is not completely secure—one can usually patch up any particular convergence problem with special tricks to speed it up—but it does weaken the initial attractions of the theory being built around simple local network interactions.

The final point is much less easy for me to express, since it rests much more than the others on unsubstantiated intuition about how the brain works. Basically, my feeling is that at these rather low levels, probabilistic approaches such as the maximum likelihood principle are not used. Partly this feeling comes from having tried to use them myself a number of times—a probabilistic approach to stereopsis yields something like gray-level correlation, and I once tried to solve some problems related to the 2½-D sketch by using this approach—and partly from the general belief that a probabilistic approach is somehow not definite enough. For a problem of any complexity, the maximum likelihood solution is always pretty improbable (in the technical sense). Yet here the answers provided by the visual system are almost always correct and, moreover, are usually accompanied by a subjective feeling of certainty, rarely of doubt—much more certain and more often right than would be indicated by a rather low probability value. In similar situations, I have usually found that better constraints are available to describe how the world is put together, and these have often led to a much firmer basis for a computational theory.

In other words, if forced to answer the question posed at the end of the section on stereopsis—namely, Does this computational theory solve the right problem?—my answer would be more equivocal than it was for stereopsis or for the other half of Ullman's theory, the structure-from-motion problem. I do not yet have any very solid alternative, but the following remarks indicate the direction of my thoughts on this problem.

A new look at the correspondence problem

One problem or two?

The heart of any computational theory of a visual process is the answer to the question, What is the process for? In Ullman's framework, the goal of the correspondence process is to establish a relation between successive frames that allows measurements of the changes that have taken place. These measurements can then provide the input for subsequent processes that can recover the structures and their motions.

No doubt this is at least part of the job of the correspondence process, but is it the whole of it? Looking ahead a little, we shall see that the recovery of structure from motion incorporates (in an internally testable way) the assumption that the moving bodies are rigid. So we may ask about the correspondence problem first of all from the point of view of an observer in a world of moving, rigid bodies.

For small time intervals, the actual correspondence problem posed by this situation is essentially equivalent to the correspondence problem in stereopsis, because moving and rotating an object a little produces the same effect as moving and rotating one eye a little. Of course, different bodies may be moving in different ways, thus being equivalent to different pairs of eye positions, but the stereopsis matching theory is a local one, and it can be applied locally, provided that its assumptions are obeyed locally. These assumptions are that surfaces are smooth locally and matching is unique, because a given position always moves to only one other, and this nearly always means only one other in the image. Of course, some visible points will become invisible, and vice versa, but this is merely the analogue of the fact that, in stereoscopic depth changes, one eye can see parts of the surface that the other eye cannot see.

What, then, about the splitting and fusion phenomena of apparent motion, in which a single element in one frame splits to match two in the next (or conversely)? These are strong and well-known phenomena in apparent motion and have caused considerable theoretical problems. How often ought they to arise in the structure-from-motion situation? We have already seen that they can arise in stereopsis, both physically, in the rare instance that two surface markings that are distinct from one eye happen to lie along the line of sight from the other, and psychophysically, in Panum's limiting case. We have even seen from Braddick's stereograms of Figure 3–19(b) that the human visual system is very catholic about accepting double matches, provided that they are unique from one eye. But the reasons there were not fundamental ones; they had to do with the implementation and arise basically because the uniqueness condition is so strongly satisfied by the physical world that the visual system can afford to assume that it holds without internally checking it.

Are the splitting and fusion phenomena of apparent motion of the same kind as in the stereo correspondence problem, or are they more fundamental? I think that if we are committed to the view that the sole function of the motion correspondence process is to solve the problems produced by rigid bodies in motion, then this problem can be solved in the same way as the stereo correspondence problem, which is equivalent. These phenomena would have to be explained away much as the examples of Panum's limiting case were in stereopsis.

However, this approach is not very satisfactory. One rather subjective reason is that the kinds of stereopsis achievable by the matching of pure texture edges are so rivalrous (see, for example, Mayhew and Frisby, 1976) and the impression of depth from them so poor that one has the feeling that "real" stereopsis is not happening at all—only some vague preliminary parts of it are (perhaps the vergence control system). In apparent motion, however, impressions are not at all as vague—such edges are clearly seen in motion with respect to one another. The matching that is obtained between pairs even as dissimilar as those in Figure 2–34 is quite clear and definite, and not at all rivalrous, as it is in stereopsis.

Another argument, which I find quite compelling, comes from a report by Ramachandran, Madhusudhan, and Vidyasagar (1973) that apparent motion can be established between subjective contours and even between disparity edges in a random-dot stereogram. This is almost a paradox from our narrow point of view, because if disparity edges have already been obtained, then we already have the three-dimensional structure, so why initiate this whole structure-from-motion process in order to obtain it?

Our narrow point of view must, I think, be inadequate—one simply cannot understand the motion correspondence process in so confined a way. How, then, is this process essentially different from the stereo correspondence process?

The crucial difference is that one is in space, and the other is in time. For rigid bodies the processes are equivalent, but for pliable surfaces they are not. The shape of an object from the left eye is always the same as its shape from the right eye at the same instant, but its shape a moment later may be different. This is not an uncommon phenomenon at all. A distant bird, for example, changes its shape and appearance very rapidly, both because it is not rigid and also perhaps because the sun catches its beating wings at one particular angle. The bird's image may be quite small and difficult to decompose into roughly rigid components. Nevertheless, although its motions may yield little or no direct clues about its structure, there is no doubt that the changing appearances are all related to one bird.

In other words, time introduces an important new factor, which is rather independent of the precise details of an object's three-dimensional structure. This factor is the *consistency of an object's identity through time,* and it is a different problem entirely. To see this difference, simply consider Ullman's (1977) example of the frog changing into the prince. This is not part of the structure-from-motion problem, because the structure changes, but it *is* part of the object identity problem.

My argument is that the theory should consider the two problems separately, because they have somewhat different computational requirements. The idea of matching disparity edges is inexplicable in the first approach but entirely explicable and almost obviously desirable in the second. For example, consider the patterns of light formed by the surface of a river playing upon a riverbed. The only constants here pertain to the geometry of the riverbed, and therefore we clearly need to be sensitive to just this, independent of its surface radiance. This type of situation may well be the real-life equivalent of Bela Julesz's random-dot "moviegrams," and this type of situation makes it quite comprehensible that we should be able to perceive them. If a fish should happen to glide leisurely by, transiently mottled by the changing patterns of light and dark falling upon it, it may be defined only by its disparity boundaries. These boundaries are moving, but it is the same fish all the time. That is a problem in object constancy.

Separate systems for structure and object constancy

Thus the problems introduced by time yield at least two rather distinct tasks for the correspondence process in apparent motion, and these are themselves distinct from the first of Braddick's two categories, which we discussed in Section 3.4. The first task is the first half of the structure-from-motion problem, and, in an environment of rigid, moving bodies, it is essentially equivalent to the matching problem in stereopsis. The only difference between the two is that a small rotation of one image is added in the motion situation, but this poses no important new problems. The aim, as in stereopsis, is to achieve a very detailed correspondence between accurately localizable items in the image, so that measurements of their position changes may be made to the (second-order) precision necessary for the structure-from-motion computations. In order to achieve this precision, one would expect the primitives used here to be rather low ones, like those in the raw primal sketch or perhaps even zero-crossings.

The goals of the second task are different, and they arise precisely because an object can change between two temporal viewpoints in a way that it cannot between two spatial viewpoints—it can change its shape and configuration (and even reflectance). Precision is not its goal; rough identity is—and this is the key to the difference between visual motion and stereopsis. There is no point to an approximate stereo correspondence by itself; it only has a point if it is a prelude to an exact match. Hence, approximate matches appear as indistinct and rivalrous perceptions. There is, however, a great deal of point to an approximate correspondence in time, since it offers a way of establishing object continuity.

My suggestion, therefore, is that two theories may be needed here, one for when the object is changing *and* moving and one for when it is only moving. The first should use everything it possibly can, including high-level primitives with catholic matching rules and any three-dimensional information that is already available. The phenomena of subjectively smooth motions may even be more concerned with the first system than with the second, since smoothness goes perceptually hand in hand with object constancy, and we know from Attneave's work that smoothness involves three-dimensional perceptual distances. The second system is at a lower level, computationally equivalent to stereopsis, and although it may not be implemented in the same way, zero-crossings may be worth looking at in this regard.

Structure from Motion

The problem

We have already seen from Ullman's (1979a) counterrotating cylinders experiment, illustrated in Figure 3–52, that both the decomposition of a scene into objects and the recovery of their three-dimensional shapes can be accomplished when the only available information is that afforded by their changing appearances as they move. Each frame in that demonstration consists of an apparently random collection of dots and is by itself uninterpretable. Only when shown as a continuous sequence does the movement of the dots create the perception of two counterrotating cylinders.

We shall therefore consider the simplified problem of how to interpret a sequence of frames, each composed of a set of random dots. In real life, the frames will contain more elaborate primitives than dots, but, just as in the case of stereopsis, the bones of the problem can be expressed in this simple form. Furthermore, we shall assume that correspondences have already been established between successive frames by the correspondence process that I discussed above. In fact, we shall need only the simpler

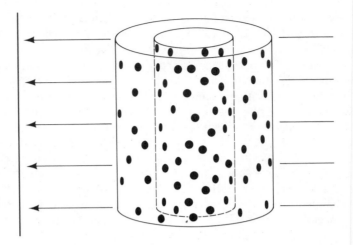

Figure 3–52.   Ullman's rotating cylinders demonstration. Dots painted on the two cylinders are projected orthographically onto a screen as indicated by the arrows, giving a sequence of frames like those illustrated in Figure 3–53. Each single frame has the appearance of a set of random dots, yet when seen as a movie, the rotating cylinders are clearly visible.

sort of correspondence process, the one for rigid objects, which we saw was computationally equivalent to the correspondence problem for stereopsis.

Thus the problem posed here is a set of data like that shown in Figure 3–53. Each frame consists of a set of labeled dots (though the labeling is not shown in the figure), where dot A in frame 1 corresponds to dot A in frame 2, and so forth. The question is, How can we make sense of these data? How should we go about a sensible three-dimensional interpretation?

The difficulty here is exactly like the one we met in the stereopsis problem, namely, that the solution is underdetermined. There are an infinite number of three-dimensional configurations that could give rise, through orthographic projection, to the images of Figure 3–53—any number of different, randomly changing snowstorms, for example. But we do not see any of these different possibilities; we see only one, and it is the correct one.

Just as in stereopsis, therefore, we must be bringing additional information to bear on the problem that constrains the solutions one finds. This additional information must at the same time be powerful and true but rather unspecific. Powerful because it forces a solution that is usually unique; true because not only does one perceive only one solution, but

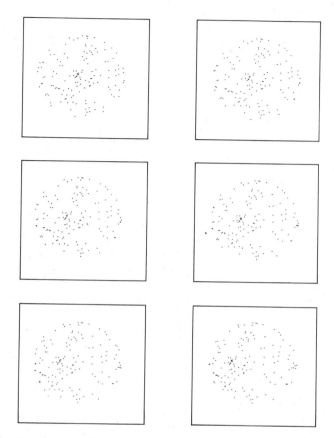

Figure 3–53.   The structure-from-motion problem. This set of frames contains three-dimensional information (see Figure 3–52). How are we to recover it?

that solution is also the correct one physically; and unspecific because the system works in unfamiliar situations, without specific a priori knowledge of the shapes to be viewed.

A previous approach

Although there have been a number of previous approaches to this problem, only one of them deserves comment. It originated with Helmholtz (1910; Braunstein, 1962; Hershberger and Starzec, 1974) and initiated the idea that motion and stereopsis are analogous: Specifically, recovering structure from motion is analogous to recovering distance from disparity.

The idea is, however, seriously flawed, because different objects in different parts of the visual field can be engaged in quite different motions. Now for the correspondence problem this does not matter, since that is essentially a local process. We have already made use of the fact that, for rigid objects and short time intervals, the two correspondence problems are in fact equivalent. We noted, however—without worrying particularly— that two different local motions would induce two different eye-pair positions to produce the equivalent stereo correspondence problem. The reason why this is not at all worrisome is that for correspondence the combination rules do not depend upon the precise position of the eyes. They have only to be close together and so have similar views. Hence, correspondence is unaffected by the fact that different portions of the visual field effectively induce different equivalent eye-pair positions.

Not so for the recovery of depth from disparity, however. As we saw, this depends critically on the effective interocular distance δ, and the induced δ's are in general different for each differently moving rigid object. There is no way of deducing their values a priori, and since they change, there is no way of comparing what is happening in one part of the visual field with what is happening in another. Hence, although this approach is actually valid for the correspondence problems in the two domains (pro-vided one restricts oneself to rigid motions and short time intervals), it is not valid at all for the recovery of three-dimensional structure.

It follows from these arguments that changes in velocity in the visual field (which are the analogues of changes in disparity) should not yield

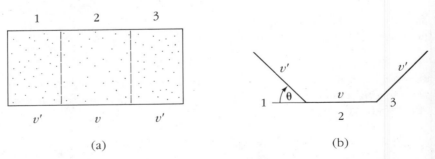

Figure 3–54. The conveyor belt demonstration. The dots in regions 1 and 2 move to the right with speed $v' = v \cos \theta$, and those in region 2 with speed v. However, the observer of (a) does not perceive the geometrical configuration (b). Instead, all of the regions appear in the frontal plane, and the dots appear to move faster in region 2. (Reprinted from Shimon Ullman, *The Interpretation of Visual Motion*, by permission of The MIT Press, Cambridge, Massachusetts, fig. 4.2. Copyright © 1979 by The Massachusetts Institute of Technology.)

direct impressions of depth, nor should common velocities be necessarily very useful for grouping. The Gestaltists, for example, had the notion of "grouping by common fate," which included grouping by common velocity, and Potter (1974) recently revived a form of this idea. However, the counterrotating cylinders demonstration includes points having the same velocity that belong to different cylinders. Evidence against the other half of the conclusion, that changes in velocity should yield changes in the impression of depth, is provided by Ullman's conveyor belt demonstration, illustrated in Figure 3–54. Dots in regions 1 and 3 have velocity v', and in region 2 they have velocity v. One does not perceive the different sections as planes at different depths or even as being arranged in the configuration of Figure 3–54(b). Instead, the dots all appear to be in the same frontal plane; they appear to speed up as they pass from region 1 to region 2 and to slow down again as they pass from region 2 to region 3.

The rigidity constraint

Most of the structures in the visual world are rigid or at least nearly so. This has been noticed by many students of motion perception (for example, Wallach and O'Connell, 1953; Gibson and Gibson, 1957; Green, 1961; Hay, 1966; Johansson, 1964, 1975), who formed the opinion that rigidity plays a special role in the problem. What they failed to realize, and what Ullman pointed out, was that searching for rigid interpretations is not merely a bias of our motion perception machinery; it enables us to solve the structure-from-motion problem unambiguously, without the need for any other constraining influence. This remarkable fact follows from a piece of mathematics that Ullman called the structure-from-motion theorem. It states that given three distinct orthographic views of four non-coplanar points in a rigid configuration, the structures and motions compatible with the three views are uniquely determined, up to a reflection where the closer points become the more distant ones. In other words, three views of four non-coplanar points suffice to determine their three-dimensional structure, provided that the correspondence problem has already been solved. Again, this result is not restricted to apparent motion; in continuous motion, what counts as three views depends solely on the resolution of the underlying systems measuring the position changes in time.

The four-points–three-views combination of the structure-from-motion theorem is the minimal combination in the following sense. With just two views, any number of points can be constructed that have no unique three-dimensional interpretation (although some combinations fortunately will), so that in general two frames are not enough. With three frames, three points are again in general too few to yield a unique solution; one needs four points.

One can give a rough plausibility argument for four points and three views based on the number of degrees of freedom involved. Suppose that we label the four points O, A, B, and C, the point O always corresponding to the origin $(0, 0, 0)$, and let us label the three views 1, 2, and 3. There are 15 variables to be determined. Nine of them determine the three-dimensional positions in view 1 of A, B, and C relative to O (three points with three coordinates for each one), and the remaining six determine the three-dimensional rotations needed to obtain views 2 and 3 from view 1. (We rule out translations by superimposing the point O in each view). It takes three variables to specify a three-dimensional rotation, two to specify the axis, and one to specify the amount.

The amount of information we gain from each view is 6 relations, the two-dimensional coordinates of each of A, B, and C. (The point O is always $[0, 0]$.) Hence, two views give us 12 relations, fewer than the 15 unknowns and so insufficient to determine the structure. Three views give us 18 relations, which exceeds 15 and so will be sufficient provided that there are not too many singularities or internal dependencies. The difficult part of the proof lies in showing that the 18 relations are in fact independent. The fact that there are 18 relations and only 15 unknowns means that there is some information left over, and this is ultimately what allows one to test internally the hypothesis of rigidity.

The rigidity assumption

In our analysis of the use of directional selectivity to infer properties of the visible surfaces, we saw that lines of discontinuity in motion direction cannot arise by accident. They have to mean the presence of a boundary between two incompatibly moving surfaces. In our analysis of the stereopsis problem, we saw that the constraints of uniqueness and continuity guarantee that a solution exists and is unique, and this theorem formed the basis for stereo analysis, since it allowed us to formulate and rely upon the fundamental assumption of stereopsis.

The same is true here. The structure-from-motion theorem, together with the general truth that most things in the world are locally rigid, allows us to formulate the fundamental assumption for the recovery of structure from motion. It was called the *rigidity assumption* by Ullman (1979a), and it states: *Any set of elements undergoing a two-dimensional transformation that has a unique interpretation as a rigid body moving in space is caused by such a body in motion and hence should be interpreted as such.*

The structure-from-motion theorem tells us that if a body is rigid, we can find its three-dimensional structure from three frames (up to a reflection, because we are dealing with the orthographic projection). If it is not rigid, the chances of there being an accidental rigid interpretation are vanishingly small, so in practice, the method will fail. The method is therefore self-verifying, and we know that if we can find a three-dimensional structure that fits the data, it is unique and correct. The proof of the theorem is constructive and enables one to formulate a set of equations whose solution yields the three-dimensional structure if it exists.

It is easy to implement this scheme, because it requires only four points as input data and so can be run in parallel independently throughout the visual field. This makes the scheme a particularly attractive candidate for understanding how human motion perception works. However, the particular algorithms suggested by directly applying the methods used in the proof of the theorem are not biologically plausible. They do not, for example, satisfy all the guidelines that I set out in Section 3.1—in particular, the principle of graceful degradation. Simply setting up the equations and solving them provides an algorithm that is far too rigid. If the data are inaccurate or if the viewed object is not quite rigid, this method will fail and give no help.

What is wanted is an algorithm that degrades gracefully in at least two senses. First, if the data are noisy but more than three views are available, the algorithm should be able to deliver an account of the structure that is at first rather rough but which becomes increasingly accurate as more views and hence more information are presented. And second, if the viewed object is not quite rigid, the algorithm should be able to produce the not-quite-rigid structure, perhaps again at the price of needing more points or more views to work on. Algorithms with this kind of robustness are being developed at our laboratory.

Until a particular algorithm has been developed as a candidate for the one actually used by our visual systems, and until the consequent psychophysical and neurophysiological experiments have been carried out, we shall not know for sure whether this approach to motion perception is appropriate. One thing, however, is certain; we now know what the important experimental questions are. Until Ullman took a computational approach to the problem, we did not know.

A note about the perspective projection

It is thought that algorithms for decoding the perspective, rather than the orthographic projection, are not part of the human visual system. The underlying reason is probably that the changes between frames are usually small already, and the differences between the changes seen by the two projections are usually very small indeed. The psychophysical evidence is that receding motion, which gives rise to changes only in the perspective and not in the orthographic projection, does not yield a clear perception of three-dimensional structure in the way that other motions do (Ullman, 1979a). The structure-from-motion scheme is an essentially local one, however, since it operates off nuclei of only four points. Even the perspective projection is locally orthographic, so there are no practical difficulties involved in using orthographic deprojection techniques like Ullman's scheme even in the real-life perspective case.

Optical Flow

J. J. Gibson has long believed that "the fundamental visual perception is that of approach to a surface. This percept always has a subjective component as well as an objective component, *i.e.* it specifies the observer's position, movement, and direction as much as it specifies the location, slant and shape of the surface" (1950). Sixteen years later he enunciated similar opinions and illustrated them with Figure 3–55 (1966, fig. 9.3).

The mathematics of this situation began to be studied quite soon, but only for special cases or for particular aspects of the general case (Gibson, Olum, and Rosenblatt, 1955; Lee, 1974; Clocksin, 1978). Nakayama and Loomis (1974) showed how depth contours may be extracted from a representation of the retinal velocity field induced by motion of the observer. Only recently, however, has a general treatment of the problem appeared (Longuet-Higgins and Prazdny, 1980).

The optical flow problem, as I shall employ the term, is the use of the retinal velocity field induced by motion of the observer to infer the three-dimensional structure of the visible surfaces around him. These visible surfaces are assumed to be stationary. The principal difference from Ullman's approach is that the optical flow effects rely on the polar projection, whereas the structure-from-motion approach is inherently orthographic. Thus, the optical flow approach can in principle deal with planar surfaces, on which the structure-from-motion approach necessarily fails.

The input representation

The information, called optical flow, on which our analysis is to operate can be thought of as the instantaneous positional velocity field (Gordon, 1965), which associates with each element on the retina the instantaneous velocity of that element. As usual, these elements are to be thought of as having some physical meaning.

This information is by no means as simple to acquire as optical flow devotees sometimes seem to assume. We have already seen in Section 3.4

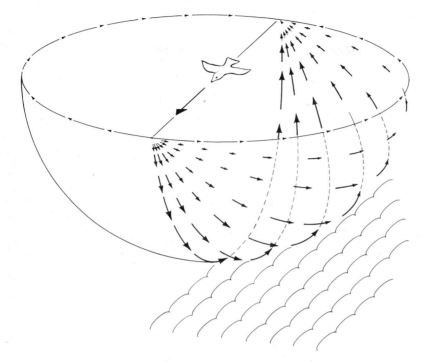

Figure 3–55.   Gibson's example of flow induced by motion. The arrows represent angular velocities, which are zero directly ahead and behind. (Reprinted from J. J. Gibson, *The Senses Considered as Perceptual Systems,* Houghton Mifflin, Boston, 1966, fig. 9.3. Copyright © 1966 Houghton Mifflin Company. Used by permission.)

that local measurements alone can give little more than the direction of movement because of the aperture problem. In fact, fully specifying the optical flow is equivalent to solving the simpler of the two correspondence problems in apparent motion, since knowing the flow field enables one to establish the correct correspondences between two frames photographed in sufficiently rapid succession. Hence if optical flow analysis is carried out by our visual systems, it must rely on an input of the same sort that feeds the structure-from-motion computations.

Mathematical results

If an observer is approaching a stationary surface on a linear trajectory, the point of impact is the singularity in the optical flow field, and the time to impact depends only on the angular velocities in the field (Koenderink and van Doorn, 1976). It is doubtful whether these facts are much used by our visual systems, since Johnston, White, and Cumming (1973) simulated optical expansion during approach to a surface and showed that human observers could reliably locate the focus of expansion only immediately prior to apparent impact with the surface. When teaching a pupil to land an airplane, a flying instructor will spend some time explaining that the current estimated landing point is the focus of expansion. This requires concentration and learning, for it is not a natural reflex. So Gibson's (1958) hypothesis that the center of optical expansion plays a major role in the control of locomotion is probably false for humans, although it may be more relevant to birds.

An authoritative account of the mathematics of optical flow has appeared only recently (Longuet-Higgins and Prazdny, 1980; Prazdny, 1979). It showed that from a monocular view of a rigid, textured, curved surface, it is possible in principle to determine the gradient of the surface at any point, the motion of the eye relative to that surface from the velocity field of the changing retinal image, and the field's first and second spatial derivatives. The relevant equations are redundant, thus providing a test of the rigidity assumption.

There is an interesting contrast between this result and Ullman's structure-from-motion theorem. In Ullman's scheme, four points are sufficient provided that the observer waits long enough to obtain at least three distinct views of them. Longuet-Higgins and Prazdny's scheme makes a slightly different trade-off; only two frames are required, so the time needed to acquire the measurements can be shorter. (Two frames suffice here because shape recovery is based on the perspective, not orthographic projection.) On the other hand, the local spatial neighborhoods involved in the computation are not just points, as in Ullman's scheme; they have to be large enough to give reliable estimates of the first and second spatial derivatives of the velocity field.

This analysis is another example of how computational theory can help empirical investigation. By solving the mathematics of the problem— and this was surely long overdue—Longuet-Higgins and Prazdny have provided a framework within which to inquire whether we humans actually do make use of optical flow, as Gibson suggested, and if we do, how. It is already clear that there are some ways in which we might have made use of it but actually do not. Attributing importance to the focus of expansion of retinal flow is one thing we could do but apparently do not. Another example is Ullman's conveyor belt demonstration, illustrated in Figure 3–54. We do not see regions 1 and 3 as having a different geometry from region 3, whereas most optical flow theories would say that we should.

Nevertheless, we could still use optical flow in some form, perhaps only weakly and more in peripheral than in central vision. That is, after all, where we might expect precision of measurement to be too low for a system based on Ullman's structure-from-motion scheme, yet it is also where we would expect to find the most evident optical flow. It remains to be seen whether optical flow is used in human vision.

3.6   SHAPE CONTOURS

As we discussed in Chapter 2, when we inquired into the physical basis for the primal sketch, there are four basic ways in which contours can arise in an image. They are (1) discontinuities in distance from the viewer, (2) discontinuities in surface orientation, (3) changes in surface reflectance, and (4) illumination effects like shadows, light sources, and highlights. Earlier in this chapter, we saw how different aspects of the primal sketch can be used as the input representation for processes based on stereopsis or on motion that are capable of finding boundaries from the differences between two or more images of a scene. We turn now to the more difficult case of a single, monocular image and ask how its contours can convey unambiguous information about shape. The mystery that needs explaining is that contours in an image are two-dimensional, yet we often see them in three dimensions. The question is how and why we make this three-dimensional interpretation.

I call the contours that we shall examine *shape contours,* because they are all two-dimensional contours that yield information about three-dimensional shape. I shall not discuss at all how to find them in an image—we spent long enough on that task in Chapter 2. Nevertheless, it is worth pointing out that although the physical origins of contours can be divided into the four categories mentioned, these origins give rise to a wide range of detectable changes in the image and hence a wide variety of ways in which a particular type of contour may be defined in the image.

For example, consider the possible effects of a discontinuity in depth. This can cause a simple intensity change—in fact, since our visual systems incorporate a predisposition for seeing brighter things as nearer, we would expect this brightness versus depth relation to be generally true of the visual world. If the surface characteristics are the same on both sides of a depth change, then a density- or size-induced texture boundary will be formed. If the two surfaces are not the same object, their textures will usually be very different and so many criteria will yield the boundary.

If the discontinuity is a change in surface orientation, intensity is likely to change and so is any density measure that the surface reflectance function may happen to support. Any clear orientation organization on the surface will also probably be shifted, and perhaps also some length measures.

If the surface reflectance is organized in any of a number of ways—for example, if it contains parallel lines—then it can convey valuable shape information to the viewer. And so forth.

The main point, then, is that contours can be defined on a surface in many ways, and they should be detected in the initial analysis and representation of the image. Some of these contours are more likely to have been caused by one kind of change than another—a discontinuity in orientation, for example, is more likely to be due to a change in surface orientation than to a change in depth—but the rules are not hard and fast. The important fact is that very many such contours can and do tell us about three-dimensional shape, and when one reflects upon it, this is actually quite an amazing fact. Such shape contours form the focus of our interest in this section.

Some Examples

The power and vividness with which contours can depict shape is not in doubt. Figure 3–56 shows some examples, and I think the reader will agree that for sheer three-dimensional realism, Figures 3–56(b) and (c) approach the effects achieved by means of stereopsis or motion. Very much more in doubt than in these other cases is precisely how these examples create this realism. Contours in an image can arise from several distinct physical causes. Some, as in Figure 3–56(a), are occluding contours—contours that arise at a discontinuity in depth, here at the edge of the viewed objects. Other contours arise from changes in surface orientation, texture boundaries, changes in reflectance and pigmentation, or from shadows falling on a surface. Most vivid and puzzling are the contours in Figures 3–56(b) and 3–56(c). To what do these correspond in nature? After all, we rarely come upon objects created by deforming a rectangular wire grid, as in Figure 3–56(b). Why then are we so good at seeing the shape of the wire room depicted there? Is it the same reason why we can see Figure 3–56(c) so well? Is there just one basic trick involved here or the happy coincidence of several that conspire and are jointly responsible for the vividness of the percept?

These, then, are the questions that we shall be studying here. Unfortunately, because we do not yet know whether one phenomenon or several are operating in cases like Figures 3–56(b) and (c), we are not in so strong

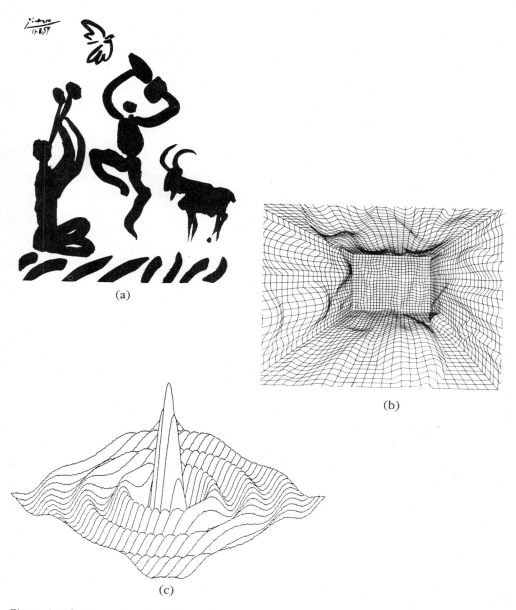

(a)

(b)

(c)

Figure 3–56.   Examples of two-dimensional contours in an image that impart three-dimensional information to the viewer. (a) *Rites of Spring* by Picasso, an example of shape information from silhouettes. (b) A "wire room." (c) A portrayal of the curve sin *x*. (b) and (c) are especially vivid. (Part (a) Copyright © SPADEM, Paris/VAGA, New York 1981. Part (b) courtesy of the Carpenter Center for the Visual Arts, Harvard University.)

a position as we were with stereopsis and motion. Psychophysics has not yet told us what the modules are, so we are still stuck in something of the linguist's predicament of not yet having a clear decomposition of language into relatively independent modules.

Nevertheless, some progress has been made. It is convenient to divide our discussion into three categories: (1) contours that occur at discontinuities in the distance of the surface from the viewer (occluding contours), (2) contours that follow discontinuities in surface orientation, and (3) contours that lie physically on the surface. This third type of contour can be due to surface markings or to shadow lines, for example. The important point is that they lie along the surface, and therefore I call them surface contours. Remember that contours in each category can be detected in several ways in an image. In all cases, our principal question is, Why and how can such contours in a single two-dimensional image convey to us unambiguous and often quite detailed information about three-dimensional shape?

Occluding Contours

An occluding contour is simply a contour that marks a discontinuity in depth, and it usually corresponds to the silhouette of an object as seen in two-dimensional projection. I became interested in occluding contours from the observation—which is almost a paradox—that when we look at the silhouettes in Picasso's *Rites of Spring* (reproduced here in Figure 3–56a), we perceive them in terms of very particular three-dimensional shapes, some familiar, some less so. This is quite remarkable, because the silhouettes could, in theory, have been generated by an infinite variety of three-dimensional shapes, which, from other viewpoints, would have no discernible similarities to the shapes that we perceive. It takes only a little imagination and moderate mischief to concoct a quite bizarre three-dimensional shape to demonstrate this point. We might, for example, arrange spikes and protuberances in a highly baroque style that happen to combine unexpectedly to produce the silhouette of a man or a goat when viewed from one special direction.

Yet we never think of such things when we are faced with these silhouettes. One can perhaps attribute part of the phenomenon to a familiarity with the depicted shapes, but not all of it, because we can use a silhouette to convey an unfamiliar shape, and because even with considerable effort it is difficult to imagine the more bizarre three-dimensional surfaces that could have given rise to the silhouettes in Picasso's painting. The paradox, then, is that the bounding contours in *Rites of Spring* apparently tell us more than they should about the shapes of the figures. For example, neighboring points on the bounding contours here could arise from widely separated points on the original surface, but our perceptual interpretation usually ignores this possibility.

This situation is so reminiscent of ignoring the many possible snowstorm interpretations of random-dot stereograms or the two-cylinder, random-dot moviegrams that one is almost forced to draw the obvious conclusion: Somewhere buried in the perceptual machinery that can interpret silhouettes as three-dimensional shapes, there must lie some source of additional information that constrains us to see the silhouettes as we do. Probably, but perhaps slightly less certainly than in the analyses of motion and stereopsis, these constraints are general rather than particular and do not require a priori knowledge of the viewed shapes.

If these constraints are general, then there must be some a priori assumptions in the way we interpret silhouettes that allow us to infer a shape from an outline. These assumptions must pertain to the nature of the viewed shape. Moreover, if a surface violates these implicit assumptions, then we should see it wrongly. Our perceptions should deceive us in the sense that the shape we assign to the contours will differ from the shape that actually caused them. One common instance is the shadowgraph, where the appropriate arrangement of the hands can, to the surprise and delight of a child, produce the shadow of an objectively quite different three-dimensional shape, like a duck, rabbit, or ostrich.

Constraining assumptions

The question we have to ask is, What assumptions are reasonable to make— that we unconsciously employ—when we interpret silhouettes like those of Figure 3–56(a) or Figure 3–57(b) as three-dimensional shapes?

Three seem to be important (Marr, 1977a). The first is that *each line of sight from the viewer to the object should graze the object's surface at exactly one point.* In other words, each point on the silhouette (Figure 3–57b) should correspond to one point on the viewed surface (Figure 3–57a). The reason for assuming this is that even if this correspondence did not exist, we could not possibly tell that it did not, and it would usually happen only as a result of an accidental alignment of two parts of the object along the line of sight.

This assumption allows us to speak of a particular curve on the object's surface called the *contour generator,* illustrated in Figure 3–57(d). It is the set of points on the surface that projects to the boundary of the silhouette in the image, and I shall use the letter Γ to denote it.

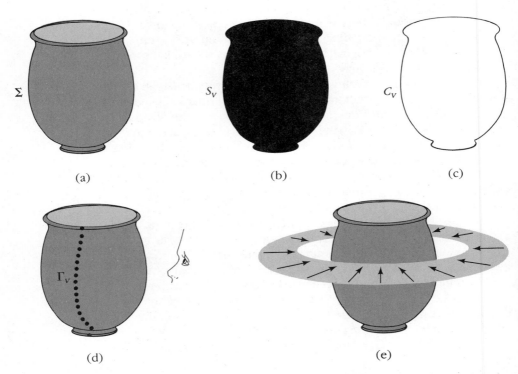

Figure 3–57.   Four structures of importance in studying the a priori conditions that we bring to bear on the analysis of an occluding contour. (a) A three-dimensional surface, Σ. (b) Its silhouette S_v as seen from viewpoint V. (c) The contour C_v of S_v. (d) The set of points Γ_v that project onto the contour. (e) A condition for the theorem discussed in the text. In particular, the meaning of "all distant viewing positions in any one plane" is shown.

The second assumption says that, except possibly in a very few instances, points that appear to be close together in the image actually are close together on the object's surface. The illustration in Figure 3–58(a) helps to explain this assumption. Think of a and b as being two hills, with the contour generators that give rise to a and b following the skyline on the top of each hill. If the dashed portion of b happens to be invisible, then at point P the visible contour generator leaps from one hill to the next—it is discontinuous. The sharp concavity at P, in fact, hints of this discontinuity, and so we half expect it. In the body of a and b, however, we do not expect it to happen, and in fact we assume it does not. This is our second assumption, and it says that *nearby points on the contour in an image arise from nearby points on the contour generator on the viewed object.*

The last assumption is a little more sophisticated, for it pertains to the

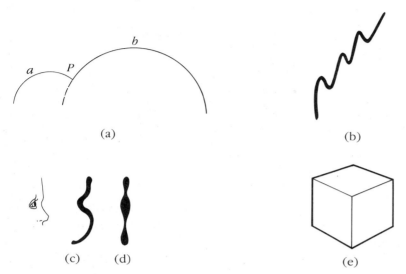

(a)   (b)

(c)   (d)   (e)

Figure 3–58.   (a) The second assumption, that nearby points on a contour arise from nearby points on the contour generator, essentially says that there are no points like *P* on the contour. If the dotted portion of *b* were invisible, the contour generator would leap from *a* to *b*, causing a discontinuity at *P*. (b) A typical piece of contour. The only features we can hope to make use of are its convexities and concavities, that is, its points of inflection, and these must be properties of the surface and not of the imaging process. For example, if a viewer is close to a snake (c), the convexities and concavities in the image (d) arise not because of properties of the snake, but because of variations in its distance away. (e) If the occluding contour shown with thick lines is present on its own, one perceives a hexagon. The interior lines change it into a cube, since they suggest that the occluding contour is not planar.

type of clue that an image contour might give about shape. Suppose, for example, that we have been presented with a piece of contour like that shown in Figure 3–58(b). The previous two assumptions allow us to think of this contour as coming from a contour generator on the surface, and we can safely assume that adjacent points on the contour come from adjacent points on the contour generator. Because the imaging process is what it is, we cannot rely on any measurements that we make on the contour in the image, and so the only remaining straightforward feature is that sometimes the contour bends one way and sometimes the other. In other words, there is a qualitative distinction between convex and concave segments, which, provided that the surface is sufficiently smooth, rests in turn on the notion of an inflection point. In general, of course, points of inflection in a contour need have no significance for the surface. The contour generator could weave around in an arbitrary and complex way, or it could move directly toward and then away from the viewer. The latter case might, under the perspective projection, give rise to convexities and concavities rather in the way illustrated by Figures 3–58(c) and (d). So our next question has to be, How exactly should we formulate an assumption saying that points of inflection in a contour are significant, that they somehow reflect real properties of the viewed surface and not artifacts of the imaging process?

Our previous two assumptions allow us to think of the contour generator as a piece of wire bent in three-dimensional space. If inflection points on the contour are to reflect genuine inflections on this piece of wire, however, two mathematical conditions must be satisfied:

1. The transformation due to the imaging process that produces the contour from the wire must be linear. This rules out the perspective transformation and restricts the validity of our theory to distant views—the object must be small relative to its distance from the viewer.

2. The curve on which the transformation acts must lie in a plane. In other words, the convex–concave distinction in the image can be meaningful only for distant views and only if the bent wire that is the contour generator lies in a plane. This gives us our third assumption, that *the contour generator is planar*.

This third assumption is a strong one that sharply delimits the class of surfaces whose shapes can be interpreted by silhouette. However, it seems unavoidable if we wish to distinguish convex and concave segments in the interpretation process. Fortunately, however, the results of using this assumption are very robust—if the contour generator is not quite planar but nearly so, then the surfaces are usually only a little misbehaved. And interestingly, the planar condition is actually embodied in much modern design. All of the outlines drawn in mechanical engineering diagrams satisfy the condition, so it has its uses even outside the study of vision. If the condition is violated, we do seem to get the shape wrong. The occluding contour in Figure 3–58(e), for example, is marked with thick lines and, if shown on its own, gives the appearance of a two-dimensional hexagon. With the additional information provided by the interior lines, however, it takes on a quite different interpretation. As a cube, the occluding contour is no longer planar.

Implications of the assumptions

In order to see what these assumptions really mean, we have to understand how they constrain the geometry of the surfaces being viewed. Clearly, some surfaces will satisfy the assumptions and some will not. What about a surface makes it satisfy them? To answer this question we should reformulate our assumptions as restrictions on the geometry of the viewed surface and then see what their consequences are. To remind ourselves of these restrictions, I restate them here:

1. Each point on the contour generator projects to a different point on the contour.

2. Nearby points on the contour arise from nearby points on the contour generator.

3. The contour generator lies wholly in a single plane.

We need one more idea before we can formulate the critical result—the idea of a *generalized cone*. This idea was introduced by T. O. Binford (1971) as a way of representing shapes in a computer program, and it is illustrated in Figure 3–59. A generalized cone is the surface created by moving a cross section along an axis. The cross section may vary smoothly in size, getting fatter or thinner, but its shape remains the same. Thus a football is a generalized cone and so is a pyramid or, roughly, a leg or an arm, or a snake, or a tree trunk, or a stalagmite. In fact, we can think of a horse as being composed of eight generalized cones, one for each leg and one each for the head, neck, body, and tail.

We are now ready for the basic result, and I hope the reader finds it as surprising as I did:

If the surface is smooth (for our purposes, if it is twice differentiable with a continuous second derivative) and if restrictions 1 through 3 hold for all distant viewing positions in any one plane, as illustrated in Figure 3–57(e), then the viewed surface is a generalized cone. The converse is also true; if the surface is a generalized cone, then restrictions 1 through 3 will be observed.

This means that if the convexities and concavities of a bounding contour in an image are actual properties of a surface, then that surface is a generalized cone or is composed of several such cones. In brief, the theorem says that a natural link exists between generalized cones and the imaging process itself. The combination of these two must mean, I think, that generalized cones will play an intimate role in the development of vision theory.

Stated baldly, this result means that, in general, shape cannot be derived from occluding contours alone unless that shape is made from generalized cones and is viewed from a position from where its axis is not

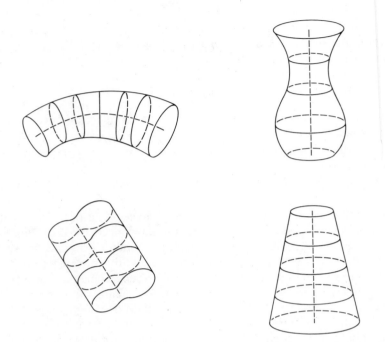

Figure 3–59.   The definition of a generalized cone. As used in this book, the term *generalized cone* refers to the surface created by moving a cross section along a given smooth axis. The cross section may vary smoothly in size, but its shape remains constant. We here show several examples. In each, the cross section is shown at several positions along the trajectory that spins out the construction.

foreshortened (foreshortening would occur in Figure 3–57(e) if the vantage point was from above or below). If there is no foreshortening, however, and even if the viewed shape is constructed of several different generalized cones like the silhouette of a man or of a horse, then the shape can be at least partially reconstructed. Perhaps the most important thing, as we shall see later in the book, is that the *axes* of the cones can be recovered from the image, because this helps to establish an object-centered coordinate system in the viewed shape. I shall say more about this in Chapter 5 and will briefly illustrate an algorithm for decomposing silhouettes into their constituent generalized cones. (See Marr, 1977a, for the theorems behind the algorithm.)

     For now, however, it is enough to note that the use of occluding contours requires the three restrictions that we formulated, and they hold if and only if the viewed shapes are generalized cones. The principal implication of these restrictions is that the surface goes in and out where the contour goes in and out. Not much more can be said from the occluding contours alone.

Surface Orientation Discontinuities

Surface orientation contours mark the loci of discontinuities in surface orientation. For example, they follow the creases on a surface, like the interior lines of Figure 3–58(e) or the longitudinal peaks and troughs of Figure 3–60. With regard to recovering the geometry of the surface, the most important question about such a contour is whether it corresponds to a convexity or a concavity on the surface. In Figure 3–58(e), all the interior contours represent convexities, whereas in Figure 3–60 convexities and concavities alternate, sometimes in an interestingly confusing way.

Unfortunately, it is often difficult to distinguish convexities and concavities from purely local cues in a monocular image. We have a predisposition to see such contours as convex (see Figure 3–61b), but even examples that are loaded one way can be made to alternate (compare Figures 3–61a and c).

There are certain things to be said about combinations of such contours—for example, Waltz-like (1975) constraints, of the form illustrated in Figure 1–3, apply, which specify that one cannot have two concave and

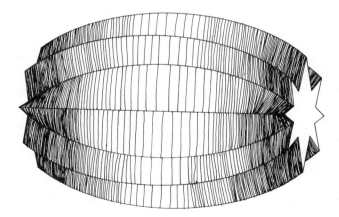

Figure 3–60.   A sketch of a generalized cone showing its silhouette (the circumscribing contour) and fluting (the contours spanning its length). The fluting marks lines of discontinuity in surface orientation.

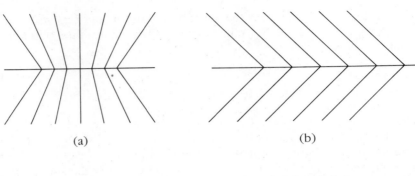

(a)   (b)

Figure 3–61.   More examples of the
portrayal of discontinuities in surface
orientation.

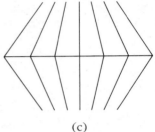

(c)

one convex contour meeting at a point. However, these are not properties
purely of the isolated contours, and I shall discuss that class of more complicated phenomena in Chapter 4. The only knowledge currently available
for helping to distinguish isolated convex and concave contours is due to
Horn (1977). He showed that, at least for the visual world of matte white
prisms, the intensity profiles across different types of edge are characteristically different. If the intensity profile across the edge is a step change
or very sharp peak, the edge is probably convex. If the intensity profile is
roof shaped, the edge is probably concave. However, there is no evidence
yet that the human visual system uses these cues in classifying edges.

Surface Contours

Surface contours arise for various reasons in the image of smooth surfaces,
and they yield information about the three-dimensional shape of the surface, in the manner illustrated by Figure 3–62. The question of interest, of
course, is how this is done, and it has recently been explored in some
detail by Stevens (1979). The underlying observation is that we do not

Figure 3–62.   The undulating surface is suggested by a family of sinusoids. The curves are naturally interpreted as surface contours, that is, the images of markings on a physical surface. What constraints can be brought to bear in making this three-dimensional interpretation? (Reprinted by permission from K. Stevens, "Surface perception from local analysis of texture and contour," Ph.D. thesis, Department of Electrical Engineering and Computer Science, Massachusetts Institute of Technology, 1979.)

perceive Figure 3–62 as purely two-dimensional; there is no doubt that what we see is a smooth, undulating surface. As we have seen many times now, this means that we are bringing some a priori assumptions to bear on our analysis of such images.

Once again, the fundamental computational questions are What are these assumptions, Why do we use them, and How do they enable us to recover three-dimensional surface orientation information from a single two-dimensional image? In this discussion of Stevens' work, I shall maintain the distinction between an image contour and its corresponding contour generator on the surface, which we met first in our analysis of occluding contours, illustrated in Figure 3–57. The difference here is that the contour generators are no longer restricted to just the silhouette boundaries of an object but may arise within the silhouette because of internal surface markings or various kinds of illumination effects. For example, the contours of Figure 3–62, are naturally interpreted as the image of markings on the surface, and we shall call these markings the contour generators of the image contours. These contours may, of course, be quite abstract objects, perhaps created by rows of dots, but we take the machinery and represen-

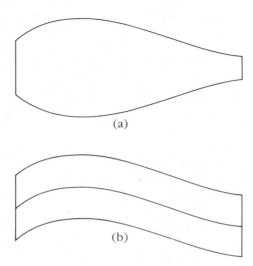

Figure 3–63. The curves in (a) are interpreted as occluding contours, and the underlying surface is seen as a generalized cone—in this case, a vaselike object. Such contours were studied in Section 3.5 and are further considered in this discussion. Those in (b) are interpreted as surface contours, and the surface appears like a gently curved flag or ruled sheet of paper. (Reprinted by permission from K. Stevens, "Surface perception from local analysis of texture and contour," Ph.D. thesis, Department of Electrical Engineering and Computer Science, Massachusetts Institute of Technology, 1979.)

tation abilities of the full primal sketch for granted here. We shall call such contours *surface contours*. Note that occluding contours are almost never surface contours (see Figure 3–63).

The puzzle and difficulty of surface contours

What makes the issue of surface contours so extremely difficult to analyze satisfactorily is that there is no obvious physical source of surface contour regularity that our perceptual machinery can use to such advantage. The world really seems to have less structure than diagrams like Figure 3–62, and I remain deeply puzzled about why we can interpret such figures so vividly.

Stevens (1979), in a useful first approach to these issues, divided the problem into two halves; inferring the shape of the contour generator in three-dimensional space and then determining how the surface itself lies in relation to the contour generator. The first step is that of discovering the

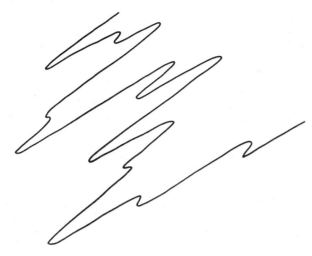

Figure 3–64. This curve appears to have a specific three-dimensional shape, as if planar and foreshortened by the slant of the plane relative to the viewer. Why and how is this interpretation derived? (Reprinted by permission from K. Stevens, "Surface perception from local analysis of texture and contour," Ph.D. thesis, Department of Electrical Engineering and Computer Science, Massachusetts Institute of Technology, 1979.)

shape of a piece of wire bent in three-dimensional space so that it lies along the contour generator and has the correct appearance in the image. The second step can then be thought of as gluing a ribbon along the wire so that it follows faithfully the strip of surface that lies directly under the contour generator.

Determining the shape of the contour generator

When we observe a single contour, the curve appears to have a specific three-dimensional shape and to lie in a plane. The impression gained from Figure 3–64, for example, is of a planar curve whose plane has a definite, if somewhat weakly specified, slant and tilt. The assumption that the contour generator is planar greatly simplifies the problem, but it is difficult to be confident of such an assumption, although shadow boundaries cast by straight edges and certain types of surface reflectance organizations will often produce planar contour generators on a surface.

There are other assumptions that one might make. Stevens (1979) pointed out that much can be done if symmetry, even of only a rough or skewed kind, is detected in the figure (see also Marr, 1977a). Witkin (1978) suggested that it is sometimes useful to assume that the real-life contour generator has the minimum possible curvature, the visible curvature of the image contour being derived in part from the imaging process. But these ideas are still ad hoc and disorganized.

The effects of more than one contour

The weakness of our perception of single contours like that of Figure 3–64 is probably related to the unsatisfactory lack of any realistic interpretive assumptions that one might bring to bear upon such perceptions. If there are several contours, however, the vividness of our perception is much enhanced, as in Figure 3–62. Except in very rare and accidental situations, if surface contours are parallel in the image, their contour generators are parallel on the surface.

That the contour generators are parallel so that one can be shifted across the surface onto its neighbor, leads to quite a powerful idea about how to recover surface orientation from surface contours. Parallel contour generators essentially mean that we can locally ignore the curvature of the surface in the direction of the shift. Technically, the surface is then devel-

Figure 3–65. The wavy lines represent visible contours in the image, and the straight lines, which have zero curvature, make explicit the parallel relationships between adjacent wavy lines. Such a surface is locally a cylinder, because one of its curvatures (and hence its Gaussian curvature) is zero. (Reprinted by permission from K. Stevens, "Surface perception from local analysis of texture and contour," Ph.D. thesis, Department of Electrical Engineering and Computer Science, Massachusetts Institute of Technology, 1979.)

opable. This means that the surface can be thought of locally as a cylinder, which is a surface with two principal curvatures, one of which is zero—the surface is flat in that direction.

The idea is illustrated by Figures 3–65 to 3–67. Figure 3–65 shows a surface in which two types of contours are visible—the wavy ones, which are the family of parallel contour generators that we suppose are in fact present in the image, and the orthogonal set of straight lines, which have zero curvature and represent the correspondence between the locally par-

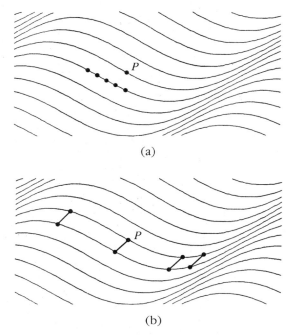

(a)

(b)

Figure 3–66. Usually, of course, the correspondences between adjacent parallel surface contours will not be explicit in the image, as they were in Figure 3–65. However, the correspondence can be found, even in the less straightforward cases. For example, if the surface contours are straight for a portion of their length, as in (a), the tangent to a point *P* on one contour may be parallel to various tangents on the adjacent contour; however, only one choice would result in a correspondence line that is parallel to the other correspondence lines between curved portions of adjacent contours, as in (b). (Reprinted by permission from K. Stevens, "Surface perception from local analysis of texture and contour," Ph.D. thesis, Department of Electrical Engineering and Computer Science, Massachusetts Institute of Technology, 1979.)

Figure 3–67. Although, strictly speaking, the assumptions and techniques illustrated in Figures 3–65 and 3–66 require that the surface be cylindrical, in practice they can be used assuming that they hold only locally, since the parallel correspondence need be established only between adjacent contours. Hence the local cylinder restriction allows us to interpret surfaces whose global structure is not cylindrical. (Reprinted by permission from K. Stevens, "Surface perception from local analysis of texture and contour," Ph.D. thesis, Department of Electrical Engineering and Computer Science, Massachusetts Institute of Technology, 1979.)

allel contour generators. In identifying the correspondence with straight lines, we are assuming that the surface is locally of a peculiarly simple sort, with one of its curvatures being zero. Once both the wavy lines and the correspondence lines are available, surface orientation is quite well constrained, because we know that in three dimensions these two types of lines are perpendicular.

Usually, of course, the correspondence contours will not be visible in the image, but Figure 3–66 illustrates how they may be recovered, even in apparently ambiguous situations (some details are given in the legend). Finally, one can extend the idea to quite general surfaces, as explained by Figure 3–67, because the fundamental assumption on which the interpretation is based must hold only locally—in this case, between adjacent surface contours. Figure 3–67 shows an example of how this basic requirement—that one of the curvatures vanishes—holds only locally and approximately. The structure of the surface depicted there can be recovered by using methods based on these ideas, even though globally it is certainly not a developable surface.

Stevens pointed out one other interesting fact, namely, that if a highlight appears along a continuous curve on a surface, then the curve is planar (assuming that the light source and vantage points are distant from the surface). This contour is like one of our correspondence contours, along which one of the principal curvatures of the surface is zero. In this case the surface normal coincides with the normal to the plane containing the gloss contour, just as in Figure 3–65 the surface normal lies perpendicular to both the straight (correspondence) and wavy lines. So the conditions that Stevens suggested for the recovery of surface orientation from surface contours do actually occur in real life.

In summary, then, the recovery of surface orientation from surface contours remains an intriguing and unsolved problem. On the other hand, Stevens' main suggestions—the planarity of the contour generator and of the locally developable assumption—seem to be powerful ingredients for achieving the recovery, and I shall be surprised if they are not used by us in practice in some form.

3.7   SURFACE TEXTURE

The notion that surface texture may provide important information about the geometry of visible surfaces has attracted considerable attention in the last 30 years. Perhaps the main impetus for this interest was the hypothesis formulated by Gibson (1950), which states that texture is a mathematically and psychologically sufficient stimulus for surface perception. By this he meant that there is sufficient information in the monocular image of a textured surface to specify uniquely the distance to points on the surface and to specify the local surface orientation. Furthermore, he claimed that the human visual system can and does use this information to derive such surface information.

In an ideal world, where the surfaces are smooth and regularly and clearly marked and exhibit sufficient density of detail so that gradients in an image can be measured quite precisely, Gibson's claim would have much to recommend it. Unfortunately, however, the world is a much rougher place, in which uniformity and regularity are the exception or only an approximation rather than the rule, so my own view is that we should be surprised when something can be done rather than when it cannot. In addition, as Stevens (1979) has pointed out, much of the rather simple mathematics associated with these questions has had a somewhat flawed presentation in the past. We shall therefore be wise to take a critical and skeptical attitude to the supposed power of texture perception except when it can be demonstrated beyond doubt that the human visual system is using it.

The Isolation of Texture Elements

The first problem, and one that has hardly been addressed at all, is how to extract from an image the uniform texture elements on which subsequent analysis must rest. A full answer to this would include a complete understanding of the full primal sketch and of the selection by similarity, whose business it is to classify items by origin and whose importance we have already encountered (for example, Figure 2–3). Let us, however, take this for granted, and assume that the world's surfaces are covered with regular and sufficient markings, and that we are capable of discovering them from our early representations of the image.

Surface parameters

As we have already seen several times, there are two ways in which a surface may be specified relative to the viewer: We can either specify the distance to local pieces of it, or we can specify the surface orientation relative to the viewer. Surface orientation itself is naturally split into two components, which we have called slant and tilt. Slant is the angle by which the surface dips away from the frontal plane, and tilt is the direction in which the dip takes place.

Mathematically, of course, distance and surface orientation are almost equivalent, being related by an integration (see Chapter 4). For the nervous system, the question is a quite different one—Which of these quantities, distance, slant, or tilt, is actually extracted directly from measurements of variations in texture? In his recent study of this question, Stevens (1979) concluded as follows:

1. Tilt is probably extracted explicitly.
2. Probably distance is also extracted explicitly.
3. Slant is probably inferred by differentiating estimates of scaled distance made in accordance with point 2.
4. In particular, measurements of texture gradients, which are closely associated mathematically with slant, are probably not made or used, perhaps because of the inaccuracies inherent in the measuring process.

We look now at the reasons for his conclusions.

Possible measurements

Stevens observes that even very different looking textures pose the same computational problems, and that one must be careful not to postulate more mechanisms than the problems require. Figure 3–68 shows an

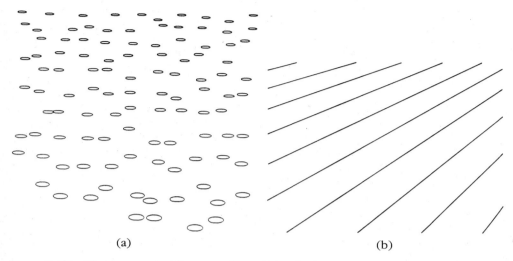

Figure 3–68.   These two types of texture, although they look very different, in fact pose the same computational problem. In (a), the ellipses vary in width, eccentricity, and density exactly as if they were produced by the perspective projection of equal-sized circles lying in a plane that slants away from the viewer. A number of measurements could be made from such an image and used to help determine the geometry of the plane, and a large part of our discussion will concern which of these measurements are likely to be used. In (b), the converging lines suggest a slanted surface ruled with parallel, equally spaced straight lines. Although it has been suggested that different processes are required to interpret (b) than are required for (a), this is not necessarily true, since measurements of spacing, separation, and so forth can be made in both. In fact, the apparent superiority of the converging contours in (b) over the more random textures of (a) could be due solely to the greater precision in image measurements that is allowed by patterns like (b). There is no a priori computational reason to invoke separate mechanisms.

example of this; although the two patterns look very different, similar measurements of spacing and size can be made in both. Our first question is, Which of the many possible measurements are in fact yielding the perceptual clues that give us the impression of a slanted surface? In Figure 3–68(a) are they the sizes of the ellipses, their distances apart, their density, or their density gradients?

In Figure 3–69, all the information that appeared in Figure 3–68(a) except the density gradient has been removed, and three types of tokens have been used to mark the positions of the ellipses. In all cases, although the density gradients are plainly visible and their directions clearly delineated, there is little or no impression of slant.

Surface tilt, on the other hand, does seem to be obtained quite directly from an image, although it is worth noting that it can be done in two ways

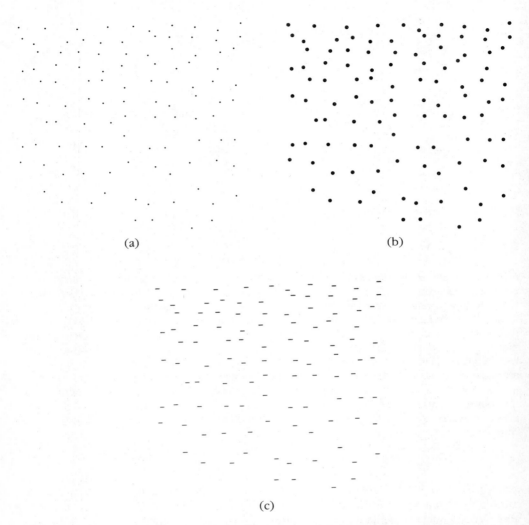

(a)   (b)

(c)

Figure 3–69.   One of the possible measurements for inferring surface slant in Figure 3–68(a) is the gradient of the density of the ellipses. Texture gradient measures, in fact, have several mathematically attractive properties. In this figure, however, the exact gradient present in Figure 3–68(a) has been reproduced using three different types of local texture element. In every case the density gradient is obvious, but the impression of a slanted surface is absent, even under the best viewing conditions. An impression of slant can sometimes be obtained using very high density gradients, but the values involved are not physically plausible. Examples like these call into question the matter of whether our own visual systems actually use texture gradient measures to infer the slant of a textured surface. (Reprinted by permission from K. Stevens, "Surface perception from local analysis of texture and contour," Ph.D. thesis, Department of Electrical Engineering and Computer Science, Massachusetts Institute of Technology, 1979.)

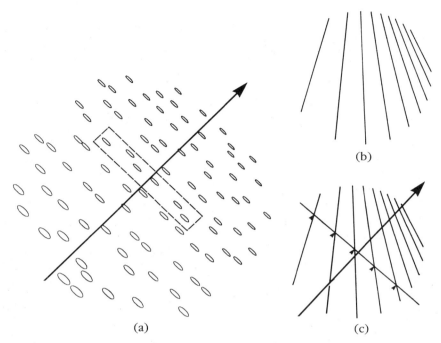

(b)

(a)   (c)

Figure 3–70.   The tilt of a surface is the direction in which it is slanted away from the viewer. If the surface bears a uniform texture, the projection of the axis of tilt in the image indicates the direction in which the local density of the texture varies most, or equivalently, it is perpendicular to the direction in which the texture elements are most uniformly distributed. Either technique can be used to recover the tilt axis, as illustrated in this figure. Interestingly, however, the tilt axis in situations like (b) can probably be recovered most accurately by using the second method, that is, searching for the line that is intersected by the perspective lines at equal intervals. This method is illustrated in (c). (Reprinted by permission from K. Stevens, "Surface perception from local analysis of texture and contour," Ph.D. thesis, Department of Electrical Engineering and Computer Science, Massachusetts Institute of Technology, 1979.)

(see Figure 3–70). We can detect either the direction in which the local density of the texture varies most or, equivalently, the line perpendicular to the direction in which the texture is most uniformly distributed. Interestingly, in cases like Figure 3–70(b), the second method probably provides the more accurate measurement. It is necessary only to search for the direction shown along line l in Figure 3–70(c), which the lines of perspective intersect at equal intervals. It is also known that the human visual system can detect equal intervals to within only a few percent.

Estimating scaled distance directly

Stevens' final demonstration appears in Figure 3–71, and it provides his reason for believing that we directly measure the size of the texture element from which we infer distance and then obtain an internal estimate of slant by a process akin to differentiation (see Chapter 4).

When viewed as a lighted display in a darkened room, Figure 3–71(a) gives the appearance of a slanted plane scattered with uniform-sized spheres. One possibility is that a texture gradient measure is being used to infer slant—for example, the gradient in the width of the circles. Figure 3–71(b), however, also appears strikingly three-dimensional under the same viewing conditions, yet there is no gradient here. The larger circles appear to be nearby, and the smaller ones further away. Both cases are explained by assuming that the circles correspond to uniform-sized spheres and that the different sizes in the image arise because of their different distances away, according to the simple geometrical rule that measured diameter varies as $1/r$. Therefore, the human visual system may not measure slant directly, preferring instead to estimate relative depth from size and perhaps brightness changes and then to infer slant from this.

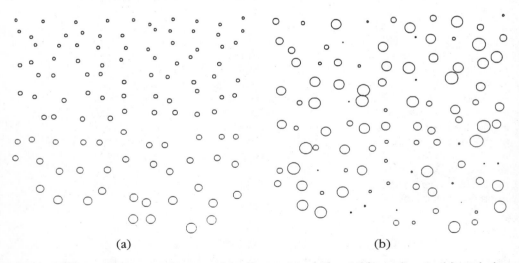

(a)   (b)

Figure 3–71. Are texture gradients used in texture vision? The visible gradient in (a) might be responsible for the apparent slant, but under suitable viewing conditions (b) appears just as three-dimensional. It could therefore be that the size or brightness of the circles is actually being used to determine slant. (Reprinted by permission from K. Stevens, "Surface perception from local analysis of texture and contour," Ph.D. thesis, Department of Electrical Engineering and Computer Science, Massachusetts Institute of Technology, 1979.)

Summary

The analysis of texture is another topic that lies in a somewhat unsatisfactory state. The mathematics is easy, but the psychophysics is not, nor is it at all obvious to what extent the vagaries of the natural world allow the visual system to make use of the possible mathematical relations. In addition, unhappily little is yet known about the later stages of the full primal sketch, where the basic texture elements are actually found. Once more is known of this matter, however, empirical studies can be conducted on a variety of natural images. Probably only then shall we ever actually understand why the human visual system handles texture information in the rather peculiar and limited way in which it appears to operate.

3.8   SHADING AND PHOTOMETRIC STEREO

The importance of makeup in the theater, and the widespread use of makeup in everyday life suggest that the human visual system incorporates some processes for inferring shape from shading. It seems likely, however, that the power of these processes is only slight, perhaps deriving from the combination of shading cues and information from occluding contours. On its own, shading acts as only a weak determiner of shape, and one of the most interesting problems in the theory of human early vision, along with color, is exactly what and how much information we are able to recover from shading.

From a purely theoretical point of view, the shape-from-shading problem was one of the very first to receive a careful analysis, and in his doctoral thesis (summarized as Horn, 1975), B. K. P. Horn showed how the differential equations relating image intensity to surface orientation could be solved provided that the illumination was simple and the surface reflectance known and uniform.

Since then Horn (1977) has reformulated his work in terms of the gradient space, which makes it much simpler to understand. The main use of his work has been in the development of methods for analyzing hill shading. Suppose, for example, one knows the terrain in a part of the Swiss Alps; the question is, How would it appear at 10 AM on a sunny summer's day? Or at 4 PM? Figure 3–72 shows that Horn's methods can answer these questions. By comparing the predicted image with an actual satellite photograph one is able to extract information about the reflectance properties of the land surface without being confused by the shading due to the particular terrain and illumination characteristics.

A mathematical understanding of the shape-from-shading problem is

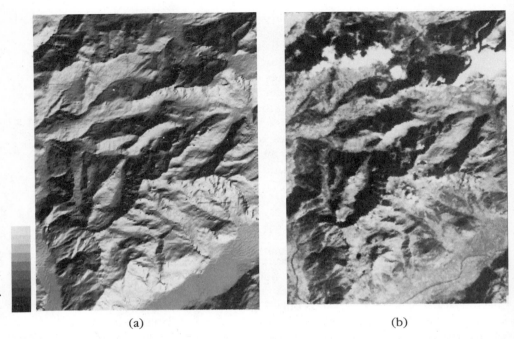

          (a)   (b)

Figure 3–72. Comparison of the predicted and actual appearance of a portion of the Swiss Alps. (a) Computed by Horn's methods from a knowledge of the terrain map and the reflectance map for that time of day. (b) is a photograph taken from a LANDSAT satellite.

probably a prerequisite for any serious study of the human capacity for recovering shape from shading, so I have outlined the important ideas here. The interested reader should consult Horn (1977) for more details, as my account will not be very technical.

Gradient Space

The first thing necessary when discussing shape from shading is a sensible way of talking about surface orientation. For this, we borrow the representation popularized in a slightly different context by Huffman (1971) and Mackworth (1973).

    Suppose we have a surface of some kind, as illustrated in Figure 3–73(a). Provided the surface is smooth, a given point on the surface will have a local tangent plane—that is, there will be a plane that is locally tangential to the surface at that point—and a local surface normal, which

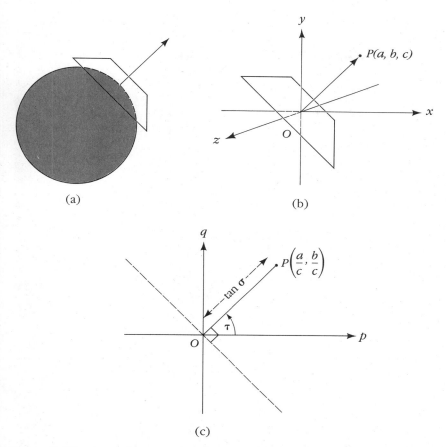

Figure 3–73. An explanation of gradient space. The local normal to the surface
(a) can be represented as a vector (a,b,c), as in (b). Since we are interested only
in the vector's direction, this can be reduced to $(a/c, b/c, 1)$, which can be represented as the two-dimensional vector $(a/c, b/c)$, as in (c). The quantity a/c is usually
denoted by p, and b/c by q.

is the outgoing normal to the tangent plane at that point. Now take the
same tangent plane, move it to the origin of the coordinate system, and
draw in its normal OP, as in Figure 3–73(b). Suppose the coordinates of
P happen to be (a, b, c). It clearly does not matter how long OP is, since
only its direction matters, so we could just as well use the point P' at $(a/c,
b/c, 1)$. But now we can represent P' by just two numbers, $(a/c, b/c)$—that
is, by just the two-dimensional point P in Figure 3–73(c). This is the *gradient space* representation of surface orientation.

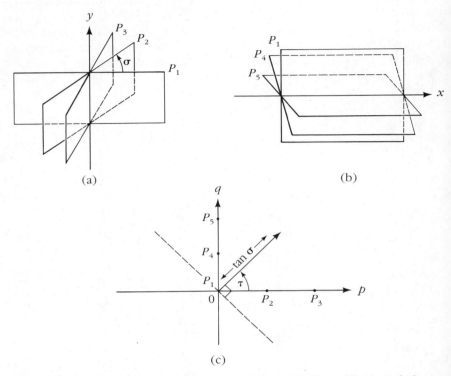

Figure 3–74.   Understanding gradient space. The orientation of the frontal plane P_1 corresponds to the origin in *(p,q)* space. As the plane is rotated about the vertical (a), the corresponding point in gradient space moves along the *p*-axis (P_2, P_3) as shown in (c). If the plane rotates about the horizontal *x*-axis, as in (b), its representation in gradient space moves out along the *q*-axis (P_4, P_5). Similar arguments apply to rotations about intermediate axes. The depicted angle τ is called the tilt, and the angle σ the slant of the plane.

Gradient space is an elegant way of representing surface orientation. A few examples will help to make its properties clear. For a frontal plane, with the surface normal coming directly toward the viewer, $a = b = 0$ and the point *P* is at the origin *O* in Figure 3–73(c). Now imagine rotating the plane clockwise about the vertical axis as in Figure 3–74(a). Then *P* moves gradually to the right along the *p*-axis, as shown in Figure 3–74(c), and the distance from *O* equals the tangent of the angle of slant. If instead we rotate the plane about the horizontal axis, as in Figure 3–74(b), *P* moves along the *q*-axis, as shown in Figure 3–74(c), again by an amount equal to the tangent of the angle of slant. If we rotate about some intermediate axis, shown dotted in Figure 3–74(c), then P moves out at right angles to it along the direction τ from the p axis, as shown in Figure 3–74(b). This angle τ is what in the psychophysical literature is referred to as the tilt of the plane, and the angle between it and the frontal plane is usually called its slant, and sometimes its dip. I shall use the letter σ to denote slant. The distance between the point P and the origin is tan σ.

The reader might like to take a few moments to play with a piece of paper and understand gradient space fully, because it is an important and useful idea. In particular, he might prove to himself that the length of OP is equal to tan σ.

Surface Illumination, Surface Reflectance, and Image Intensity

The study of shape from shading is concerned with finding ways of deducing surface orientation from image intensity values. The problem is complicated because intensity values do not depend on surface orientation alone; they depend on how the surface is illuminated and on the surface reflectance function. In the real world, the prevailing illumination is often complex, especially indoors. Outside is more straightforward—the sun is nearly a distant point source, and the ground illumination that is produced by thick cloud cover is nearly uniform, so these two situations are quite simple. A partly cloudy day can sometimes be treated as a combination of the two. But at ground level, the situation is often made very complex by secondary illumination effects—light bouncing off one surface onto another and thence into our eyes. These effects are almost impossible to treat analytically.

Just like the echo effects in acoustics, secondary illumination becomes especially important for indoor scenes, where light from a ceiling fixture can reach the coffee table top either directly or after reflecting off the ceiling or walls. The ceiling will help to illuminate the walls, and these in turn will reflect light back, helping to illuminate the ceiling—a condition called mutual illumination. The combined complexity introduced by all these effects makes the analysis of shape from shading extremely difficult, and no real progress has yet been made with the problem except in the very simple illumination condition of one distant point source. Horn, however, has effectively solved this situation, and we shall shortly look at how he did it.

The second factor that profoundly influences the shape-from-shading problem is the surface reflectance function. The fraction of light reflected

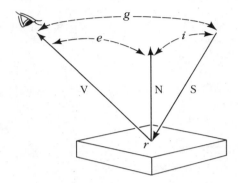

Figure 3–75.   The definitions of the angles *i*, *e*, and *g*.

toward the viewer from a source depends on the microstructure of the reflecting surface, and this is usually described as a function of the three angles shown in Figure 3–75—the *angle of incidence i* between the source and the surface normal, the *angle of emittance e* between the line of sight to the viewer and the surface normal, and the *phase angle g* between the incident and emitted rays. The *reflectance function* $\phi(i,e,g)$ is the fraction of incident light reflected per unit surface area per unit solid angle in the direction of the viewer. Intuitively, this says that the amount of light incident on a surface patch that will be reflected to a detector depends directly on the area of the illuminated patch, the value of $\phi(i,e,g)$, and the angular size of the detector.

There are many kinds of reflectance function. A perfect Lambertian surface—a pure matte surface—looks equally bright in all directions and has the simple reflectance function $\phi(i,e,g) = \cos i$. The surfaces of rocky, dusty objects that are viewed from great distances have another interesting type of reflectance function; for fixed-phase angle g, ϕ depends only on $\cos i/\cos e$. This relationship applies to the material in the maria of the moon—and for observation from the earth, g is indeed constant. This has greatly helped the study of lunar topography.

A polished metallic surface has a particularly simple reflectance function ϕ; it is 1 when $i = e$ and $g = i + e$, the properties of a pure mirror. If the surface is not quite so polished, then ϕ is smudged a little around this value, often by convolution with a Gaussian. This smudged property is particularly interesting because many everyday surfaces have a reflectance function that combines two components, one matte and one specular.

The reflectance function of glossy white paint is made up of such a combination. For example, if s is the fraction of light reflected specularly, its reflectance function might have the form

$$\phi(i,e,g) = \frac{s(n+1)(2\cos i \cos e - \cos g)^n}{2} + (1-s)\cos i$$

where the first term is the specular component, and the second the matte. The number n determines how sharp the specular peak is; a typical value for a glossy paint might be $n = 16$ (see Horn, 1977).

The Reflectance Map

The best way of understanding the shape-from-shading problem is to understand the reflectance map, which is a way of relating image intensities directly to surface orientation.

Suppose we take a particular type of surface with a known reflectance function ϕ. Suppose we take distant source and viewing positions, so that the phase angle g is constant, and suppose that we take just a single source, so that the problem is expressed in its very simplest form. Then each surface orientation will produce a particular intensity in the image, which we can plot in the (p,q) gradient space map. In fact, let us choose to plot our reflectance map in a particularly simple way—let us draw in the contours of constant reflected intensity, normalized to some scale lying between 0 (for darkness) and 1 (the maximum possible intensity of light that could be found in the image). Then if the measured intensity at a given point is, say, 0.8, we know that the surface orientation (p,q) must lie on the 0.8 contour in the reflectance map.

Figures 3–76 to 3–79 show some examples. Figure 3–76 is the reflectance map for a pure matte (Lambertian) surface illuminated from a source that is by the viewer. In Figure 3–77, the surface is the same, but the source is in a different direction (actually in direction $p = 0.7, q = 0.3$). Notice here there is a shadow line—the line of surface orientations at which the surface becomes self-shadowed from the source. Figure 3–78 shows the peculiar reflectance map characteristic of the maria of the moon, and Figure 3–79 shows the reflectance map for our glossy white paint. The very closely spaced circular contours correspond to values of intensity that change very rapidly with any change in surface orientation, and so they correspond to the specular component. The rest of the map looks more like Figure 3–77 and corresponds to the matte component.

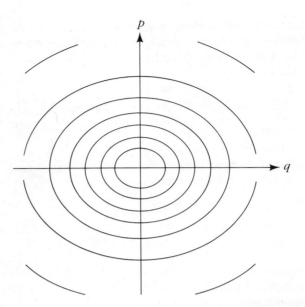

Figure 3–76. Contours of constant cos *i*. Contour intervals are 0.1 unit wide. This is the reflectance map for objects with Lambertian surfaces when there is a single light source near the viewer.

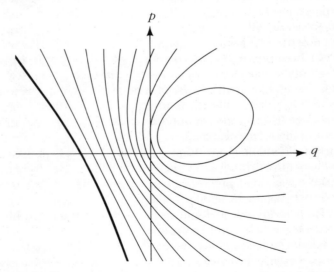

Figure 3–77. Contours of constant cos *i*. Contour intervals are 0.1 unit wide. The direction to the source is $(p_s, q_s) = (0.7, 0.3)$. This is a typical reflectance map for objects with Lambertian surfaces when the light source is not near the viewer.

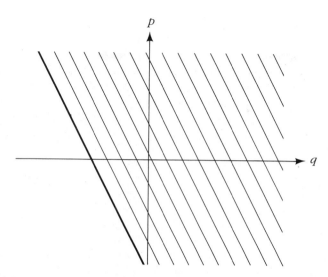

Figure 3–78. Contours of $\phi(i,e,g) = \cos i/\cos e$. Contour intervals are 0.2 unit wide. The reflectance function for the material in the maria of the moon is constant for constant $\cos i/\cos e$.

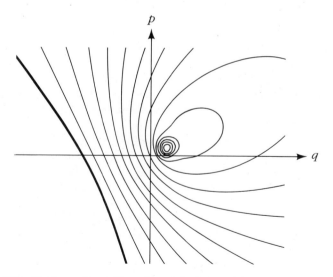

Figure 3–79. Contours for $\phi(i, e, g) = 0.5s\,(n + 1)(2 \cos i \cos e - \cos g)^n + (1 - s) \cos i$. This is the reflectance map for a surface with both matte and specular components of reflectance when the surface is illuminated by a single point source. Glossy white paint can produce such a ϕ.

Recovery of Shape from Shading

The fundamental problem with recovering shape from shading is that, even with all the simplifying assumptions that enable us to use a reflectance map, it is still very difficult. Knowing the intensity value places one on a particular isoluminance contour in the reflectance map—for example, it might tell us that the surface orientation lies on the 0.8 contour—but it does not tell us where. Unless we have additional information, one position on the contour is just as good as any other.

However, the problem can be solved. The extra condition we need is to assume that the surface is smooth and that surface orientation varies smoothly (that is, is differentiable). Essentially this says that if you are at one point in the image and know the surface orientation there and how it changes locally, then if you move in one direction across the image, you can tell from the new intensity value what the new local orientation is.

This is an amazing fact, because one would not think that smoothness constrains the answer enough. But it does because of a beautiful mathematical trick (Horn, 1977), which I am unfortunately unable to reduce to succinct English. So from a mathematical point of view, the problem is soluble. However, from a biological point of view, this type of solution, even given the major simplifications on which Horn's approach rests, is still far too complicated to be used. To solve the shape-from-shading equations for a general reflectance map requires successive integration along paths in the image whose loci can be determined only as the integration proceeds. Solving these equations in a simpler, more parallel way appears quite hopeless unless we are prepared to introduce other constraints.

A number of other approaches have therefore been tried. Woodham (1977) combined constraints on surface orientation—like minimizing local curvature—and constraints from shading to produce a local iterative approach to determine surface orientation. Brady (1979) suggested restricting the type of surface as well, for example, to generalized cones, and showed how one can then determine the direction of the light source.

However, I think it is fair to say that none of these approaches has yet thrown much light on the use of shading information by the human visual system. The difficulty is probably that we do not use this information very well. The human visual processor seems to use only coarse shading information, often but not always correctly, which is probably why shading is easily overridden by other cues. Situations where the human visual system does not perform well always cause trouble because knowing how the problem ought to be solved mathematically may throw very little light on how we ourselves approach it. Unfortunately, the same may be true for color, as we shall see. Nevertheless, we do make some use of shading, so there is definitely something here to be understood.

Photometric Stereo

Finally, there is a technique for recovering shape from reflectance maps that cannot possibly have any biological significance, but which is so elegant that I cannot resist mentioning it. The idea was introduced by Woodham (1978) and developed by Horn, Woodham, and Silver (1978), and it rests on the following idea. Given an image and a reflectance map for one position of the light source, suppose that we measure image intensity at one particular point. As we have seen, we may then deduce that the corresponding surface orientation lies on a particular contour in gradient space—the 0.8 contour was our example in the previous section—and I have illustrated it in Figure 3–80(c). The problem is that we do not know where along this contour the correct surface orientation (p,q) is.

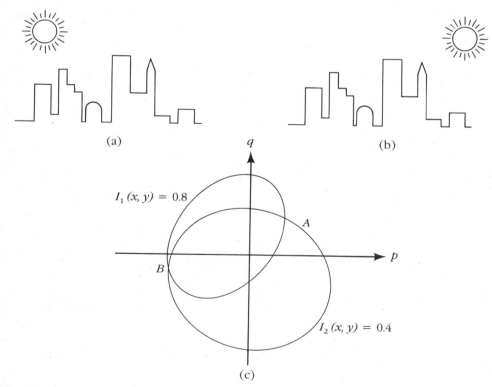

Figure 3–80.   The idea behind photometric stereo. Images I_1 and I_2 are taken of the same scene under two different lighting conditions, and so two different reflectance maps are employed. From the first, image intensity measurements may place a particular point in the image on the 0.8 contour (a); from the second, on the 0.4 contour (b). Hence the true surface orientation (p,q) corresponds to either point A or point B in (c), the intersection points of the two contours.

Suppose, however, that we now move the light source—or, in an outdoor scene, we wait until later in the day—and then take a second image from the same viewpoint. The surface geometry relative to the viewer is all the same, but the reflectance map changes. For example, the situation may change to look like Figure 3–80(b), and the intensity measurement at the same point in the image puts us on the 0.4 contour in the reflectance map, as shown in Figure 3–80(c). Then the true surface orientation is narrowed down to just two possibilities—the two points at which the first 0.8 contour and the second 0.4 contour intersect, points A and B in Figure 3–80(c). This essentially solves the problem, since the choice between A and B can usually be made easily by using continuity information or by taking a third picture with yet another lighting position.

This type of scheme may be of practical use, since we can usually construct a reflectance map even for complicated lighting conditions, although we usually have to measure the reflectance map empirically because it is too difficult to compute. Provided that the lighting and surface characteristics are the same everywhere in a scene, the sole determiner of image intensity is surface orientation.

3.9   BRIGHTNESS, LIGHTNESS, AND COLOR

All the processes that we have considered so far have used the image of reflectance and illumination changes on a surface to recover information about the geometry of the surface. Nothing has been said about the nature of the surface itself. Yet the reflectance of a surface—whether it is light or dark, whether it reflects red light well or poorly, and so forth—carries information that often has important biological significance. For example, we can tell just by looking whether a fruit is ripe, whether a branch is strong enough to bear one's weight, whether a leaf is green and supple, whether an insect is likely to be poisonous, and many other things.

The business of recovering surface reflectance, then, is important, and we are actually quite good at it. It is surprising how much perceived color depends upon the reflectance of a surface and how little it depends on the spectral characteristics of the light that enters our eyes. According to Helson (1938), an illuminant may be up to 93% chromatic, but provided it contains at least 7% "daylight", surfaces with uniform spectral reflectance—that reflect equally at all wavelengths—will remain achromatic. The opposite aspect of the problem is by how wide a range of stimuli we can be fooled into saying that brightness differences exist where they objectively do not—from the Hering grid and Benussi ring on the one hand to the phenomenon of subjective contours on the other. Some examples appear in Figure 3–81.

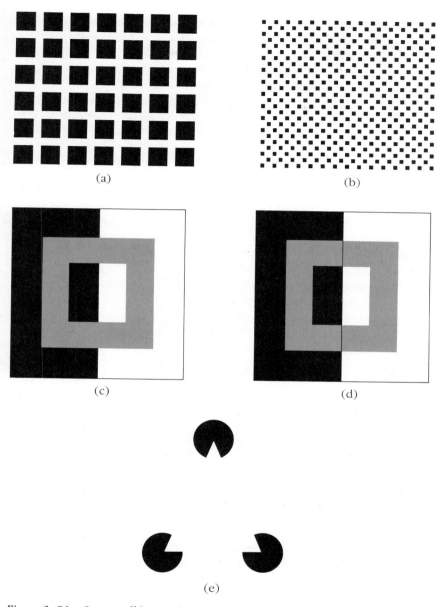

Figure 3–81. Some well-known brightness illusions. (a) The Hering grid. (b) An illusion by Robert Springer that provokes the appearance of faint diagonal lines. (c), (d) The Benussi ring; notice how the simple addition of a contour in (d) can cause the two gray regions to look different. (e) The Kanizsa triangle.

The theory of color vision is in an unsatisfactory and interesting state. On the one hand, we have for a long time had a fairly adequate phenomenological description, due to Helson (1938) and Judd (1940). Their equations can be used to predict the colors that will be perceived by a subject about as accurately as the subject is able to describe them, and they can, without modification, account for Land's (1959a, 1959b) famous two-color projection demonstrations in which images produced with only two colors gave full color percepts (Judd, 1960; Pearson, Rubinstein,, and Spivack, 1969). As Helson and Judd themselves commented, however, there are probably many other equations that describe color perception just as well; in fact, Richards and Parks (1971) proposed a simpler model that is nearly as accurate.

The problem is that these formulations are *descriptions* of color vision, not *theories* of it. The researchers do not say why their equations are good at separating the effects of the illuminant from the effects of surface reflectance. Of course, there may be no real theory of color vision, and these descriptions may be as close as we can get—but I hope not. The only attempt at a true theory of color vision is Land and McCann's (1971) retinex theory. This theory has been criticized for explaining nothing that the Helson–Judd formulation cannot account for, and this is probably true. But that comment misses what, from this book's perspective, is the most important difference between these two theories, namely, that the Helson–Judd formulation is a phenomonological description, whereas the retinex idea is a computational theory that is based on particular assumptions about the physical world. To bring these points out, let us look in more detail at the two formulations.

The Helson–Judd Approach

The basis for Helson and Judd's approach to color vision is the time-honored view that object color depends on the ratios of light reflected from the various parts of the visual field rather than on the absolute amounts. Helson and Judd tried to construct a formula that predicts what color a given piece of paper will appear to have under different illumination conditions and against different backgrounds. Thus they were not so much interested in color constancy as in quantifying the extent to which constancy is violated as the illumination and background are changed.

Their formulation is based on two steps. First, find out what "white" should be for the conditions prevailing in the scene; second, compute what color the paper should have by referring to this estimate of white. The basic idea behind finding the white is (1) to take the standard daylight white, which by a suitable choice of coordinates we can denote as (r_w, g_w); (2) to measure the "average" color of the whole visual field, which we denote by (r_f, g_f); and (3) to assume that the current white (r_n, g_n) lies somewhere between these two. For example, we might write

$$r_n = r_f - k(r_f - r_w)$$

$$g_n = g_f - k(g_f - g_w)$$

according to which the current white lies on the straight line joining daylight white to the average over the current visual field.

This basic idea is then modified by incorporating various empirical observations that Helson and Judd made to produce a complex expression that is no longer linear. In other words, the modifications push the current white off the line joining daylight white to the current average, so as to account for the various odd effects that Helson and Judd found empirically. The most important modification comes about because of a notion they had called *adaptation reflectance,* which is essentially a shade of gray that depends on the scene. Papers that are lighter than this gray take on the hue of the illuminant, whereas darker papers take on the complementary hue. Of course, linear formulas cannot account for this effect. Other modifications arise because adaptation effects increase in power as we move away from white, peculiar effects occur if the blue component of the illuminant is intense, and so forth. The result is a long and complicated formula, adding to the basic equations above a number of second-order, nonlinear terms, each justified by a particular aspect of the experimental findings. The second part of the scheme, assigning color relative to this estimate of white, has a simple formulation. To determine the hue to be associated with the point (r, g), we simply examine the orientation of the line joining it to the current white (r_n, g_n); the length of this line determines the saturation.

The interesting thing about this approach is that these assumptions lead to a successful predictor of perceived color. What is missing is an explanation of why we can make these assumptions and why they lead to valid color perception under such a wide range of circumstances.

Retinex Theory of Lightness and Color

Land and McCann (1971), on the other hand, base their theory firmly on assumptions about the physical world. It applies to the planar world of so-called Mondrians, which, as we saw in Chapter 2, consist of rectangular

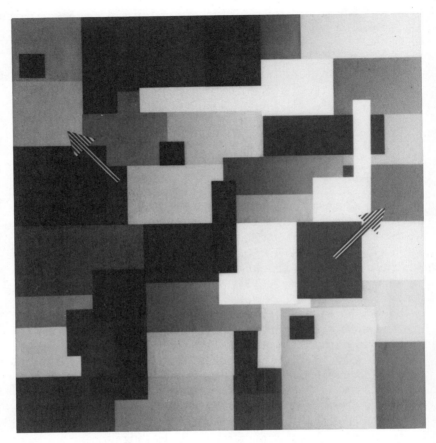

Figure 3–82.   The two marked squares have the same luminance, yet one is perceived as being much darker than the other. (Reprinted by permission from E. H. Land and J. J. McCann, "Lightness and retinex theory", *J. Opt. Soc. Am. 61* (1971), 1–11, fig. 3.)

patches affixed to a large board that can be illuminated in various ways (see Figure 2–30). The first part of the theory, concerned with what Land and McCann called lightness, deals with monochromatic images of just this kind. The central problem, as they state, is to separate the effects of surface reflectance from the effects of the illuminant, because as has long been known, what we perceive as the color of a surface is much more closely connected with spectral characteristics of its reflectance function than with the spectral characteristics of the light falling upon our eyes.

How can these effects be separated? What critical characteristics might enable us to separate the effects due to changes in illumination from the effects due to changes in reflectance? Land and McCann proposed the following: Changes due to the illuminant are on the whole gradual, appearing usually as smooth illumination gradients, whereas those due to changes in reflectance tend to be sharp. This dichotomy is certainly true in the Mondrian world that they studied, and hence if we can separate the two types of change, we can separate effects of changes in the illuminant from the effects of changes in reflectance in these images.

An example of what Land and McCann mean appears in Figure 3–82. This shows the image of a monochromatic Mondrian lit from above. The two patches marked with arrows have exactly the same intensity, yet one appears to be darker than the other. If one removes the effects of the illumination gradient, one patch would actually become much darker than the other. The argument is that this computation is essentially what our visual systems do, and it is called the retinex computation.

Algorithms

The retinex computation has been implemented in at least two ways. Land and McCann themselves used the one-dimensional approach illustrated in Figure 3–83(a). If we trace the image intensities along any path from A to B as shown, they will have the form shown in the first graph, portions of slow changes interspersed with large jumps at the reflectance boundaries. By applying a threshold, we can remove the effects of the slow changes, thus arriving at the curve in the second graph, which describes the effects of the reflectance changes only. Since the system is conservative, it does not matter which path from A to B is used—the resulting assignments of reflectance will always be the same. Land and McCann used this technique together with a sufficient number of randomly chosen paths across the image to cover all locations adequately.

Horn (1974) derived a two-dimensional analogue of this algorithm, illustrated in Figure 3–83(b) and consisting essentially of the same three steps. The first step is to take a differencing operator, here having a two-dimensional center–surround form. Then we ignore small values and accept only large ones, which correspond to the reflectance changes. Finally, using only the large changes, we reconstruct the image to get a two-dimensional analogue of the second graph in Figure 3–83(a). For this, Horn suggested an interesting iterative algorithm based on nearest-neighbor interactions in order to implement the equations shown in Figure 3–83(b).

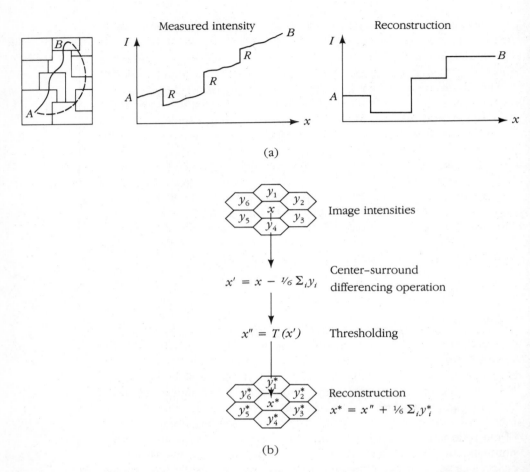

Measured intensity

Reconstruction

(a)

y_6　y_1　y_2
y_5　x　y_3　　Image intensities
　　y_4

$x' = x - \frac{1}{6}\Sigma_i y_i$　　Center–surround differencing operation

$x'' = T(x')$　　Thresholding

y_6^*　y_1^*　y_2^*
y_5^*　x^*　y_3^*　　Reconstruction
　　y_4^*　　$x^* = x'' + \frac{1}{6}\Sigma_i y_i^*$

(b)

Figure 3–83. Diagrams illustrating retinex algorithms. (a) Land and McCann's one-dimensional algorithm. (b) Horn's two-dimensional version. In both, the idea is to ignore smooth changes in intensity, taking account only of discontinuities. See text for details.

Extension to color vision

The operations diagrammed in Figure 3–83 show the retinex operating monochromatically. In order to apply the operation to color, Land and McCann require that it be performed independently in each of the red, green, and blue channels. What then emerges from each, they hope, are signals that depend not on the illuminant but solely on the surface reflectance. These can be combined to give a percept of color that happily rests solely on properties of surface reflectance and not on the vagaries of its particular, present illuminant. Of course, there is still the need to calibrate the signals in the three channels relative to one another, but Land and McCann suggest that this can be done by calling the brightest point in the scene white.

McCann, McKee, and Taylor (1976) have recently published comparisons between the results predicted by such an algorithm on their Mondrian stimuli and the psychophysical estimates of color made by subjects who viewed them. They found that the agreement between their subjects and their predictions was as good as the agreement among their subjects.

Comments on the retinex theory

To me, the positive aspects of Land and McCann's work seem to be threefold. First, they have attempted to construct a real theory of color vision, as opposed to a description of color perception. Second, they have drawn attention to the importance of boundaries and described one way in which boundary effects may propagate across an image. Such effects had been known for a long time—for example, the Craik–Cornsweet illusion and the Benussi ring—but boundary effects do not appear explicitly in the Helson–Judd equations. Third, Land's earlier work formulated an interesting principle considered important by Judd, namely, that when the colors of the patches of light making up a scene are restricted to a one-dimensional variation of any sort, the observer usually perceives the objects in that scene as essentially without hue.

The case against the retinex theory seems to consist of one major and several minor arguments. The major argument is that there is more to simultaneous contrast than is present in the retinex theory. That is, formulations like Helson and Judd's that are based on the idea of simultaneous contrast may be able to explain Land and McCann's effects, but the gradient-eliminating retinex theory cannot explain all of simultaneous contrast, because these effects occur perfectly well in situations of uniform illumination, where there are no illumination gradients. In addition, Land and McCann apparently did not always pay adequate attention to the effects of simultaneous contrast in their displays. For example, in Figure 3–82, one of the squares has darker neighbors than the other, so one might expect them to appear different just on these grounds. In any event, brightness perception and color perception appear to involve at least some effects that are not predicted by Land and McCann's approach.

One possible explanation is that these extra effects are introduced by aspects of the problem that Land and McCann did not consider. For example, their theory applies only to planar surfaces, and these other effects may be introduced only to deal with the added complications of having different surface orientations in different parts of the visual field. This, however, is unlikely. Although there certainly are three-dimensional effects on brightness perception, they are probably not very large. Gilchrist (1977) recently claimed that perceived orientation could affect brightness perception by factors of up to 30%, but, in repeating his experiments, Ikeuchi (1979) was unable to obtain factors much greater than 5%–10%.

The first of the minor arguments against the retinex idea is computational: The theory involves a threshold (the level of gradient at which the cutoff occurs), but it does not say what that threshold should be. It is a matter of unhappy experience that whenever we have to set a threshold in an image-processing task, we usually have problems—which is one reason why the idea of zero-crossings is so attractive. The problem is that if the threshold is too low, it will not remove the illumination gradient; but if it is too high, it will remove valuable shading information. Gradual changes in surface orientation also produce gradual changes in intensity across an image, and these might be too valuable to throw away cavalierly. And gradual changes in surface coloration can also be important. After all, we can see a rainbow, even one that has been enlarged by binoculars. The color changes are not thresholded out.

The second minor argument arises from neurophysiological observations. According to the retinex theory, the red, green, and blue channels are processed independently, each in the manner of Figure 3–83, and combined only afterward. This, however, is not the observed situation. Neural processing seems to be based on an opponent-color approach—where the output depends on the difference between two color channels—right from the start. Even in the retina, most color-sensitive cells have an opponent-color organization (DeValois, 1965), and DeValois and his associates have found an impressive correlation between the psychophysics of color discrimination and the observed neurophysiological properties of lateral geniculate color-opponent cells.

These findings do not disprove the notion that the retinex function is being computed in the visual pathway. One could argue, as Horn (1974) pointed out, that the retinex can be carried out on any three linear combinations of red, green, and blue just as well as on the original channels themselves, and this adjustment might make the retinex theory compatible with the neurophysiological observations. But this argument is not very convincing, especially since the theory provides no very good reason why one should want to operate on linear combinations rather than on the original signals.

Some Physical Reasons for the
Importance of Simultaneous Contrast

It is a widespread and time-honored view, going back at least to Ernst Mach, that object color depends upon the ratios of light reflected from the various parts of the visual field rather than on the absolute amount of light reflected. Of course, this must be because although the illumination of a scene, which greatly influences the spectral distribution in its image, changes drastically from time to time and from place to place, we are relatively immune to the variation. The range of color constancy is, of course, bounded—when we buy clothing we insist on seeing the items in daylight or under tungsten illumination if the store's lighting is fluorescent. But the important point is that although our perceptions may only approximate the objective reflec-tances, they do this much more accurately than they reflect the spectral qualities of the light falling upon the retina.

Even within a single scene, the intensity of illumination can change drastically, from sunlight to shadow, for example, or from near the lights in a large hall to the dim recesses of the furthest-flung corners. The spectral characteristics can also change, although usually not by so much. The light becomes greener under a tree than in the open; in the mouth of a cave it can turn browner. So even though the main fluctuations in spectral content occur over time, they can still occur in a single scene, and this does not much affect us.

How can we deal with such a wide range of effects? What the simul-taneous-contrast phenomena* seem to be drawing attention to is an argu-ment of the following kind. Suppose you pass an embankment where a yellow or blue flower happens to be growing amid a background of green grass and clover. Although the absolute spectral characteristics of the light coming from the flower cannot at all be relied upon as a clue to its surface reflectance characteristics, either in the matter of its lightness or of its spectral properties, nevertheless its characteristics relative to other nearby surfaces probably are reliable. If the flower appears lighter than the grass, this is probably due to a characteristic of the flower and not of the illu-mination (though the head of the flower could be just catching the sun). If the flower looks bluer than the grass, then it probably really is. If it looks yellower, then, again, it probably really is.

Furthermore, what is so amazing about simultaneous-contrast effects— even as simple as those in Figures 3–81(b) and (c)—is that the visual

---

*The tendency for color or brightness of one area to affect neighboring areas.

system seems to take them so seriously. That is, we get what looks like wrong answers in situations as simple as the Bernussi ring (Figure 3–81c), where we would think that almost any sensible scheme would give an answer reflecting the objective truth of the situation. I find this so striking that I am tempted to believe that relative observations may be *all* one relies on.

Even so, to make a success of a scheme based only on relative measurements, we have to make a basic distinction between changes in the image due to changes in reflectance (like the difference between a flower and grass) and those due to changes in illumination (like the shadow of a nearby tree). The fact is that shadowed lawn looks darker than unshadowed lawn, and a daisy in the sun looks brighter than a daisy in the shade, but the shadowing does not much affect the color of the lawn or the daisy. The sunlit daisy and the shadowed daisy both look white, and (critically) the shadowed daisy does not look gray.

We naturally consider the sunlit daisy brighter than the shadowed one. This suggests that brightness is a subjective quality related to the intensity of the prevailing illuminations. The reflectance of the surfaces, on the other hand; is more closely related to the qualities of lightness and color. Changes in lightness are ideally pure scalar changes in a surface's reflectance involving no changes in the surface's spectral characteristics (detectable through the three color channels), whereas changes in color refer ideally to changes in the spectral characteristics of a surface and may be described by the two components hue and saturation. Helson (1938) and Judd (1940) defined the terms *brightness, lightness,* and *color* purely psychophysically, but I think that to regard them as perceptual approximations to illumination intensity and to the value and spectral distribution of surface reflectance is consistent with their definitions (see Judd, p. 3).

The computational problem, therefore, is how to formulate in a reasonable way the physical basis for estimating brightness, lightness, and color from an image. The first point to note is that surface orientation can influence brightness (according to our definition) but not usually a surface's lightness or color, because at some orientations a surface will be more directly illuminated than at others. The final solution to the computation of brightness will therefore have to await an estimate of the surface's orientation. As we have noted, however, the effects of 3-D interpretation on perceived brightness are still not fully established.

The major source of brightness changes is shadows, and again, as we saw in Section 2.4, these can be detected autonomously by using the ideas behind the operator $\nabla I/I$. These two phenomena, surface orientation changes and shadows, provide the main sources of discontinuity in brightness, so provided that they are taken adequately into account, we can be

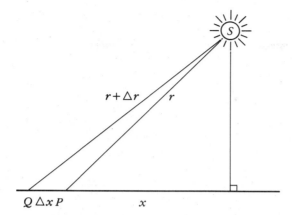

Figure 3–84.   Gradients in intensity that are due solely to illumination are usually small and almost linear. *S* is a source that illuminates the plane containing *P* and *Q*. The most significant terms in the difference between the intensities at *P* and *Q* depend on △*x*, the distance between the two points.

fairly sure that the remaining changes in the illuminant are smooth rather than sharp.

Our next observations are (1) that locally measurable illumination gradients on a flat surface can occur only if the light source is not very far away, (2) that they are small unless the source is very near, and (3) that they are approximately linear except perhaps directly under the source. This can be seen from Figure 3–84. The illumination at *P* is I/r^2, and at *Q* nearby it is $I/r^2 - 2x\triangle x/r^4 + O(1/r^4)$. If $\triangle x/x$ is small, the change from *P* to *Q* varies approximately with $-2\triangle x/x$. This is essentially linear in $\triangle x$, the distance between *P* and *Q*, provided that $\triangle x$ is small compared with x. In other words, illumination gradients are almost always small and linear. This may be one reason why the human visual system is insensitive to small linear changes in intensity (see Brindley, 1970, p. 153).

Hypothesis of the Superficial Origin of Nonlinear Changes in Intensity

These observations suggest that the following approach to the physical basis of color vision may be profitable: *In the absence of sharp changes in brightness, detectable as shadow boundaries or changes in surface orientation, all nonlinear changes in intensities may be assumed to be due to properties of the surface—either its orientation or its reflectance.* In other words, in the absence of obvious illumination effects like shadows, measurable nonlinear local differences in either image intensity or spectral distributions are due to changes in the lightness or color of the surface. This assumption allows us to ignore small linear changes and provides a basis for the idea that lightness and color may be recovered from measurements of nonlinear local changes in intensity and spectral distribution made, for example, by comparing their values at each point with their values in the surrounding neighborhood.

Implications for Measurements on a Trichromatic Image

According to physiological descriptions, some opponent-color cells in the retina of the monkey have receptive fields with rather mixed properties, like a red center and green surround (Gouras, 1968; de Monasterio and Gouras, 1975). There seem to be no internal reasons for doubting these reports; nevertheless, I find such cells extremely difficult to understand in general and impossible to fit into the $\nabla^2 G$ framework that we developed in Chapter 2.

The reason for the difficulty is that a cell with such a receptive field, illustrated for convenience in Figure 3–85(a), signals a complex mixture of spatial and chromatic information. It signals neither a pure $\nabla^2 G$ function for a single chromatic channel, like the red $\nabla^2 G$ receptive field illustrated in Figure 3–85(b), nor a purely chromatic message about the relative strengths of signals in the two channels at one point in the image, as would the receptive field illustrated in Figure 3–85(c). In fact, Figure 3–85(a) is not even a zero-mean operator—it is not like a second derivative, and its zero-crossings are meaningless. To use it, we have to pay special attention to *changes* in its value—for example, if its green-center, red-surround analogue looks at a lawn, it will fire everywhere over it, slightly harder for the more saturated greens. This seems to me not only poor engineering but also a contradiction to the experience we have about how the nervous system likes to code changes rather than pure values; in other words, it violates Barlow's (1972) second dogma about the economical neural encoding of stimulus information.

In order to make a reasonable concrete suggestion about what these cells are signaling, I would like to combine two pieces of information. The first is that the $\nabla^2 G$ style of analysis requires that the spectral characteristics of the center and of the surround be essentially the same, related to one another by a minus sign. This is necessary for zero-crossings to be useful.

The other piece of information is the idea that lightness and brightness

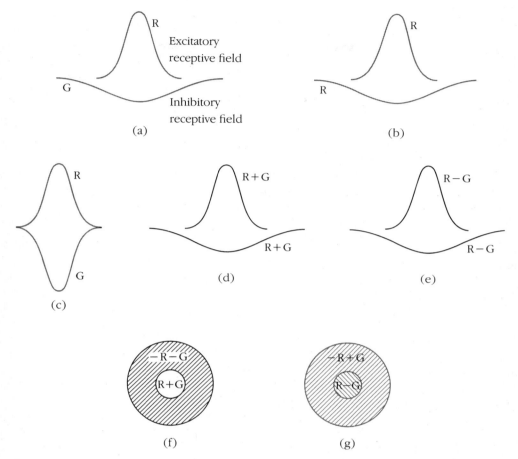

Figure 3–85.   Various possible organizations of chromatic receptive fields. The organizations are assumed to be composed spatially of the difference of two Gaussian distributions. (a) So-called red–green opponent-color receptive field. (b) A red-center, red-surround receptive field. (c) A red–green opponent-color receptive field in which the spatial distribution of the two are identical. (d) A pure luminance (red + green) receptive field. (e) A pure color-difference receptive field (red − green). (f) The two-dimensional receptive field corresponding to (d). (g) The two-dimensional receptive field corresponding to (e). R = red; G = green.

should be separated from color. Luminance boundaries correspond effectively to change in the summed contributions of the red and green channels, which we can write $(R + G)$. To detect these boundaries requires a $\nabla^2 G$ operator running on this sum, as illustrated in Figure 3–85(d). To detect changes in color, on the other hand, our hypothesis of the last section tells us to detect *relative* changes in the amounts of red and green light. That can be done by a $\nabla^2 G$ operation on $(R - G)$, the difference between red and green signals, as illustrated in Figure 3–85(e).

Now the first type of cell, whose receptive field is illustrated in Figure 3–85(f), will not be very color selective, since its maximal stimulus will be a white central spot, and it can be turned off by any combination of red and green in the center and in the surround. The only criterion is that the effective luminances should balance.

The second type of cell is quite different, however. Its optimal stimulus would be a red center accompanied by a green surround, and it would therefore look like a color-opponent cell. Such a cell would respond best to changes in color and it should not respond at all to a pure white spot at its center, provided that the red and green are appropriately balanced in the white. Such a cell should respond to color boundaries but not to other boundaries. In order for such a cell to be insensitive to nonwhite lightness boundaries, like the boundary between two reds that differ only in the fraction, not the quality, of the light they reflect, the quantities R and G would have to be in logarithmic units. Such a cell would then act as a pure detector of changes in color. $\nabla^2 G$ operators are also insensitive to linear gradients.

Summary of the Approach

The main ideas of this approach, then, are to separate brightness from lightness and color and then to separate the estimation of lightness (average percent reflectance) from color (spectral distribution). Local changes may be recovered from zero-crossings in the lightness $(R + G)$ image and in the color image based on $(R - G)$ and $(B - G)$ (where B = blue).

The principal neurophysiological consequences are that no receptive fields should mix color and spatial variations in the manner of Figure 3–85(a); instead, receptive fields should exist as shown in the configurations of Figure 3–85(d) (for changes in lightness and brightness) and Figure 3–85(e) (for changes in color). Zero-crossing segment detection can subsequently occur in a similar way on each type of measurement yielding luminance contours from the first type, and color change contours from the second.

3.10   SUMMARY

In this chapter we have seen some of the quite striking variety of ways in which surface information is encoded in images, and we have explored as

Table 3–2.   Processes producing surface information from image information and their probable input representations.

Process	Probable input representation
Stereopsis	Mainly ZC with eye movements helped by FPS
Directional selectivity	ZC
Structure from motion	FPS for correspondence; perhaps only RPS for detailed measurements
Optical flow	FPS(?) if process is used at all
Occluding contours	RPS, BC
Other occlusion cues	RPS
Surface orientation contours	RPS, BC
Surface contours	RPS, IC, GT
Surface texture	RPS, GT
Texture contours	BC
Shading	IC, RPS; possibly others

Note: BC = boundary contours created by discrimination processes and curvilinear aggregation of tokens; FPS = full primal sketch = RPS + GT + IC + BC. GT = group tokens, created by grouping processes in the full primal sketch; IC = illumination contours (shadows, highlights, and light sources); RPS = raw primal sketch (edges, blobs, thin bars, discontinuities, and terminations); ZC = zero-crossings, discontinuities, and terminations.

far as is presently possible how such information may actually be recovered. At the moment, the different processes appear to use slightly different input representations; the simplest, like directional selectivity, is driven by the zero-crossings, and the more elusive, like surface texture, probably involves the most complex aspects of the full primal sketch. I have summarized the discussion in Table 3–2.

Another interesting aspect of all these processes is that, in addition to using slightly different input representations, they all involve slightly different assumptions about the world in order to work satisfactorily. As we have seen, in each case the surface structure is strictly underdetermined from the information in images alone, and the secret of formulating the processes accurately lies in discovering precisely what additional information can safely be assumed about the world that provides powerful enough constraints for the process to run—for example, uniqueness and continuity in stereopsis, rigidity in motion, and so forth. Much of the art of formulating these processes lies in the precision and accuracy with which these additional constraints are expressed, and our survey has included some processes that I find satisfactorily formulated and others that remain puzzling and rather ill defined. The constraints assumed by the various processes have been set out roughly in Table 3–3, but the reader should bear in mind that few of these are certain. Thus the table should be regarded more as a guide to current thinking than as a definite statement of what allows these processes to run.

Finally, a few words about research strategy in this area. As we have seen, there are striking differences in the clarity and precision with which we have been able to formulate the different processes. Some are straight-forward and clean, like stereopsis, structure from motion, and directional selectivity, whereas others, like visual texture and surface contour analysis, seem to be inherently muddy. That is not because the first kind are intellectually easier—on the whole they are not. For example, the mathematics associated with stereopsis or with structure from motion is not as easy as that associated with visual texture. Rather, the analytical difficulties arise from deciding what can be safely assumed about the world in order to help the processes interpret images of it. Where this can be done cleanly, more or less by inspection of the real world, we have on the whole been able to develop a clean theory. But where it cannot, I think there is no hope of understanding the processes properly until some other means have been found for determining what is safe to assume about the world and what is not, together with the related question of the reliability of the different kinds of information.

In the end, these are empirical questions, not so much about our visual systems (although the answers will be reflected in their structural design), as about the statistical structure of the visual world. I think that one will have to accept this, taking more of an engineering point of view when trying to answer them. As our knowledge of how to implement these early processes improves, we shall have to build fast machines that can run these processes in real time and acquire in that rather direct way a more detailed knowledge of which tricks pay off in practice and which do not. Studying vision is a mixture of studying processes and studying the world from this rather specialized point of view—something that natural evolution has been doing for a long time.

The first step is to build a unified system that employs all the processes that we currently understand, but much remains to be done before even this limited goal should be attempted. For one thing, processes like the construction of the raw primal sketch require a great deal of computational

Table 3–3.   Guide to additional assumptions implicit in processes deriving surface information from images.

Process or representation	Implicit assumptions
Raw primal sketch	Spatial coincidence
Full primal sketch	Various assumptions about spatial organization of reflectance functions
Stereopsis	Uniqueness; continuity
Directional selectivity	Continuity of direction of flow
Structure from motion	Rigidity
Optical flow	Rigidity
Occluding contours	Smooth, planar contour generator
Surface contours	Surface locally cylindrical; planar contour generators
Surface texture	Uniform distribution and size of surface elements
Brightness and color	Only local comparisons reliable
Fluorescence	Uniform light source

power. Even the fastest general purpose machines are several orders of magnitude too slow for real-time vision, and although the emerging very large scale integration (VLSI) technologies will eventually provide the necessary power, the sensors and technology are not yet available and will not be for several years. And then, of course, there is the question of what one would do with the output of a machine that could run a set of processes like the ones described in this chapter. It is to this question that we now turn our attention.

CHAPTER 4

The Immediate Representation of Visible Surfaces

4.1 INTRODUCTION

In this chapter, we shall discuss the issues and problems surrounding the idea of the 2½-D sketch, whose acquaintance we have already made in Section 3.3. The central point is a simple one—that the 2½-D sketch provides a viewer-centered representation of the visible surfaces in which the results of all the processes described in Chapter 3 can be announced and combined. The construction of the 2½-D sketch is a pivotal point for the theory, marking the last step before a surface's interpretation and the end, perhaps, of pure perception.

The idea that such a representation might exist and that its construction can be regarded as the goal of early visual processing will probably strike the reader as unsurprising, especially since this book is written within precisely such a framework. But when we started out we had no such framework, and in trying to find a way of understanding what vision was, we were confused, having to grapple with almost philosophical difficulties concerning what perception was for. The reader who cares to examine Marr (1976) closely, for example, will find no explicit statement of what the primal sketch was for. He will find it more or less defined, justified on general grounds, and closely tied to physical reality. But the idea that the purpose of early vision is to recover explicit information about the visible surfaces was only implicit.

In fact, at that point, much of computer vision was in considerable disarray, because, with the exception of Horn's (1975) work, the idea that the main point of vision was to tell the shapes of things had not yet been taken seriously. And although perceptual psychologists like Gibson had the notion that surfaces are important, the idea of an internal representation obtained by certain processes was foreign to their thinking. In retrospect, our lines of thought and the kind of questions we asked at that time were rather muddled; inquiry had to do with feature-based recognition, how to separate figure from ground, how to extract and interpret a "form" or "figure," how much analysis could be done in a data-driven or bottom up way, and how much needed top–down influences. In addition, we had no coherent framework that allowed us to see how processes like stereopsis, shading, or motion perception could combine with one another and with the rest of vision to create what we call seeing.

All this type of thinking was dramatically swept away by the idea of the 2½-D sketch, which simultaneously resolved these and many other issues. It told us what the goals of early vision were, it related them to the notion of an internal representation of objective physical reality that *preceded* the decomposition of the scene into "objects" and all the concomitant difficulties associated with object recognition. At the same time, it hinted at the limits of what one might call pure perception—the recovery of surface information by purely data-driven processes without the need for particular hypotheses about the nature, use, or function of the objects being viewed. And finally, it provided the cornerstone for an overall formulation of the entire vision problem—the framework that this book has been written to explain and that has since enabled us to structure our research in a rational and strategic way.

For all these reasons, the emergence during the autumn of 1976 of the idea of the 2½-D sketch, which first appeared in Marr and Nishihara (1978, fig. 2) and was developed at length a little later (Marr, 1978, sec. 3), was for me the most exhilarating moment of the whole investigation. Its first positive consequence was the theory of stereo vision (Marr and Poggio, 1979) which was formulated during the first half of 1977. The reformulation of early visual processing was begun later that year, and of course, the 2½-D sketch ultimately led to the overall framework that we now have (Marr, 1978).

4.2   IMAGE SEGMENTATION

Perhaps the best way to introduce the whole question of the 2½-D sketch is to describe in some detail the impasse that it was intended to resolve. The neurophysiologists' and psychologists' belief that figure and ground constituted one of the fundamental problems in vision was reflected in the attempts of workers in computer vision to implement a process called *segmentation*. The purpose of this process was very much like the idea of separating figure from ground, the idea being to divide the image into regions that were meaningful either for the purpose at hand (which for computer vision might be assembling a water pump) or for their correspondence to physical objects or their parts.

Despite considerable efforts over a long period, the theory and practice of segmentation remained primitive for two reasons. First, it was well-nigh impossible to formulate precisely in terms of the image or even of the physical world what the exact goals of segmentation were. What, for example, is an object, and what makes it so special that it should be recoverable as a region in an image? Is a nose an object? Is a head one? Is it still one if it is attached to a body? What about a man on horseback?

These questions show that the difficulties in trying to formulate what should be recovered as a region from an image are so great as to amount almost to philosophical problems. There really is no answer to them—all these things can be an object if you want to think of them that way, or they can be a part of a larger object (a fact that is captured quite precisely in Chapter 5). Furthermore, however these questions were answered in a given situation did not help much with other situations. People soon found the structure of images to be so complicated that it was usually quite impossible to recover the desired region by using only grouping criteria based on local similarity or other purely visual cues that act on the image intensities or on something like the raw primal sketch. Regions that have "semantic" importance do not always have any particular visual distinction. Most images are too complex, and even the very simplest, smallest images like one depicting just two leaves (Marr, 1976, fig. 13) often do not contain enough information in the pure intensity arrays to segment them into different objects.

Despite the lack of any precise formulation of what it meant, the notion of segmentation continued to be investigated with increasingly complex techniques. It had been a long-standing view that visual perception was analogous to problem solving and should therefore involve the testing and modifying of hypotheses about the viewed object. This idea was common in computer vision (for example, see Minsky, 1975), and it had its counterpart in the psychology of vision (as exemplified by Gregory, 1970). The critical difference between this idea and the use of constraints as described in Chapters 2 and 3 is that, in the problem-solving approach, the additional knowledge or hypothesis that is brought to bear is not general but particular and true only of the scene in question and others like it. Instead of using things like rigidity, we make inferences such as: A black blob at desk level has a high probability of being a telephone.

Naturally, because of their specificity, any very general vision system must command a very large number of such hypotheses and be able to find and deploy just the one or two demanded by the particular situation. This prospect casts a whole complexion on the vision problem, in which the main questions to be addressed concern how to manage vast amounts of information in an efficient way. That is why so much effort was expended on the design of efficient program control structures* for deploying visual knowledge. Incidentally, for this type of reason people in other branches of artificial intelligence believe the problem of control to be an important one.

The main thrust of the then-current ideas was, therefore, to invoke specialized knowledge about the nature of the scene being viewed to aid segmentation of the image into regions that corresponded roughly to the objects expected in the scene. Tenenbaum and Barrow (1976), for example, applied knowledge about several different types of scene to the segmentation of images of landscapes, an office, a room, and a compressor. Freuder (1974) used a similar approach to identify a hammer in a simple scene. If this approach had been correct, then a central problem for vision would have been arranging for the availability of the right piece of specialized knowledge at the appropriate time during segmentation. Freuder's work, for example, was almost entirely devoted to the design of what was called a heterarchical control system that made this possible. A little while later, the constraint relaxation technique of Rosenfeld, Hummel, and Zucker (1976) attracted considerable attention for just this reason—it appeared to be a technique whereby constraints drawn from disparate sources could be applied to the segmentation problem while making the control processes required to manage the information only slightly more complex. Our own work on cooperative algorithms was also slightly colored by thoughts that they could perhaps be used to combine constraints from disparate sources, and this provided one of the motivations for trying to develop precise methods of analyzing the convergence of such algorithms (Marr, Palm, and Poggio, 1978).

---

*The interaction among subprocesses in a computer program.

4.3   REFORMULATING THE PROBLEM

What was wrong with the idea of segmentation? The most obvious flaw seemed to be that "objects" and "desirable regions" were almost never visually primitive constructions and hence could not be recovered from the primal sketch or other similar early representations without additional specialized knowledge. Edges that ought to be significant are either absent from an image or almost so (see, for example, Figure 4–1), and the strongest changes in an image are often changes in illumination and have nothing to do with meaningful relations in a scene. Given a representation like the primal sketch and the many possible boundary-defining processes that are naturally associated with it, which of all the possible boundaries should one attend to, and why? In order to answer these questions, it was necessary to discover precisely what information we should try to recover from an image and then to design a representation for expressing it.

In order to find the answer, it was necessary to go back to first principles, to return to the physics of the situation. As we have seen several times, the principal factors that determine the intensity values in an image are (1) the illumination, (2) the surface geometry, (3) the surface reflectance, and (4) the vantage point. At some stage, the effects of these different factors are separated.

The main argument was, therefore, as follows: Most early visual processes extract information about the visible surfaces directly, without particular regard to whether they happen to be part of a horse, or a man, or a tree. It is these surfaces—their shape and disposition relative to the viewer—and their intrinsic reflectances that need to be made explicit at this point in the processing, because the photons are reflected from these surfaces to form the image, and they are therefore what the photons are carrying information about. In other words, the representation of the visible surfaces should be carried out before knowing whether the surface belongs to a horse, man, or tree. As for the question of what additional knowledge should be brought to bear, general knowledge must be enough—general knowledge embedded in the early visual processes as

---

Figure 4–1. (opposite)   This image of two leaves is interesting because there is not a sufficient intensity change everywhere along the edge inside the marked box to allow its complete recovery from intensity values alone, yet we have no trouble perceiving the leaves correctly. The table shows the actual intensity values within the box. However, the surface is clearly discontinuous within the box. Consistency-maintaining processes operating in the 2½-dimensional sketch may be partially responsible for this.

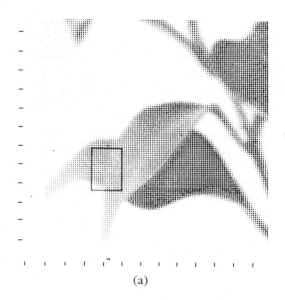

(a)

X =	34	35	36	37	38	39	40	41	42	43	44	45	46	47	48	49
Y
58	171	169	167	167	166	165	166	164	167	171	171	174	174	175	173	171
57	168	168	168	167	166	167	167	165	169	168	174	176	175	175	175	172
56	168	167	167	165	166	166	167	167	168	170	178	177	176	174	174	173
55	168	168	165	169	167	168	167	165	168	175	177	177	175	175	172	171
54	169	170	167	169	169	168	163	166	172	169	174	173	175	178	173	173
53	171	169	170	168	169	168	169	168	168	170	175	173	175	177	178	176
52	172	171	170	168	169	169	167	168	173	172	173	177	174	175	178	176
51	172	174	171	170	166	168	167	168	172	172	172	177	179	172	175	175
50	171	167	176	169	170	169	168	169	171	172	174	174	173	173	174	178
49	174	172	173	173	173	174	171	171	172	174	172	172	172	169	173	173
48	173	173	173	176	178	172	171	174	174	173	175	175	175	173	173	171
47	173	175	178	173	173	171	171	175	175	177	178	175	174	173	175	178
46	178	175	174	169	173	175	177	175	177	177	174	175	176	177	177	174
45	173	175	173	174	172	173	174	175	174	171	173	174	175	174	172	171
44	177	174	175	175	172	171	172	176	172	173	172	172	173	170	170	175
43	173	171	174	168	176	172	173	173	173	174	171	174	175	173	174	174
42	175	173	171	172	170	171	176	175	178	172	174	175	175	175	175	172
41	181	179	177	172	170	170	169	179	175	174	175	174	172	175	174	175
40	188	184	179	178	176	176	176	174	172	178	172	174	173	172	174	173
39	195	191	188	186	185	183	180	177	178	175	174	176	175	174	176	176
38	200	199	197	193	190	187	185	180	176	175	180	177	175	175	176	177
37	202	202	199	202	199	194	187	180	175	179	177	176	174	175	176	173

(b)

general constraints, together with the geometrical consequences of the fact that the surfaces coexist in three-dimensional space.

Was there any chance that such an idea might work? In order to explore it, we needed to look at three questions. First, what might it mean to represent the visible surfaces? In order to answer this, we needed to preview the general classification of shape representations, which we shall spend more time on in the next chapter. Second, we needed to look at the information provided by psychophysics, both about the early processes that we studied in the last chapter and about whether there is any evidence that such processes are combined before the visible shapes are interpreted as objects. Third, we needed to look at the computational aspects of the problem. In what form do these early processes deliver information about the visible surfaces, and how might one combine all the different resources?

Part of our task in formulating the problem of intermediate vision is to examine ways of representing and reasoning about surfaces. We start our inquiry by discussing the general nature of shape representations. What kinds are there, and how may one decide among them? Although formulating a completely general classification of shape representations is difficult, we had already set out the basic design choices that have to be made when a representation is formulated. Three characteristics of a shape representation are largely responsible for determining the information that the representation makes explicit. The first is the type of coordinate system the representation uses—whether it is defined relative to the viewer or to the object being viewed; the second concerns the nature of the shape primitives used by the representation, that is, the elements whose positions the coordinate system is used to define. Are they two- or three-dimensional, in what sizes do they come, and how detailed are they? And the third characteristic is concerned with the organization a representation imposes on the information in a description—is it, for example, flat like an image intensity array, or does it have a hierarchical structure, like the full primal sketch of Chapter 2?

The first question about the coordinate system and the second about the shape primitives both have fairly straightforward answers. The coordinate system must be viewer centered, and the shape primitives must be two-dimensional and specify where the local pieces of surface are pointing. Briefly stated, the reason for this is that the information delivered by all the early visual processes of Chapter 3 depends upon aspects of the imaging process—for example, measures of depth, or surface orientation are obtained relative to the viewer, and so fall naturally into a viewer-centered coordinate frame. The second point is that all these processes tell about the visible surfaces, usually only locally, and so it is this information that needs representing, usually only locally. It is worth going into these points more deeply.

4.4   THE INFORMATION TO BE REPRESENTED

Vision, as we have already seen, provides several sources of information about shape. The most direct are stereopsis and motion, but surface contours in a single image are nearly as effective, and we have seen several examples of other, less effective cues. It often happens that some parts of a scene are open to inspection by some of these techniques and other parts by others. Yet different as the techniques are, they have two important characteristics in common: They rely on information from the image rather than on a priori knowledge about the shapes of the viewed objects, and the information they specify concerns the depth or surface orientation at arbitrary points in an image, rather than the depth or orientation associated with particular objects.

When viewing a stereo pair of a complex surface, like a crumpled newspaper or the "leaves" cube of Ittelson (1960), which is a box with leaves attached to the sides and pointing nearly at the viewer, we can easily state the surface orientation of any piece of the surface and whether one piece is nearer to or further from the viewer than its neighbors. Nevertheless, memory for the shape of the surface is poor, despite the vividness of its orientation during perception. Furthermore, if the surface contains elements lying nearly parallel to the line of sight, their apparent orientation when viewed monocularly can differ from the apparent surface orientation when viewed binocularly.

The reader can check this in a room with a textured ceiling: If you look at it with one eye through a narrow tube, any portion you see through the tube will soon come to be oriented apparently at a right angle to your line of sight. This impression persists despite the certainty of one's knowledge that it is false.

From these observations, we may draw some simple inferences:

1. There is at least one internal representation of the depth, surface orientation, or both associated with each surface point in a scene.

2. Because surface orientation can be associated with unfamiliar shapes, its representation probably precedes the decomposition of the scene into objects.

Table 4–1.   Forms in which early visual processes would deliver information about surface geometry changes most naturally.

Process	Natural output form
Stereopsis	Disparity, hence δr, Δr, and s
Directional selectivity	Δr
Structure from motion	r, δr, Δr, and s
Optical flow	? r and s
Occluding contours	Δr
Other occlusion cues	Δr
Surface orientation contours	Δs
Surface contours	s
Surface texture	Probably r
Texture contours	Δr and s
Shading	δs and Δs

Note: r = relative depth (in orthographic projection); δr = continuous or small local changes in r; Δr = discontinuities in r; s = local surface orientation; δs = continuous or small local change in s; Δs = discontinuities in s.

3. Because the apparent orientation of a surface element can change, depending on whether it is viewed binocularly or monocularly, the representation of surface orientation is probably driven almost entirely by perceptual processes and is influenced only slightly by specific knowledge of what the surface orientation actually is. Our ability to perceive the surface much better than we can memorize it may also be connected with this point.

4. In addition, it seems likely that the different sources of information can influence the same representation of surface orientation.

In order to make the most efficient use of these different and often complementary sources of information, they need to be combined in some way. The computational question is, How best to do this? The natural answer is to seek some representation of the visual scene that makes explicit just the information that these processes can deliver.

Fortunately, the physical interpretation of the representation that we seek is clear. All these processes deliver information about the depth or orientation associated with surfaces in an image, and these are well-defined physical quantities. We therefore seek a way of making this information explicit, of maintaining it in a consistent state and perhaps also of incorporating into the representation any physical constraints that hold for the values which depth and surface orientation take over the kinds of surface that occur in the real world.

Table 4–1 lists the types of information that the different early processes can extract from images. The interesting point here is that although processes like stereopsis and motion are in principle capable of delivering depth information directly, they are in practice more likely to deliver information about local *changes* in depth, for example, by measuring local changes in disparity. Surface contours and shading provide more direct information about surface orientation. In addition, occlusion and brightness and size clues can deliver information about discontinuities in depth. The main function of the representation we seek is therefore not only to make explicit information about depth, local surface orientation, and discontinuities in these quantities but also to create and maintain a global representation of depth that is consistent with the local cues that these sources provide. We call such a representation the 2½-D sketch, and the next section describes a particular candidate for it.

4.5   GENERAL FORM OF THE 2½-D SKETCH

In order to provide an example of a representation as a basis for a more thorough discussion about the details of its composition, I will describe first the original proposal for a viewer-centered representation (this is the force of the word *sketch*) that uses surface primitives of one (small) size. It includes a representation of contours of surface discontinuity, and it has enough internal computational structure to maintain its descriptions of depth, surface orientation, and surface discontinuity in a consistent state.

Depth may be represented by a scalar quantity r, the distance from the viewer of a point on a surface. Surface discontinuities may be represented by oriented line elements. As we have seen, surface orientation may be represented as a vector (p,q) in two-dimensional space, which is equivalent to covering the image with needles. The length of each needle defines the slant (or dip) of the surface at the point, so that zero length corresponds to a surface that is perpendicular to the vector from the viewer to that point, and the length of the needle increases as the surface slants away from the viewer. The orientation of the needle defines the tilt, that is, the direction of the surface's slant. Figure 4–2 illustrates this representation; it is like having a gradient space at each point in the visual field.

In principle, the relation between depth and surface orientation is

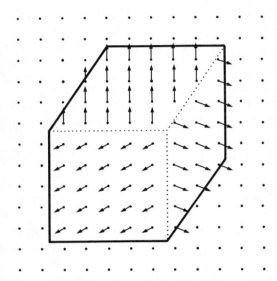

Figure 4–2. Another example of a 2½-dimensional sketch, this time of a cube. The surface orientation is again represented by arrows, as explained in the text and in the legend to Figure 3–12. Occluding contours are shown with full lines, and surface orientation discontinuities with dotted lines. Depth is not shown in the figure, though it is thought that rough depth is available in the representation.

straightforward—one is simply the integral of the other, taken over regions bounded by surface discontinuities. It is therefore possible to devise a representation with intrinsic computational facilities that can maintain the two variables of depth and surface orientation in a consistent state. But note that in any such scheme surface discontinuities acquire a special status (as curves across which integration stops). Furthermore, if the representation is an active one, maintaining consistency largely through local operations, curves that mark surface discontinuities (for example, contours that arise from occluding contours in the image) must be filled in completely, so that the integration cannot leak across any point along an object boundary. It is interesting that subjective contours have this property and that they are closely related to subjective changes in brightness often associated with changes in perceived depth. If the human visual processor contains a representation that resembles the 2½-D sketch, it would be interesting to ask whether subjective contours occur within it.

In summary, then, the argument is that the 2½-D sketch is useful because it makes explicit information about the image in a form that is closely matched to what early visual processes can deliver. We can then formulate the goals of early visual processing as being primarily the construction of this representation. For example, specific goals would be to discover the surface orientations in a scene, which contours in the primal sketch correspond to surface discontinuities and should therefore be represented in the 2½-D sketch, and which contours are missing in the primal sketch and need to be inserted into the 2½-D sketch so that it is consistent with the structure of three-dimensional space. This formulation avoids all the difficulties associated with the terms *figure* and *ground, region* and *object*—the difficulties inherent in the image segmentation approach; for the gray-level intensity array, the primal sketch, the various modules of early visual processing, and finally the 2½-D sketch itself deal only with discovering the properties of surfaces in an image.

This outline raises many questions of detail, and we shall examine some of them in the next few sections. The reader, however, should be warned not to expect very precise answers. Our knowledge from here on is much less detailed than it has been up to this point. Unfortunately, I cannot provide much more than a framework within which to ask questions. Nevertheless, this has its value, even though denying the satisfaction of permanent answers. Thus, it is worth setting this description out with a little more precision than our discussion of the 2½-D sketch has had hitherto.

4.6 POSSIBLE FORMS FOR THE REPRESENTATION

There has not yet been any determined psychophysical assault on the 2½-D sketch, so we know very little about it or even whether it in fact exists in the sense suggested by our approach to vision. The main questions, however, are not difficult to formulate: What precisely is represented and how? What precisely is the coordinate system?—even saying that it must be viewer centered leaves one with several options. And perhaps most difficult, what kinds of internal computations are carried out within the representation either to maintain its own internal consistency or to keep it consistent with what is allowed by the three-dimensional world?

The first question is, Exactly what kind of surface information is made explicit? Are both depth r and surface orientation s represented, for example, or is only r actually carried in the representation, surface orientation being computed on demand by local differentiation? Or alternatively, is only surface orientation carried explicitly, depth being obtained somehow by local integration?—a more difficult possibility to accept but definitely different from the first alternative.

The best argument for the explicit representation of some function like distance from the viewer comes from the theory of stereopsis. The maximum range of disparities that are simultaneously perceivable without

(a)

(b)

Figure 4–3. A selection of large-disparity stereograms. The reader can test for himself what is the largest disparity for which he can simultaneously fuse both foreground and background. When viewed from 20 cm, these stereograms have disparities of (a) 2°, (b) 2.25°, (c) 2.5°, and (d) 2.75°.

diplopia is the same under four rather different conditions. First, in stabilized-image conditions,* Fender and Julesz (1967) obtained a figure of about 2° for a random-dot stereogram. Second, in the absence of any stabilization—that is, under normal viewing conditions—about the same range is obtained. When the complex stereograms given by Julesz (1971; for example, fig. 4.5–3) are viewed from about 20 cm, they give rise to

---

*Images are held fixed on the retinas so that eye movements have no effect.

(c)

(d)

Figure 4–3 (continued).

disparities of about the same order; if one views them from much closer, one cannot "see" all of them at once. Third, it seems at present unlikely that the maximum range of simultaneously perceivable disparities is much affected by their distribution. The reader can see for himself from Figure 4–3 that the figure of about 2°, which holds for stabilized-image conditions and for freely viewed stereograms with continuously varying disparities, also applies to stereograms with a single disparity. And fourth, if you experiment informally, using your fingers and real-world surfaces, you will arrive at a similar figure.

These examples suggest that the figure of about 2° for the maximal range of simultaneously perceivable disparities has a rather general validity (provided there is enough surface at the extreme disparities) and that the figure is independent of eye movements. It is difficult to see how a memory buffer that stored only surface orientation could impose such a restriction, so I would conclude that depth is held in some form, perhaps only roughly, and that the amount that is being held corresponds to 2°–2¼° of disparity.

The second set of arguments concerning why depth should be represented explicitly in some form has to do with the importance of discontinuities in depth. Several early visual processes can yield information about such discontinuities, some of them in only a qualitative way. The most striking are probably occlusion cues, certain texture boundaries, disparity boundaries, and also directional selectivity (see Table 4–1). The perceptual vividness of subjective contours testifies to their importance. And subjectively, if two surfaces lie at very different depths, we seem to be very aware of this fact, even if they have the same surface orientations.

Both kinds of arguments suggest that some form of depth representation exists, and one interesting question is whether the range of simultaneously perceivable depths from apparent motion is commensurate with what we can see stereoscopically. But neither argument forcefully requires that depth information be held very accurately, as it would have to be if it formed the primary representation. Very locally we can easily say from motion or stereopsis information whether one point is in front of another. But if we try to compare the distances to two surfaces that lie in different parts of the visual field, we do very poorly and can do this much less accurately than we can compare their orientations.

This casts doubt, then, on the idea that depth is the basic represented variable, that it is stored accurately over a particular range of values, and that it is differentiated on demand to give surface orientation. There are better arguments against this possibility, too, which come from the fact that many of the processes listed in Table 4–1 yield information about surface orientation directly rather than via information about depth. The most obvious are surface contours, shading, and contours that deal with discontinuities in surface orientation. But in fact, stereopsis and structure from motion are both best suited to delivering information about how things are changing locally rather than about absolute depth—stereopsis because the brain rarely seems to know the actual absolute angle of convergence of the two eyes, dealing instead only with variations in it, and structure from motion because the analysis is local and orthographic, thus yielding only local changes in depth. There is therefore a strong sense in which both processes are very well suited to delivering surface orientation information, and it is probably more accurate to think of them in this way than as if they were primarily concerned with distance from the viewer.

Finally, we can judge surface orientation very accurately, to within a degree or two over the entire range of possible orientations (Stevens, 1979, app. B). This is not on its own conclusive evidence that we represent it explicitly, but taken in conjunction with our poor depth-judging abilities, I think that it is a significant fact which would require explanation if we did not represent it.

My conclusion from these arguments is that we likely represent both quantities s and r internally, but that although we may represent s quite accurately, we represent r only roughly. We may also have facilities for representing local differences in depth more accurately, which would be in addition to our representation of surface orientation.

4.7 POSSIBLE COORDINATE SYSTEMS

Perhaps we should next address the question of a coordinate system. We have already observed that it must be centered on the viewer, but this still leaves several possibilities. The first and most conspicuous point is that all the processes we have discussed are naturally retinocentric, as illustrated in Figure 4–4(a). Relative depth and surface orientation are obtained along and relative to the line of sight, not any external frame. So at least initially,

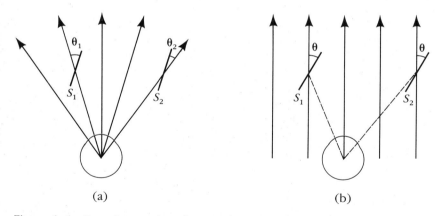

(a)   (b)

Figure 4–4. In retinocentric polar coordinates, the natural angle to measure a surface's orientation is that formed between the surface and the line of sight. Hence, as in (a), two parallel surfaces S_1 and S_2 are associated with different angles θ_1 and θ_2, respectively, which here have opposite signs. A much more convenient representation is to refer all angles to the direction straight ahead, as illustrated in (b). It is then easy to tell whether two surfaces are parallel and whether they are flat, convex, or concave.

we are almost forced to expect a retinocentric frame within which to express the results of each process.

On the other hand, it must be remembered that coordinates referring to the line of sight are not very useful to the viewer. Decisions about whether two surfaces have the same orientation or whether a surface is flat are not easily made from specifications in such a frame. One must continually allow for the angle of the line of sight, as illustrated in Figure 4–4(a)— a difficulty that is compounded by the effects of eye movements.

The second point, which follows from the first, is that although most early visual processes that deliver surface orientation information do so relative to the line of sight, each process may do so in its own way. In stereopsis, as we saw, there is a natural preference for specifying the components of surface orientation in the vertical and horizontal directions separately, simply because the horizontal positioning of the two eyes distinguishes these two directions. Surface contour and texture information prefer a slant-and-tilt representation of the sort discussed in Sections 3.6 and 3.7. Structure-from-motion information is probably like surface contour information in this respect.

To summarize, then, there are several different ways of representing surface orientation in a retinocentric coordinate frame, and the different early visual processes may use slightly different ones in which to express their own first guesses at what the surface orientation actually is.

The third point is that we have a fovea. Different parts of the visual field are analyzed at very different resolutions for a given direction of gaze. An important consequence of this is that the amount of memory or buffer space necessary to record the results of early visual processes varies widely in the visual field, being much greater for the fovea than for the periphery. This provides another reason for expecting a retinocentric frame, because if one used a frame that had already allowed for eye movements, it would have to have foveal resolution everywhere. Such luxurious memory capacity would be wasteful, unnecessary, and in violation of our own experience as perceivers, because if things were really like this, we should be able to build up a perceptual impression of the world that was everywhere as detailed as it is at the center of the gaze.

The final general point involves the question of consistency. We have already observed that the early visual processes can run independently to a large extent, and that some parts of the visual field will be accessible to some processes, and other parts to other processes. Therefore, the question of maintaining consistency among the different types of information will arise, as well as the question of assigning priorities that accurately reflect the reliabilities of the different processes, that is, assigning priorities so that the best source is believed when different sources are in conflict.

This question of consistency should clearly be resolved as early as possible, because until it is, all the information cannot be reduced to just one representation.

These four observations lead to two conclusions. First, information from the different sources is probably checked for consistency and combined in some kind of retinocentric frame. This is because the information is all delivered in this form and because such a representation, containing among other things an enlarged foveal capacity, best matches the capabilities of the preceding processes.

Second, some conversion of the coordinate frame probably takes place at this point in order to express information from the different processes in a standard form and probably also to allow for the angle of gaze. An example of a suitable conversion is illustrated in Figure 4–4(b), where all angles are referred to the direction straight ahead instead of to the local line of sight. Such a conversion would (1) facilitate the computation of predicates like flat, convex, or concave; (2) allow easy comparison of the orientation of surfaces in different parts of the visual field; and (3) prepare the way for the business of allowing for eye movements.

4.8 INTERPOLATION, CONTINUATION, AND DISCONTINUITIES

The issues I wish to discuss next are based on three different types of psychophysical observation. The first is the observation, first studied in detail by White (1962), that one "sees" even a low density (2%–3%) random-dot stereogram as portraying a continuous surface, not as a set of isolated dots. The reader may confirm this for himself by looking at the 5% stereogram in Figure 3–8. The impression of a solid surface is strong. We are aware that the dots all lie at the same depth—they are clearly markings on an otherwise transparent sheet, which is flat and whose surface orientation is clearly apparent. This phenomenon is not altogether surprising in view of the theory of stereopsis described in Section 3.3, because the zero-crossings at which disparity is assigned do not cover the image—most of its area has no zero-crossings at all (examine Figure 3–14, for example)—so the notion that some kind of filling-in has to be carried out is to be expected. Notice, incidentally, that in the cooperative stereo algorithm of Figure 3–7, the filling-in process is incorporated into the algorithm, and this indeed was one of its initial attractions for us.

Eric Grimson (1979) has studied the filling-in or interpolation problem from a psychophysical and a computational point of view and has found that the visual system is very conservative in the amount of filling-in

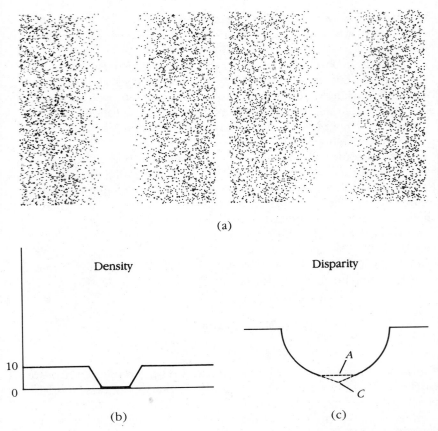

Figure 4–5.   The stereogram (a) has the density distribution given in (b) and the disparity distribution indicated by the solid lines in (c). Such a stereogram can be used to explore psychophysically whether and how we interpolate across the gap. The dotted lines in (c) illustrate two interpolation possibilities.

it allows without additional evidence. He created various stereograms like the ones depicted in Figure 4–5, in which the density and disparity both decrease toward the center, as shown. The question is, How, if at all, does the observer fill in across the region where there are no dots? Two of the three possible candidates are shown in Figure 4–5(c): Candidate *A* fills in straight across with constant disparity; candidate *B* (not shown) produces some smooth interpolation that connects the two surfaces without any discontinuity in surface orientation; and candidate *C* continues the surfaces linearly until they intersect.

What the viewer perceives can be determined by putting a probe spot

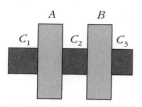 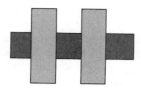

Figure 4–6.   In this stereo pair, C_2 is seen at the same depth as C_1 and C_3, despite the fact that there are no disparity cues to the depth of C_2.

in the intermediate region at various disparities and asking the viewer whether it lies above or below the place "where the surface goes." Grimson found that the percept is unfortunately not a vivid one in these circumstances; although the subjects confidently exclude possibilities A and C, they are vague about the position of B. They never report any discontinuities in surface orientation. He concluded that although there seems to be some interpolation, the matter is not straightforward. I shall look at the computational side of the problem a little later.

The second aspect of the problem is what I shall call continuation, which is best illustrated by a stereo pair of Andrew Witkin's, shown in Figure 4–6. This stereogram is perceived as two rectangles A and B occluding a continuous rectangle containing C_1, C_2, and C_3. The curious thing about this demonstration is that the information about stereo disparity can come from only the vertical lines in the figure. Thus, regions A, B, C_1, and C_3 contain points at which the disparity is defined, and the fact that we see each as a whole surface is a problem only in interpolation. But for region C_2 there are no such cues. The fact that it is assigned the same depth as C_1 and C_3 must therefore be the result of some continuation process operating "behind" the occluding planes A and B. It is critically important for the demonstration that lines like the horizontal edges of C_1, C_2, and C_3 be in good alignment. It is as though their accurate alignment in the two-dimensional image allows them to be viewed as evidence of the same surface discontinuity in three dimensions, which then allows surface C_2 to be seen at the same depth as surfaces C_1 and C_3. A similar inference may perhaps be made from some experiments by Naomi Weisstein (1975), who displayed a drifting grating, occluded a central rectangular patch of it, and yet found adaptation effects occurring even within this patch.

These experiments suggest that the viewer-centered representation of surfaces may be capable of representing more than one surface at once. It may also be significant that in suitably constructed random-dot stereograms, like that given in figure 3–19(b), one can simultaneously and vividly see two surfaces. I personally cannot see three at once (compare Julesz, 1971, fig. 5.7–1), although there may be people who can.

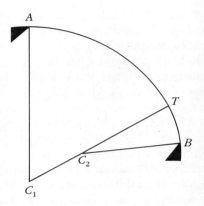

Figure 4–7. The shape of curved subjective contours. They are composed of two circles with centers C_1 and C_2, one emanating smoothly from each initiating point, *A* and *B,* that are joined smoothly (at *T*). Of the infinite number of pairs of circles with these properties, subjective contours follow the pair having minimal curvature.

   Finally, there is the question of discontinuities in depth and in surface orientation. We have already mentioned discontinuities in depth in relation to the kind of continuation necessary for the phenomenon of Figure 4–6 and also in relation to the phenomenon of subjective contours. In both cases, continuity and smoothness (minimum curvature) seem to be important criteria. Ullman (1976a) examined phenomenologically the shape of curved subjective contours and found that it could be described accurately by two circles, one emanating from each source point, that are joined smoothly, as illustrated in Figure 4–7. Of the infinite family of pairs of circles with this property, one selects the pair yielding minimal curvature. Ullman also described a local network capable of generating this shape.
   Although the shape of these contours is quite well understood, little is known about the conditions that cause their formation except the rather general notions that evidence of occlusion is required, together with rather direct monocular cues as to the exact position of the discontinuity. The Kanisza triangle (Figure 3–81e), the radial sun (Figure 2–25b) and the 5% random-dot stereogram (Figure 3–8), in which the dots themselves contain short, vertical edge segments, all provide both kinds of information in slightly different ways. The topic needs further psychophysical study.

4.9   COMPUTATIONAL ASPECTS
OF THE INTERPOLATION PROBLEM

From a computational point of view, two problems need to be understood before planning detailed psychophysical experiments. The first is the notion of discontinuity, and the second, the different possibilities for interpolation.

Discontinuities

Although the distinction between a continuous and a discontinuous change over a continuum is a clear one, where the sample space is discrete the distinction is more elusive. We have already met this problem twice, once in detecting discontinuities in the orientation of zero-crossings where, strictly speaking, they cannot occur, and again in Land and McCann's (1971) lightness algorithm. In both cases one has to set a threshold. In the first case it was based on the point at which a "real" underlying discontinuity can no longer be discriminated from a very high curvature change. This point depends upon the receptive field size associated with the channel, so that what the smaller channels might "see" as smooth, the larger ones might "see" as discontinuous.

In an absolute sense, the resolution of the sample space does impose restrictions on what can be considered a continuous change. For example, in the one-dimensional case, suppose that the underlying representation consists of values specified a distance δ apart. Then by the sampling theorem, the representation cannot contain complete information about frequencies higher than, say, $\pi/\delta = \Omega$. Thus, the representation is effectively band limited by frequency Ω.

Now, although a signal that is band limited by frequency Ω can be represented completely by samples at intervals of δ, there is no guarantee that such a signal can accommodate all sample points at which one places arbitrary numbers. In other words, if the sample values change too fast, the overall signal may exceed the bandwidth of the representation. If this occurs, then the representation is forced to attribute the change to a discontinuity, since it is simply not rich enough to accommodate the changes that are actually occurring. This point is captured precisely by a theorem due to Bernstein, which says that the derivative of a band-limited function cannot get too large compared with the value of the function. If $f(x)$ is a function that is band limited by Ω, and if $f'(x)$ is its derivative, then the theorem states that

$$\sup |f'(x)| \leq \Omega \sup |f(x)|$$

That is, the largest value of $|f'(x)|$ over all x's is not bigger than the largest value of $\Omega|f(x)|$.

This constraint is a fundamental one that applies whenever we try to represent information on a discrete grid, and it is of particular interest here that the human visual system appears unable to represent sine waves in depth whose frequencies exceed 3–4 cycles per degree at the fovea

(Tyler, 1973). For example, the constraint may help to explain why subjective contours that do not appear or that are not very strong when we look at them directly appear much more vivid if we look at them indirectly. Presumably, the resolution of the representation also decreases with eccentricity, so that what can be represented foveally as a very steep gradient must, when presented more eccentrically, be represented as a discontinuity.

As we saw in Section 3.3, stereopsis can sometimes provide clear evidence for a surface discontinuity; if, for example, the horizontal rate of change of disparity, which we shall call d', reaches 1 in either eye, there is a discontinuity in depth as seen from the other eye. But in sparsely featured images, there is often not enough information to decide even this. Perceptually one may be left with a vague feeling that the disparity does change but no exact impression of where. In a sparse random-dot stereogram, if two squares happen to line up along a disparity boundary, vivid subjective contours are formed and the boundary is clearly delineated; however, if the squares in the stereogram are replaced by blurred dots, for example, the perception of the discontinuity is much less vivid.

Although these observations are little more than suggestions, they do hint that the interpolation process is conservative and that the visual system is reluctant to insert contours of discontinuity in either depth or surface orientation unless the image itself provides reasonable evidence of their positions. A contour may not be evident all along its length, but it is unlikely that direct visual evidence of it will be lacking everywhere along it. Eric Grimson (1979) enshrined this view in a dictum, which states that *places of no information are actually places of information.* In other words, one cannot hide discontinuities, and conversely, if the image provides no evidence at all about the presence of a discontinuity, not even an edge fragment anywhere along where one is expected, then such a discontinuity may not be assumed. Hence, in contrived situations where direct evidence is deliberately removed, as in Figure 4–5, we neither insert contours nor interpolate the surfaces in a definite way, and we are thus left with a vague and unsettled perception.

Interpolation Methods

Three principal interpolation methods deserve notice: (1) linear interpolation in depth r, (2) linear interpolation in surface orientation, and (3) "fair surface" interpolation, which is a method used by car makers to give car bodies a smooth shape. Very roughly, method 1 is similar to the inverse transform we met in Horn's (1974) algorithm for the retinex. It tries to minimize the value of the Laplacian operator ∇^2 on the surface. Method 2 approximately minimizes the first curvature of the surface in any given concave or convex region. (This follows from the facts that the first curvature $J = -\text{div } n$, where div n is the divergence of n, n is the surface normal, and that locally averaging n almost minimizes div n.) The objection to both methods 1 and 2, implemented on a grid, is that convergence rates are slow—quadratic, in fact, with the distance between fixed points of the computation. I have already stated my reservations about the use of iterative methods in perceptual computations (see Sections 3.2 and 3.5).

The third possibility, favored over the other two by Grimson, involves the notion of a fair surface, which is a surface whose first and second derivatives vary continuously but which allows discontinuities in the third and higher derivatives. One-shot methods are available for filling-in between neighboring triplets of points and knitting them together along the seams, so as to preserve smoothness in arbitrarily high-level derivatives. Choice of the second derivative as the cutoff point rests on the empirical observation of car designers that customers notice discontinuities in the first and second derivatives of a surface but not in the third. Figure 4–8 illustrates the result of applying a filling-in method of this kind to the output derived from a stereo pair. It gives a smooth and pleasing appearance.

As for the connection between these computational ideas and the truth of how we ourselves find discontinuities or fill in surfaces—to the limited extent that we do these things—these are questions for the future.

4.10   OTHER INTERNAL COMPUTATIONS

The notion of surface continuity may, as we have seen, give rise to various active computations in the 2½-D sketch, including filling-in and the smooth continuation of discontinuities. We would expect other local constraints to be embedded there in a similar way—for example, consistency relations concerning the possible arrangements of surfaces in three-dimensional space, such as the constraints made explicit by Waltz (1975; recall Figure 1–3). Such constraints may eventually form the basis for an understanding of phenomena like the reversal of the Necker cube. From this point of view, it is natural that many illusions concerning the interpretation of three-dimensional structure (the Necker cube, subjective contours, the Muller–Lyer figure, the Poggendorff figure, and so on) should take place after stereoscopic fusion (see Julesz, 1971; Blomfield, 1973). Illusions like the reversing bucket of Figure 5–9 should also have part of their cause here, since the continuity of the bucket's surface plays a critical role in keeping its appearance consistent. The interesting questions here concern how much

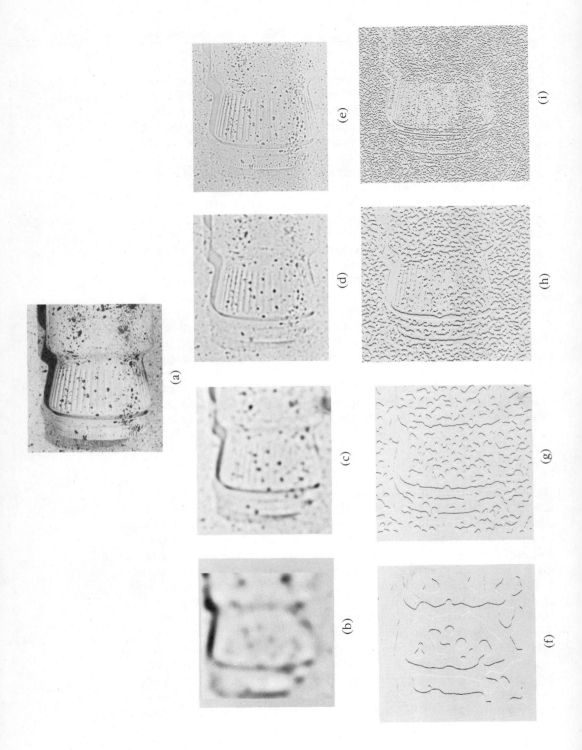

(a)

(b)

(c)

(d)

(e)

(f)

(g)

(h)

(i)

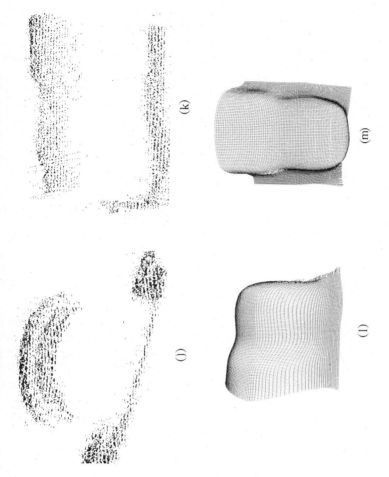

(j)

(k)

(l)

(m)

Figure 4–8. (a) Shows one of the images from a stereo pair. (b)–(e) Show its convolution with $\nabla^2 G$ filters of four sizes, and (f)–(i) display the zero-crossings thus obtained. (j) and (k) show two different views of the disparity map obtained after stereo matching, and (l) and (m) show the surface obtained from this information by Eric Grimson's interpolation algorithm.

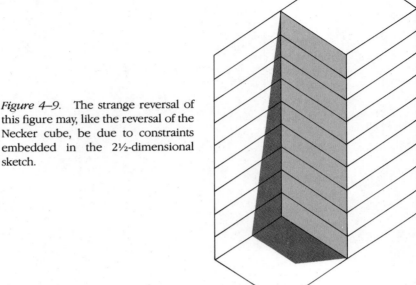

Figure 4–9. The strange reversal of this figure may, like the reversal of the Necker cube, be due to constraints embedded in the 2½-dimensional sketch.

is done in the 2½-D sketch proper and how much occurs as this immediate representation is computed into a three-dimensional representation of the kind that we remember (see the next chapter). Examples like the Penrose triangle, many of Escher's figures, and even Figure 4–9 probably depend on a mixture of effects, some local in the 2½-D sketch, and other effects due to a failure to construct an overall, consistent three-dimensional inter-pretation from a set of local views.

One final point that might be thought puzzling. Why should the Necker cube reversal occur when depicted in a random-dot stereogram? It might be argued that since stereopsis definitely assigns the edges all to a plane, the figure should be seen in two-dimensions and not in three. I think it is best to regard all contours in the 2½-D sketch as trying for a three-dimen-sional interpretation. The fact that the contours are put there by stereopsis rather than by, say, the primal sketch is unimportant.

CHAPTER 5

Representing Shapes
for Recognition

5.1 INTRODUCTION

We come now to the final and perhaps most fascinating of the steps in our overall program, the transformation of shapes from a representation that is matched to the processes of perception into a representation that is suitable for recognition. There are many issues to be explored here, and this chapter, which rests heavily on Marr and Nishihara (1978), touches only the surface of some of them. Nevertheless, the main ideas are once more clear in outline, and I shall emphasize exactly what creating a shape representation that is suitable for recognition entails. This involves us in a discussion of what recognition is and how it comes about.

The single most important point is that we must now abandon the luxury of a viewer-centered coordinate frame on which all representations discussed hitherto have been based because of their intimate connection with the imaging process. Object recognition demands a stable shape description that depends little, if at all, on the viewpoint. This, in turn, means that the pieces and articulation of a shape need to be described not

295

relative to the viewer but relative to a frame of reference based on the shape itself. This has the fascinating implication that a canonical coordinate frame* must be set up within the object *before* its shape is described, and there seems to be no way of avoiding this. For some shapes, like a cigar, it will be easy to do this, and for others, like a crumpled newspaper, it will not.

Let us therefore look at these questions in detail. I shall reserve the term *shape* for the geometry of an object's physical surface. Thus, two statues of a horse cast from the same mold have the same shape. A *representation* for shape is a formal scheme for describing shape or some aspects of shape together with rules that specify how the scheme is applied to any particular shape. I shall call the result of using a representation to describe a given shape a *description* of the shape in that representation. A description may specify a shape only roughly or in fine detail.

5.2   ISSUES RAISED BY THE REPRESENTATION OF SHAPE

There are many kinds of visually derivable information that play important roles in recognition and discrimination tasks. Shape information has a special character, because unlike color or visual texture information, the representation of most kinds of shape information requires some sort of coordinate system for describing spatial relations. For example, the information that distinguishes the different animal shapes in Figure 5–1 is the spatial arrangement, orientation, and sizes of the sticks. Similarly, since left and right hands are reflections of each other in space, any description of the shape of a hand that is sufficient for determining whether it is left or right must in some manner specify the relative locations of the fingers and thumb.

Criteria for Judging the Effectiveness of a Shape Representation

There are many different aspects of an object's shape, some more useful for recognition than others, and any one aspect can be described in a number of ways. Although formulating a completely general classification

---

*A coordinate frame uniquely determined by the shape itself.

of shape representations is difficult, we can attempt to set out the main criteria by which they may be judged and the basic design choices that have to be made when formulating a representation.

Accessibility

Can the desired description be computed from an image, and can it be done reasonably inexpensively? There are fundamental limitations to the information available in an image—for example, regarding its resolution— and the requirements of a representation have to fall within the limits of what is possible. Moreover, a description that is in principle derivable from an image may still be undesirable if its derivation involves unacceptably large amounts of memory or computation time.

Scope and uniqueness

What class of shapes is the representation designed for, and do the shapes in that class have canonical descriptions in the representation? For example, a shape representation designed to describe planar surfaces and junctions between perpendicular planes would have cubical solids within its scope, but would be inappropriate for describing a billiard ball or a comb. If the representation is to be used for recognition, the shape description must also be unique; otherwise, at some point in the recognition process, the difficult problem would arise of deciding whether two descriptions specify the same shape. If, for example, we chose to represent shape using polynomials of degree n, the formal description of a given surface would depend on the particular coordinate system chosen. Since we would be unlikely to use the same coordinate system on two different occasions without observing some additional conventions, even the same image of a surface could give rise to very different descriptions.

   Another example would be to represent a shape by a large collection of small cubes, packed together so as to approximate the shape as closely as possible. If the cubes were sufficiently small, the shape could be approximated quite accurately so that the scope of such a representation would be quite broad. On the other hand, a small shift of, say, half the side of a $\frac{1}{8}$-in "minicube" could significantly change the representation of a shape, thus violating the uniqueness condition. If we used 1-ft cubes instead, the uniqueness problem would be greatly alleviated (a human might be represented by just six of them stacked up), but at considerable cost to other aspects of the representation.

Stability and sensitivity

Beyond the above scope and uniqueness conditions lie questions about the continuity and resolution of a representation. To be useful for recognition, the similarity between two shapes must be reflected in their descriptions, but at the same time even subtle differences must be expressible. These opposing conditions can be satisfied only if it is possible to decouple stable information that captures the more general and less varying properties of a shape from information that is sensitive to the finer distinctions between shapes.

For example, consider a stick figure representation that uses the three-dimensional arrangement and the relative size of sticks as primitive elements to describe animal shapes, as in Figure 5–1. The size of the sticks used gives one control over the stability and sensitivity of the resulting stick figure description. Stability is increased by using larger sticks; a single stick provides the most stable description of the whole shape, describing only its size and orientation. A description built of smaller sticks, on the other hand, would be sensitive to smaller, more local details, such as the extremities of an animal's limbs. Although such details tend to be less stable, they can nevertheless be important for making fine distinctions between similar shapes.

Choices in the Design of a Shape Representation

We can now relate the effects of different designs of shape representation to our three performance criteria. It is worth repeating once more that the most fundamental property of a representation is that it can make some types of information explicit, and this property can be used to bring the essential information to the foreground allowing smaller and more easily manipulated descriptions to suffice. We shall consider three aspects of a representation's design here: (1) the representation's coordinate system; (2) its primitives, which are the primary units of shape information used in the representation; and (3) the organization that the representation imposes on the information in its descriptions.

Coordinate systems

The most important aspect of the coordinate system used by a representation is the way it is defined. If locations are specified relative to the viewer, we say the representation uses a viewer-centered coordinate system. If locations are specified in a coordinate system defined by the viewed object,

Figure 5–1. These pipe cleaner figures illustrate several of the points developed in this chapter. A shape representation does not have to reproduce a shape's surface in order to describe it adequately for recognition; as we see here, animal shapes can be portrayed quite effectively by the arrangement and relative sizes of a small number of sticks. The simplicity of these descriptions is due to the correspondence between the sticks shown here and natural or canonical axes of the shapes described. To be useful for recognition, a shape representation must be based on characteristics that are uniquely defined by the shape and that can be derived reliably from images of it. (Reprinted by permission from D. Marr and H. K. Nishihara, "Representation and recognition of the spatial organization of three-dimensional shapes," *Proc. R. Soc. Lond. B 200,* 269–294.)

the representation uses an object-centered coordinate system. There are, of course, several versions of each type.

For recognition tasks, viewer-centered descriptions are easier to produce but harder to use than object-centered ones, because viewer-centered descriptions depend upon the vantage point from which they are built. As a result, any theory of recognition that is based on a viewer-centered representation must treat distinct views of an object essentially as distinct objects. Thus this approach requires a potentially large store of descriptions in memory in exchange for a reduction in the magnitude and complexity of the computations required to compensate for the effects of perspective.

Minsky (1975) has suggested that this number of descriptions might be minimized by choosing appropriate shape primitives and views to be stored in memory. Clearly much can be accomplished by this approach in some circumstances. For example, suppose squirrels need to distinguish trees from other objects but do not need to identify particular trees by their shape. They may be able to note some general characteristics of the appearance of a vertical tree trunk on the ground nearby that do not depend on the vantage point. In a representation based on these characteristics, all trees in the squirrel's environment would produce essentially the same description.

For more complex recognition tasks involving the arrangement of an object's components, however, any viewer-centered representation is likely to be sensitive to the object's orientation. For example, consider the many orientation-dependent appearances of a human hand, even if the fingers and thumb remain fixed with respect to each other. In order to distinguish a left hand from a right by using a viewer-centered representation, this problem would have to be treated as many separate cases, one for each possible appearance of a hand.

The alternative to relying on an exhaustive enumeration of all possible appearances is to use an object-centered coordinate system and thus to emphasize the computation of a canonical description that is independent of the vantage point. Ideally, only a single description of each object's spatial structure would have to be stored in memory in order for that object to be recognizable from even unfamiliar vantage points. However, an object-centered description is more difficult to derive, since a unique coordinate system has to be defined for each object, and, as I mentioned earlier, that coordinate system has to be identified from the image before the description is constructed.

Primitives

The primitives of a representation are the most elementary units of shape information available in the representation, which is the type of information that the representation receives from earlier visual processes. For instance, the 2½-D sketch is an example of a representation whose primitives carry information about local surface orientation and distance (relative to the viewer) at thousands of locations in the visual field. We can separate two aspects of a representation's primitives; the type of shape information they carry, which is important for questions of accessibility, and their size, which is important for questions of stability and sensitivity.

There are two principal classes of shape primitives, surface-based (two-dimensional) and volumetric (three-dimensional). As we have already seen, surface information is more immediately derivable from images. The simplest primitives useful for surface descriptions would specify just the location and size of small pieces of surface. More elaborate surface primitives like those used in the 2½-D sketch could include orientation and depth information as well.

On the other hand, volumetric primitives carry information about the spatial distribution of a shape. This type of information is more directly related to the requirements of shape recognition than information about a shape's surface structure, and this often means that much shorter and therefore more stable descriptions can still satisfy the sensitivity criterion. The simplest volumetric primitive specifies just a location and a spatial extent, and corresponds to a roughly spherical region in space. By adding a vector to this information, a roughly cylindrical region can be specified, whose length is indicated by the length of the vector and whose diameter is indicated by the spatial extent parameter of the primitive. A second vector could indicate a rotational orientation about the first vector, making it possible to specify a pillow-shaped region whose cross section along the first vector is thicker in the direction of the second vector. The additional vector could alternatively be used to specify the direction and magnitude of a curvature in the axis of the cylindrical region.

The complexity of the primitives used by a representation is limited largely by the type of information that can be reliably derived by processes prior to the representation. While the information-carrying capacity of primitives can be increased arbitrarily, there is a limit to the amount that is useful, since very detailed primitives will be derived less consistently by those earlier processes. In the extreme case, descriptions in a shape representation would consist of a single primitive. Such a representation would satisfy the uniqueness and stability conditions only if the information carried by the primitive was derived consistently by the processes supplying it. If this were so, however, those processes would already have accomplished shape recognition in specifying the primitive, and there would be no need for the representation.

Size is the other aspect that influences the information that the representation's primitives make explicit. In particular, information about features much larger than the primitives used is difficult to access, since it is represented only implicitly in the configuration of a larger number of smaller items. For example, consider how the arm of the human shape would be described in a surface representation like the 2½-D sketch. The representation here is essentially what one would get by covering the surface with fish scales, each specifying a local surface orientation. Only information about small patches of surface is present, so a rather sophisticated analysis of a large assembly of these patches is required to make explicit the presence of the arm shape itself. A stick figure representation, on the other hand, can specify an arm explicitly with a single stick primitive of the appropriate size. Similar arguments can be applied to the representation scheme based on small cubes, discussed earlier; larger-scale shape information is not immediately available from such a representation.

At the other end of the scale, features of a shape that are much smaller than the primitives used to describe it are not just inaccessible, they are completely omitted from the description. For example, the fingers of a human shape are not expressible in a stick figure description that uses only primitives the size of the arms and legs. And even the arms and legs would be inexpressible in terms of 1-ft cubes. Similarly, surface details much smaller than the basic surface primitives used in the 2½-D sketch would be inexpressible in that representation. Thus the size of the primitives used in a description determines to a large degree the kind of information made explicit by a representation, the information made available but not directly obtainable, and the information that is discarded.

Organization

The third design dimension is the way shape information is organized by a representation. In the simplest case, no organization is imposed by the representation and all elements in a description have the same status. The local surface representation provided by the 2½-D sketch is one such example, and another would be our pile of minicubes that approximates a three-dimensional shape.

Alternatively, the primitive elements of a description can be organized into modules consisting, for example, of adjacent elements of roughly the same size, in order to distinguish certain groupings of the primitives from others. A modular organization is especially useful for recognition because it can make sensitivity and stability distinctions explicit if all constituents of a given module lie at roughly the same level of stability and sensitivity.

5.3   THE 3-D MODEL REPRESENTATION

We have formulated the requirements for a representation for shape recognition in terms of the criteria of accessibility, scope and uniqueness, and stability and sensitivity. We concluded that the design of a suitable representation should involve an object-centered coordinate system, include but perhaps not be limited exclusively to volumetric shape primitives, and impose some kind of modular organization on the primitives involved in a description. These choices have strong implications, and a limited representation, called the *3-D* (three-dimensional) *model representation,* can be defined quite directly from them.

Natural Coordinate Systems

Our first objective is to define a shape's object-centered coordinate system. If it is to be canonical, it must be based on axes determined by salient geometrical characteristics of the shape, and conversely, the scope of the representation must be limited to those shapes for which this can be done. A shape's natural axes may be defined by elongation, symmetry, or even motion (for example, the axis of rotation); thus, the coordinate system for a sausage should be defined by its major axis and the direction of its curvature, and that of a face by its axis of symmetry. Objects with many or poorly defined axes, like a sphere, a door, or a crumpled newspaper, will inevitably lead to ambiguities. For a shape as regular as a sphere, this poses no great problem, because its description in all reasonable systems is the same. A door has four distinguished axes, defined by the directions of its length, its width, and its thickness and also by the axis on which it is hinged. Since the number of descriptions is small and doors are important, we could deal with each of the four possible descriptions of a door as a separate case. This would not be true of a crumpled newspaper, however, which is likely to have a large number of poorly defined axes.

At present, the problems we understand best are those involving the determination of axes based on a shape's elongation or symmetry (Marr, 1977a), and for the sake of simplicity we shall restrict the scope of the 3-D model representation to shapes that have natural axes of this type. One large class of shapes that satisfy this condition is the generalized cones, which we have already met and studied in Section 3.6 and illustrated in Figure 3–59. This class of shapes is important to us not because the surfaces are conveniently described—they may actually not be at all simple (Hollerbach, 1975)—but because such shapes have well-defined axes. This critical feature helps to define a canonical object-centered coordinate system, which is of course the central and most difficult task we face here.

In real life, a wide variety of common shapes is included in the scope of such a representation, because objects whose shape is achieved by growth are often described quite naturally in terms of one or more generalized cones. The animal shapes depicted in Figure 5–1 provide some examples—the individual sticks are simply axes of generalized cones that approximate the shapes of parts of these animals.

Axis-Based Descriptions

To be useful for recognition, a representation's primitives must also be associated with stable geometrical characteristics. The natural axes of a shape satisfy this requirement, and we shall therefore base the 3-D model representation's primitives on them. A description that uses axis-based primitives can be thought of as a stick figure, like those depicted in Figure 5–1, but one must be careful to think of the stick as a local coordinate axis. While only a limited amount of information about a shape is captured by such a description, that information is especially useful for recognition. We shall further limit the information carried by these primitives to pertain just to size and orientation. This will enable us to develop the 3-D model representation with a minimal commitment to inessential details. More elaborate details, such as curved axes or the tapering of a shape along the length of its axis, will not be included here.

The concept of a stick figure representation for shape is not new. Blum

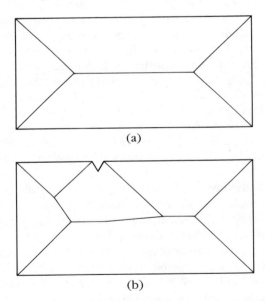

(a)

(b)

Figure 5–2. Blum's (1973) grassfire technique for recovering an axis from a silhouette. It can be thought of as lighting a fire at the boundary, the axis being defined as where two configurations meet. However, the technique is undesirably sensitive to small perturbations in the contour. (a) Shows the Blum transform of a rectangle, and (b) of a rectangle with a notch. (Reprinted by permission from G. Agin, "Representation and description of curved objects," Stanford Artificial Intelligence Project, memo AIM-173, Stanford University, Stanford, California.)

(1973), for example, has studied a classification scheme for two-dimensional silhouettes based on a "grassfire" technique for deriving a kind of stick figure from those shapes (see Figure 5–2), and Binford (1971) introduced the generalized cone for three-dimensional shapes. These representations have an important limitation, however; they do not impose a modular organization on the information they carry. For example, each part of the arm of a human shape can correspond to at most one stick in these representations; it would not be possible to have both a single stick corresponding to the whole arm and three smaller sticks corresponding to the major segments of the arm in the same description.

Modular Organization
of the 3-D Model Representation

The modular decomposition of a description used for recognition must be well defined—such a decomposition must exist and it should be uniquely determined. In the 3-D model representation as specified so far, this is best achieved by basing the decomposition on the canonical axes of a shape. Each of these axes can be associated with a coarse spatial context that provides a natural grouping of the axes of the major shape components contained within that scope. We shall refer to a module defined this way as a *3-D model*. Thus, each 3-D model specifies the following:

1. A model axis, which is the single axis defining the extent of the shape context of the model. This is a primitive of the representation, and it provides coarse information about characteristics such as size and orientation about the overall shape described.

2. Optionally, the relative spatial arrangement and sizes of the major component axes contained within the spatial context specified by the model axis. The number of component axes should be small and they should be roughly the same size.

3. The names (internal references) of 3-D models for the shape components associated with the component axes, whenever such models have been constructed. Their model axes correspond to the component axes of this 3-D model.

Each of the boxes in Figure 5–3 depicts a 3-D model with the model axis on the left and an arrangement of the component axes on the right. The model axis of the human 3-D model makes explicit the gross properties (size and orientation) of the whole shape with a single primitive. The six component axes corresponding to the torso, head, and limbs can each be associated with a 3-D model containing additional information about the decomposition of that component into an arrangement of smaller components. Although a single 3-D model is a simple structure, the combination of several in this kind of organizational hierarchy allows one to build up a description that captures the geometry of a shape to an arbitrary level of detail. We shall call such a hierarchy of 3-D models a *3-D model description* of a shape.

The example in Figure 5–3 illustrates the important advantages of a modular organization for a shape description. The stability of the representation is greatly enhanced by including both large and small primitive descriptions of the shape and by decoupling local spatial relations from

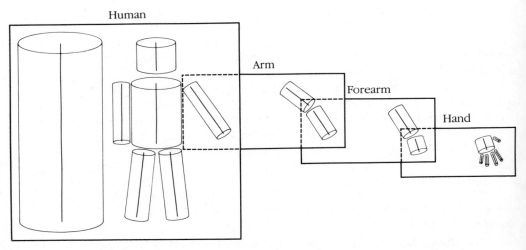

Figure 5–3. This diagram illustrates the organization of shape information in a 3-D model description. Each box corresponds to a 3-D model, with its model axis on the left side of the box and the arrangement of its component axes on the right. In addition, some component axes have 3-D models associated with them, as indicated by the way the boxes overlap. The relative arrangement of each model's component axes, however, is shown improperly, since it should be in an object-centered system rather than the viewer-centered projection used here (a more correct 3-D model is given by the table shown in Figure 5–5c). The important characteristics of this type of organization are: (1) Each 3-D model is a self-contained unit of shape information and has a limited complexity; (2) information appears in shape contexts appropriate for recognition (the disposition of a finger is most stable when specified relative to the hand that contains it); and (3) the representation can be manipulated flexibly. This approach limits the representation's scope, however, since it is only useful for shapes that have well-defined 3-D model decompositions. (Reprinted by permission from D. Marr and H. K. Nishihara, "Representation and recognition of the spatial organization of three-dimensional shapes," *Proc. R. Soc. Lond. B 200*, 269–294.)

more global ones. Without this modularization, the importance of the relative spatial arrangement of two adjacent fingers would be indistinguishable from that of the relation between a finger and the nose. Modularity also allows the representation to be used more flexibly in response to the needs of the moment. For example, it is easy to construct a 3-D model description of just the arm of a human shape that could later be included in a new 3-D model description of the whole human shape. Conversely, a rough but usable description of the human shape need not include an elaborate arm description. Finally, this form of modular organization allows one to trade off scope against detail. This simplifies the computational processes that derive and use the representation, because even though a complete 3-D model description may be very elaborate, only one 3-D model has to be dealt with at any time, and individual 3-D models have a limited and manageable complexity.

Coordinate System of the 3-D Model

There are two kinds of object-centered coordinate systems that the 3-D model representation might use. In one, all the component axes of a description, from torso to eyelash, are specified in a common frame based on the axis of the whole shape. The other uses a distributed coordinate system, in which each 3-D model has its own coordinate system. The latter is preferable for two main reasons. First, the spatial relations specified in a 3-D model description are always local to one of its models and should be given in a frame of reference determined by that model for the same reasons that we prefer an object-centered system over a viewer-centered one. To do otherwise would cause information about the relative dispositions of a model's components to depend on the orientation of the model axis relative to the whole shape. For example, the description of the shape of a horse's leg would depend on the angle that the leg makes with the torso. Second, in addition to this stability and uniqueness consideration, the representation's accessibility and modularity is improved if each 3-D model maintains its own coordinate system, because it can then be dealt with as a completely self-contained unit of shape description.

The coordinate system for specifying the relative arrangement of a 3-D model's component axes can be defined by its model axis or by one of its component axes. We shall refer to the axis chosen for this purpose as the model's *principal axis*. For the examples given here, the principal axis will be the component axis that meets or comes close to the largest number of other component axes in the 3-D model (for example, the torso of an animal shape). The location of the principal axis must also be spec-

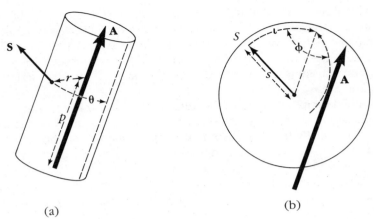

(a)   (b)

Figure 5–4. The spatial organization of a 3-D model's axes is specified in terms of pairwise relationships between those axes that we call adjunct relations. The disposition in space of one axis **S** is determined relative to another, **A**, by specifying the location of one of its endpoints in a cylindrical coordinate system (p, r, θ) about **A** as shown on the left, and its orientation and length in a spherical coordinate system (ι, ϕ, s) centered on that point and aligned with **A** as shown on the right. (Reprinted by permission from D. Marr and H. K. Nishihara, "Representation and recognition of the spatial organization of three-dimensional shapes," *Proc. R. Soc. Lond. B 200*, 269–294.)

ified relative to the model axis in order to maintain the connectedness of the distributed coordinate system.

Two three-dimensional vectors are required to specify the position in space of one axis relative to another. One way of doing this is illustrated in Figure 5–4, which represents the position of a vector **S** relative to an axis vector **A** by means of two vectors. The first vector, written in cylindrical coordinates (p, r, Θ), defines the starting point of **S** relative to **A** (Figure 5–4a); the second vector, written in spherical coordinates (ι, ϕ, s), specifies **S** itself (Figure 5–4b). We shall call the combined specification $(p, r, \Theta, \iota, \phi, s)$ an *adjunct relation* for **S** relative to **A**.

Because the precision with which 3-D models can represent a shape varies, it is appropriate to represent the angles and lengths that occur in an adjunct relation in a system that is also capable of variable precision. For instance, one might wish to state that a particular axis, like the arm component of the human 3-D model in Figure 5–3, is connected rather precisely at one end of the torso (that is, the value of p is exactly 0), but with Θ only coarsely specified and with very little restriction on ι. An example of a suitable system incorporating variable precision is illustrated in Figure 5–5.

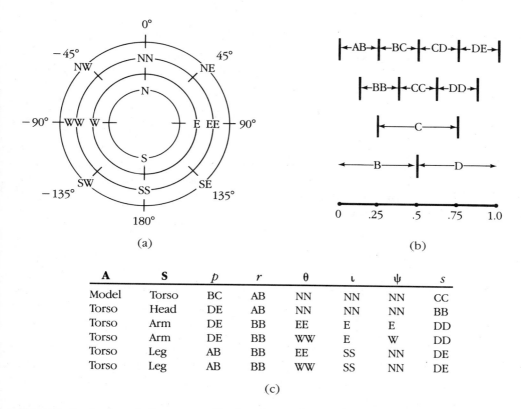

A	S	*p*	*r*	θ	ι	ψ	*s*
Model	Torso	BC	AB	NN	NN	NN	CC
Torso	Head	DE	AB	NN	NN	NN	BB
Torso	Arm	DE	BB	EE	E	E	DD
Torso	Arm	DE	BB	WW	E	W	DD
Torso	Leg	AB	BB	EE	SS	NN	DE
Torso	Leg	AB	BB	WW	SS	NN	DE

(c)

Figure 5–5. Angle and distance specifications in an adjunct relation must include tolerances so that specificity of these parameters can be made explicit in the representation. One way to do this is shown in the upper diagrams, which associate symbols with (a) angular and (b) linear ranges. An example of adjunct relations for the human 3-D model in Figure 5–3 that are expressed in these symbols is shown in the table (c). **A** and **S** identify the two axes related by the adjunct relation specified along each row of this table. If the mnemonic names listed under **A** and **S** were replaced by internal references to the corresponding 3-D models whenever they exist and left blank otherwise, this table would show essentially all the information carried by a 3-D model. (Reprinted by permission from D. Marr and H. K. Nishihara, "Representation and recognition of the spatial organization of three-dimensional shapes," *Proc. R. Soc. Lond. B 200*, 269–294.)

5.4 NATURAL EXTENSIONS

These representational ideas, perhaps best epitomized by the hierarchical scheme depicted in Figure 5–3, begin to show how the complexities of shape description may be approached. Perhaps if J. L. Austin had seen such a figure, he would not have thrown up his hands in such despair at the prospect of formulating rules for representing the shape of his cat (see Section 1.2)! Nevertheless, the ideas are still quite crude, and little work has gone into their development since 1977, mainly because we have been preoccupied with the details of early visual processing. However, questions have frequently arisen about ways of generalizing these ideas, and while answers have not yet been developed in detail, it is worth indicating briefly the most obvious directions in which the representation can be extended.

Perhaps the first point is that one can represent two-dimensional configurations just as easily as three provided, of course, that the patterns are endowed with a natural axis of elongation or of symmetry. Thus we can as easily represent a two-dimensional drawing of a face as the features and details on a real three-dimensional head. A primitive example appears in Figure 5–6. It is particularly interesting to note in this connection that the existence of symmetry in a pattern yields a canonical axis but not a canonical direction along the axis. We still have to decide which end is 0 (down) and which is 1 (up). This choice has to be made when one starts to construct a particular 3-D model, and we seem to make this final choice ourselves using the direction that we are currently taking to be up—usually it is vertically up. If you construct a detailed face description while adhering to this convention and then stand on your head, the details become completely unrecognizable, perhaps because the innate choice mechanism is now using the opposite convention! In addition, face recognition seems to be a rather accurate, specialized, and late-developing process in humans, and interested readers should consult Carey and Diamond (1980) and other works on the subject.

The second point is that the primitives of the 3-D model representation can be extended to include surface primitives, roughly of two kinds. First would be just rough, two-dimensional rectangular surfaces of various sizes, including elliptical shapes and circular ones. Not very many primitives would be needed by the average man, although presumably a sculptor like Henry Moore has a repertoire of hundreds. The second kind of primitive is the notion of something that is not solid but hollow—like a tube or cup, for example. It is not hard to see how such primitives may be organized along much the same lines as the original 3-D model representation, and Figure 5–7 illustrates some preliminary ideas about how such a vocabulary may be deployed to represent various common objects. If we also admit curved axes into the representation, much can be done to represent the more common objects we encounter in everyday life (see Figure 5–7a, and especially Hollerbach, 1975).

The other major directions in which these ideas need to be extended concern not so much the spatial arrangement of a given shape but the spatial configurations formed by several separate objects. These will need

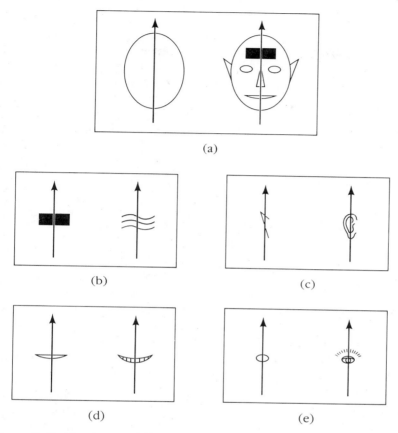

Figure 5–6. The 3-D model for a two-dimensional pattern portraying a face. (a) The overall 3-D model, with the axis determined by symmetry. (b)–(e) Possible 3-D models for the pattern's principal constituents.

at least three types of description. One is the incorporation of their positions in a standard space frame around the viewer in terms of angles and distances from him. Another is the representation of configurations of objects relative to the viewer, for example, the notion that you and two other people happen to create an equilateral triangle. The critical point here is that the position of the viewer is involved and that angular relations—the internal structure of the configuration—are made explicit. Finally, there is the representation of the relative positions of a number of external objects without particular reference to the viewer. For example,

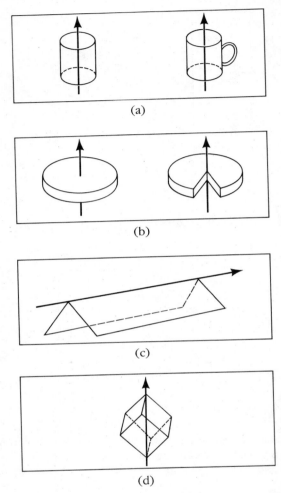

Figure 5–7. Some 3-D models for more complex shapes (a). (b) and (c) may require surface primitives in the representation. (d) Illustrates the representation of a familiar object (a cube) obtained by G. Hinton's unusual choice of axis (a diagonal from one vertex to the opposite one).

three trees might lie in a row, or four buildings form a square. The underlying problems here are exactly the same as those we have already met—how to choose an appropriate canonical coordinate frame within which to make explicit the spatial relations of the configuration.

It is already clear how to approach representational problems of this sort, and for the designers of a vision machine I do not think that these questions will raise any insurmountable difficulties. The major scientific obstacles here, it seems to me, are how to discover what systems and schemes are actually used by humans. I do not expect the answers to be very surprising, but at present I see no empirical way of approaching this type of problem. It seems to be much more difficult to design experiments to answer questions at these rather high levels of analysis than at the lower ones. In fact, perhaps we could say that at these higher levels we are beginning to face all the problems that the linguists have. Designing a successful empirical approach to such questions would represent a major breakthrough.

5.5 DERIVING AND USING THE 3-D MODEL REPRESENTATION

The advantages of modularity, which has been one of our major concerns in the design of the 3-D model representation, will become especially visible as we discuss the processes that derive and use the representation for recognition. In particular, none of the processes have to deal with the internal details of more than one 3-D model at a time even if the complete description of a shape involves many 3-D models. We begin by examining the basic problems associated with identifying a model's coordinate system and its component axes and transforming the viewer-centered axis specifications into specifications in the model's coordinate system. We then treat the task of recognizing this description as a problem of indexing into a catalogue of stored 3-D model descriptions. Finally, we consider the interaction between the process that derives a 3-D model description and the recognition process. The ambiguities introduced by the perspective projection often mean that only coarse specifications of the lengths and orientations of a shape's axes are directly accessible from its image. However, if the recognition process in conjunction with the derivation process, is conservative—so that all the information recognition recovers is reliable— the early stages of the recognition process can make additional constraints available so that a more precise description can be produced.

Deriving a 3-D Model Description

To construct a 3-D model, the model's coordinate system and component axes must be identified from an image, and the arrangement of the component axes in that coordinate system must be specified.

Even if a shape has a canonical coordinate system and a natural decomposition into component axes, there is still the problem of deriving these features from an image. At present we do not have a complete solution to this problem, but some results have been obtained for shapes that fall within the scope of the 3-D model representation. For example, we saw in Section 3.6 that the image of a generalized cone's axis may be found from the occluding contours in an image provided that the axis is not too foreshortened. An example of the decomposition formed by this method appears in Figure 5–8, and a brief description is given in the legend. Notice that the final decomposition (Figure 5–8f) was derived from the contour (Figure 5–8a) without knowledge of the three-dimensional shape apart from the assumption that it is composed of generalized cones. The method can therefore be used to find the component axes for the 3-D model of a shape that has not been seen before.

This result is somewhat limited, but so is the information it uses, namely, the contours formed by rays that are tangential to the side of a smooth surface. Interestingly, as we saw in Section 3.2, these particular contours are unsuitable for use in either stereopsis or structure-from-motion computations, because they do not correspond to fixed locations on the viewed surface. Creases and folds on a surface also give rise to contours in an image, and these have yet to be studied in detail. Similarly, much work remains in the study of how to use information about shape from shading and texture.

A major difficulty in the analysis of images arises when an important axis is obscured because it is either foreshortened or hidden behind another part of the shape. For example, although the torso-based coordinate system for the overall shape of a horse is easily obtained from a side view, it is difficult to obtain when the horse faces the viewer. There are three ways of dealing with such a situation. The first is to allow for recognition the use of partial descriptions based on the axes visible from the

Figure 5–8. (opposite) The occluding contours of simple shapes composed of generalized cones can be used to locate projections of the natural axes of the cones provided that the axes are not severely foreshortened. One algorithm for doing this is shown in this example from a program written by P. Vatan. The initial outline in (a) was obtained by applying local grouping processes to the primal sketch of an image of a toy donkey. This outline was then smoothed and divided into convex and concave sections to get (b). Next, strong segmentation points like the deep concavity circled in (c) were identified and a set of heuristic rules used to connect them with other points on the contour to get the segmentation shown in (d). The component axes shown in (e) were then derived from these. The thin lines in (f) indicate the position of the head, leg, and tail components along the torso axis, and the snout and ear components along the head axis. (Reprinted by permission from D. Marr and H. K. Nishihara, "Representation and recognition of the spatial organization of three-dimensional shapes," *Proc. R. Soc. Lond. B 200,* 269–294.)

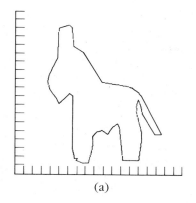

(a)

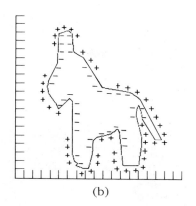

(b)

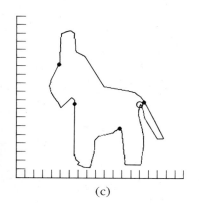

(c)

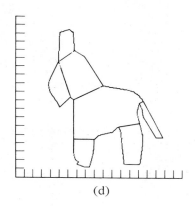

(d)

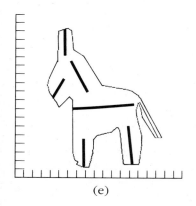

(e)

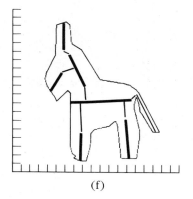

(f)

front. If this is done, the representation is slightly weakened in terms of the uniqueness criterion, but not as severely as a purely viewer-centered representation would be. Another strategy is to use a shape's visible components whenever their recognition is easy but that of the overall shape is difficult. For example, the front view of a horse usually contains an excellent view of the horse's face, which can be recognized directly and provides

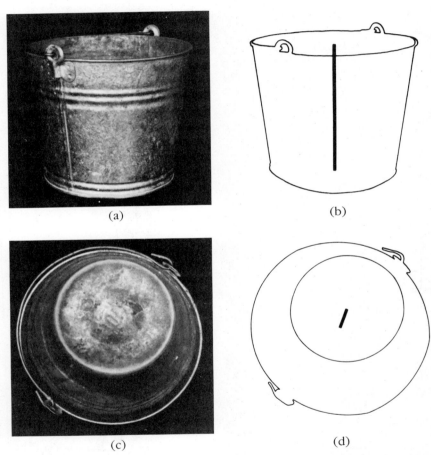

(a)

(b)

(c)

(d)

Figure 5–9.   These views of a water bucket illustrate an important characteristic of any system based on the derivation of canonical axes from an image. The techniques useful for the axis shown in (b) from the image (a) are quite different from those that are best for situations where the axis is foreshortened, as in (c) and (d). (Reprinted by permission from D. Marr and H. K. Nishihara, "Representation and recognition of the spatial organization of three dimensional shapes," *Proc. R. Soc. Lond. B200*, 269–294.)

another route by which the horse can be recognized. This strategy will be discussed further at the end of this section. Finally, a foreshortened axis can sometimes be found from an analysis of radial symmetry in the image.

A water bucket like that shown in Figure 5–9 provides an interesting example. Its principal axis and the shape about that axis are derivable by the methods discussed above for the view shown in Figure 5–9(a) but not for that in Figure 5–9(c), where the bucket's principal axis is foreshortened. An erroneous axis is likely to be established instead, perhaps going through the flanges that attach the handle to the rim. However, a failure to produce a recognizable description with this erroneous axis would suggest that the correct axis is not the most pronounced in the image, and an alternative could be sought. The two concentric circles (made by the top and bottom rims of the bucket) are strong clues that the principal axis passes through their centers. Furthermore, because they are concentric, these circles may be at widely separated locations along that axis. Considering that possibility could lead to the desired description of the bucket even though the identity of the closer rim remains ambiguous. A local surface depth map like the 2½-D sketch, computed by means of stereopsis, shading, or texture information, is likely to play an important role in interpreting images like these.

Relating Viewer-Centered to Object-Centered Coordinates

Techniques for finding axes in a two-dimensional image describe the locations of the axes in a viewer-centered coordinate system, and so a transformation is required to convert the specifications of the axes to an object-centered coordinate system. In the 3-D model representation, all axis dispositions are specified by adjunct relations, as in Figure 5–4, so a mechanism is required for computing an adjunct relation from the specification of two axes in a viewer-centered coordinate system. We shall call this mechanism the *image-space processor*.

The image-space processor can be kept very simple, since the adjunct relation is the only positional specification that has to be interpreted. An adjunct relation $(p, r, \Theta, \iota, \phi, s)$, as we have seen, is a way of specifying the position of a vector **S** relative to an axis vector **A**. What the image-space processor must do is make the coordinates of **S** available simultaneously in a frame centered on the viewer and in one centered on vector **A**, so that specifying the vector **S** in either frame makes it available in the other. This is not a difficult task (see Marr and Nishihara, 1978, for more details).

The accuracy of the adjunct relations computed by the image-space processor is limited by the precision with which vectors **A** and **S** are specified in the viewer-centered coordinate system. Since depth information is lost in the orthographic projection, the orientation specifications for axes derived from the retinal images are least precise in the amount the axes slant toward or away from the viewer. Axis slant parameters can often be reconstructed at least roughly by using stereopsis, shading, texture, structure-from-motion, and surface contour analysis. Constraints supplied by the recognition process can also be used to improve the precision of the slant specifications. We shall consider this possibility later when we discuss the interaction between the derivation process and recognition.

Indexing and the Catalogue of 3-D Models

Recognition involves two things: a collection of stored 3-D model descriptions, and various indexes into the collection that allow a newly derived description to be associated with a description in the collection. We shall refer to the above collection along with its indexing as the *catalogue of 3-D models*. Although our knowledge of what information can be extracted from an image is still limited, three access paths into the catalogue appear to be particularly useful. They are the specificity index, the adjunct index, and the parent index.

All 3-D models can be classified hierarchically according to the precision of the information they carry, and an index can be based on this classification that we call the *specificity index*. Figure 5–10 shows an exam-

---

Figure 5–10. (opposite)   If the recognition process of relating new shape descriptions to known shapes is to be a reliable source of information about the shape, it must be conservative. This diagram illustrates an organization (or indexing) of stored shape descriptions according to their specificity. The top row contains the most general shape description, which carries information about size and overall orientation only. Since no commitment about the shape's internal structure is made, all shapes are described equally well. Descriptions in the second row include information about the number and distribution of component axes along the principal axis, making it possible to distinguish a number of shape configurations (a few are shown in this example). At this point only very general commitments are made concerning the relative sizes of the components and the angles between them. These parameters are made more precise at the third level so that distinctions can be made, for example, between the horse and cow shapes. A newly derived 3-D model would be related to a model in this catalogue by starting at the top level and working downward as far as the information in the new description allows. At that point, it could branch and form a new shape category. (Reprinted by permission from D. Marr and H. K. Nishihara, "Representation and recognition of the spatial organization of three-dimensional shapes," *Proc. R. Soc. Lond. B 200*, 269–294.)

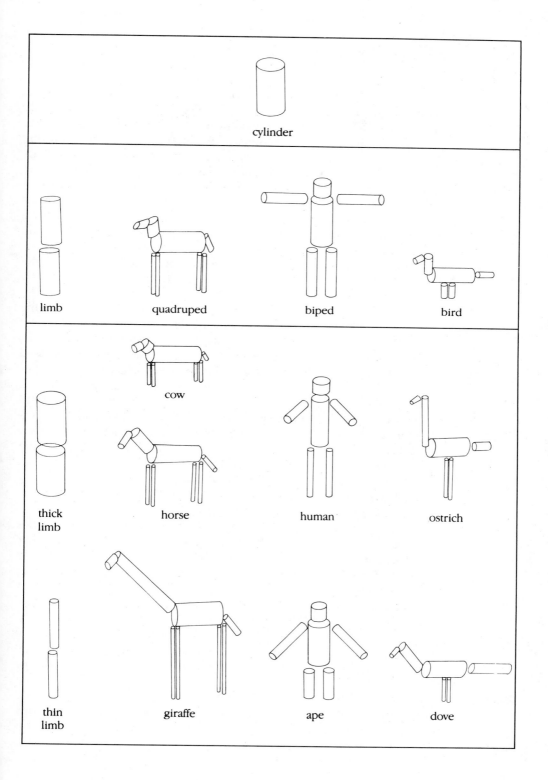

cylinder

limb   quadruped   biped   bird

thick
limb   cow   human   ostrich

horse

thin
limb   giraffe   ape   dove

ple of this organization for models of a few animal shapes. The top level contains the most undifferentiated description available, a 3-D model without a component decomposition. Only the model axis is specified, so the model describes any shape. At the next level of detail are various limbs and a general quadruped shape, a biped shape, and a bird shape. These descriptions are most sensitive to the number of component axes in the model and to their distribution along the principal axis (which is the torso for most animal shapes), while only very coarse information about the lengths and orientations of the components is available. One level lower in the hierarchy the descriptions become more sensitive to angles and lengths, so that distinctions can be made between horse, giraffe, and cow shapes, for example. A newly derived 3-D model may be related to a model in the catalogue by starting at the top of the hierarchy and working down the levels through models whose shape specifications are consistent with the new model's until a level of specificity is reached that corresponds to the precision of the information in the new model.

Once a 3-D model for a shape has been selected from the catalogue, its adjunct relations provide access to 3-D models for its components based on their locations, orientations, and relative sizes. This gives us another access path to the models in the catalogue, which we call the *adjunct index*. It tells, for example, that the two similar components lying at the front end of a quadruped model are general limb models and that, for a horse model, they are more specific horse limb models. Thus the adjunct index provides useful defaults for the shapes of the components of a shape prior to the derivation of 3-D models for them from the image. It is also useful in situations where a catalogued model is not accessible via the specificity index because the description derived from the image is inadequate (perhaps because the component has very little structure).

The third access path that we consider important is the inverse of the second, and we call it the *parent index* of a 3-D model. When a component of a shape is recognized, it can provide information about what the whole shape is likely to be. For example, the catalogue's 3-D model for a horse can be indexed under each of its component 3-D models so that the 3-D model for a horse's leg provides access to the 3-D model for a horse shape.

This index would play an important role in the situation, discussed earlier, where an important axis of a shape is obscured or foreshortened. When a horse faces the viewer, the omission of the torso and hind leg axes might cause the neck axis to be selected incorrectly as the principal axis. Unless special provision has been made to handle this case, the specificity index will fail to access a horse model in the catalogue. A reasonable strategy at this point is to apply the derivation process to the components of the image. In this example, 3-D models for the head, neck, and the two forelegs would be produced. Catalogued models for the head and legs are likely to be found using the specificity index, and each of these would indicate via the parent index that it is a component of either the quadruped or the horse 3-D model (depending on the quality of the derived component models), providing strong evidence for considering the quadruped or horse model for the whole shape.

It is important to note that the adjunct and parent indexes play a role secondary to that of the specificity index, upon which our notion of recognition rests. We shall see below that their purpose is primarily to provide contextual constraints that support the derivation process, for example, by indicating where the principal axis is likely to be when such information cannot be obtained directly from the image. They do not prevent novel composite shapes, such as a centaur, from being described faithfully and recognized (in the case of a centaur, as a horse shape with a human bust).

It may be useful to posit other indexes in the catalogue, perhaps based on color or texture characteristics (for example, the stripes of a zebra) or even on nonvisual clues, such as the sounds an animal makes, but these lie outside of the scope of this investigation.

Interaction Between Derivation and Recognition

So far, the derivation of a 3-D model has been treated separately from the process of relating that model to the stored models of the 3-D model catalogue. We view recognition as a gradual process that proceeds from the general to the specific and that overlaps with, guides, and constrains the derivation of a description from the image. After a catalogued model is selected by using one of the three indexes, we want to use it to improve the analysis of the image. There are two phases to this: First, the component axes from the image must be paired with the adjunct relations supplied by the catalogue; second, the image-space processor must be employed to combine the constraints available from the image with those provided by the model to produce a new set of derived adjunct relations that are more specific than those from the catalogue model. This last phase involves an analysis of constraints that must be satisfied by adjunct relations consistent with both the image and the information from the catalogue. The general idea of using a stored model of a shape to assist in the interpretation of an image was first used by Roberts (1965) in a computer program for producing edge descriptions of shapes built out of cubes, wedges, and hexagonal prisms from their images.

Finding the correspondence between image and catalogued model

The first phase can be thought of as a homology problem, in which the adjunct relations of a catalogue model must be related to the axes derived from an image. There may not be a complete solution. For example, the leg axes in a silhouette of a horse as viewed from the side are easily identified, but the left and right forelegs usually cannot be disambiguated without further information. Often this ambiguity may be tolerable, however, since the corresponding adjunct relations for the two legs have the same general orientation specifications (they differ only in their locations), and this is all that the following analysis makes use of.

The information available for establishing the correspondence between image and model increases as the derivation–recognition process proceeds. Initially, positional information along the principal axis of the stick figure has priority, since it is the least distorted by the perspective projection. Other clues available initially include: (1) the relative thicknesses of the shapes about the component axes (the neck of a horse is much thicker than the legs), (2) possible decompositions of component axes (the tail and legs of a horse may be roughly straight but the bust has two components that always make a large angle with one another), (3) symmetry or repetition (the legs of a horse are all the same thickness and are roughly parallel, and so have roughly the same length and orientation in the image distinguishing them from the tail), and (4) large differences in ϕ of the adjunct relation (in an image, the legs and tail of a horse usually extend to one side of the torso while the neck extends to the other). Collectively, such clues are often sufficient to relate the major components of a 3-D model to the axes derived from an image.

Homology information is also available from the adjunct and parent indexes. When a 3-D model from the catalogue is obtained by using the adjunct index, the polarity of that component's axis is automatically determined. For example, when continuing the analysis of the image of a horse to one of the legs, the polarity of the leg axis is indicated by its connection with the torso (the hoof end being distal to the junction). When the parent index is used to select a catalogue model based on the identification of some of a shape's components, the pairings for these identified components strongly constrain pairings for the remaining components. For example, in the case of a horse facing the viewer, the missing torso's location in the image can be found from the locations of the head, neck, and forelegs.

Constraint analysis

Once a homology has been established between a 3-D model and the image, we want to use the information that it makes available to constrain the possible slant angles for the axes. The basic idea is that there are often only a few combinations of the slant specifications for the projected axes in the image for which the adjunct relations derived from the image would be consistent with those supplied by the catalogue model. Or equivalently, there are often only a few orientations of the catalogue model's principal axis (relative to the viewer) for which its component axes match closely the projected axes in the image.

The combination of information from the image and the catalogue model is often sufficient to determine the axis slants uniquely up to a reflection about the image plane. For example, Figure 5–11(a) shows the locus of orientations of vector **A** (relative to the viewer) that are consistent with an inclination of 90° between **A** and vector **S**, and an angle of 47° between their projections onto the image plane; Figure 5–11(b) shows the allowed orientations for an inclination angle of 45° and a projected angle of −111°; and Figure 5–11(c) shows the intersection of these two sets. The sharpness of these constraints depends on the particular viewing angle (as indicated by the other examples in the figure) and on the particular adjunct relations in the 3-D model. Generally, the constraints are strongest when the component axes have very different orientations and when the principal axis does not lie in the image plane.

There are several algorithms that can use these constraints. Perhaps the simplest is a relaxation process that adjusts the orientation of **A** incrementally, seeking the disposition for which the projections of the angles between the component axes of the catalogue model, as computed by the image-space processor, best agree with those in the stick figure image. At this point vector **A** will indicate the orientation of the principal axis that is most consistent with all of the constraints, and the image-space processor can use its other vector, **S**, to compute the orientations of each of the component axes by using the adjuncts from the catalogue model. This hill-climbing approach converges quite efficiently when the constraints are sufficiently strong.

Alternatively, instead of relaxing the orientation of the catalogue model's principal axis, one can relax the slant angles of the sticks obtained from the image. In this case, the discrepancy measure is obtained by comparing adjunct relations derived between the sticks in the image with the corresponding adjunct relations from the catalogue. This approach is interesting because in its implementation, all of the transformations carried out by the image-space processor are in the same direction (from viewer-centered to object-centered coordinates). In a final step, improved orientation information may be used to recover more information from the image. In particular, once the orientations of the axes have been determined, their relative lengths may be computed.

The overall recognition process may be summarized as follows. We

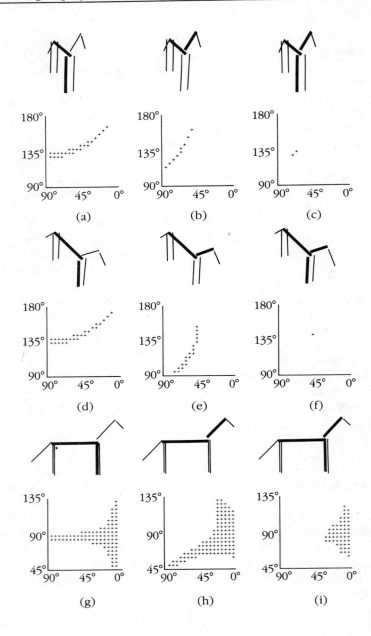

Figure 5–11. (opposite) If we know the three-dimensional inclination angle ι that vector **S** makes with axis **A**, as well as the two-dimensional projection of this angle, then the orientation of **A**'s coordinate system relative to the viewer is strongly constrained. (a) The orientations consistent with an inclination of 90° and an image angle like that between the heavy lines in the accompanying stick figure (allowing a tolerance of 5° in the image angle). The horizontal axis of the graph indicates the angle by which **A** dips out of the image plane toward the viewer. The vertical axis is the amount the coordinate system is rotated about **A**. (b) shows the set of orientations consistent with $\iota = 45°$, and the visible angle between the images of the torso and neck axes. (c) The intersection of the two sets, which is restricted to a narrow range of orientations having a dip of approximately 67° out of the image plane (there is another solution, not shown here, at $-67°$). The remaining rows show the same analyses for dips of 45° and 0°, respectively. In this way, two-dimensional information from the image and angles from the stored 3-D model can be combined to give sometimes quite accurate information about the spatial disposition of the viewed shape relative to the viewer. (Reprinted by permission from D. Marr and H. K. Nishihara, "Representation and recognition of the spatial organization of three-dimensional shapes," *Proc. R. Soc. Lond. B 200*, 269–294.)

---

first select a model from the catalogue based on the distribution of components along the length of the principal axis. This model then provides relative orientation constraints that help to determine the absolute orientations (relative to the viewer) of the component axes in the image, and with this information the image-space processor can be used to compute the relative lengths of the component axes. This new information can then be used to disambiguate shapes at the next level of the specificity index.

5.6 PSYCHOLOGICAL CONSIDERATIONS

In our study of the primal sketch and of processes capable of deriving surface information from such image representations, we were much helped by evidence from neurophysiology and psychophysics and by a careful computational examination of what can in fact be derived from the available information. Our approach rested heavily on the principle of modularity (Marr, 1976), which states that any large computation should be split up into a collection of small, nearly independent, specialized subprocesses. Our analysis was based on evidence from psychophysics and from everyday experience about what the modules were likely to be, the underlying argument being that if visual information processing is not organized in a modular way, incremental changes in its design, presumably an essential requirement for its evolutionary development, would be unable to improve one aspect of visual performance without simultaneously degrading the operation of many others.

Unfortunately, we receive little help from the biological sciences about the kinds of questions raised by the later aspects of the visual process. Virtually nothing is known about the physiological and anatomical arrangements that mediate the construction of three-dimensional visual descriptions of the world, and even the best psychological information is for the most part anecdotal and derived from neurological rather than psychophysical studies.

Nevertheless, I think it is clear in principle that the brain must construct three-dimensional representations of objects and of the space they occupy. As Sutherland (1979) has remarked, there are at least two good reasons for this. First, in order to manipulate objects and avoid bumping into them, organisms must be able to perceive and represent the disposition of the objects' surfaces in space. This gives us a minimal requirement for something like the 2½-D sketch. Second, in order to recognize an object by its shape, allowing one then to evaluate its significance for action, some kind of three-dimensional representation must be built from the image and matched in some way to a stored three-dimensional representation with which other knowledge is already associated. As we have seen, the two processes of construction and matching cannot be rigorously separated because a natural aspect of constructing a three-dimensional representation may include the continual consultation of an increasingly specific catalogue of stored shapes.

This forces us to rely, in our study of these later problems, much more on a careful consideration of the computational and representational requirements. Stated baldly, the strong constraints come from what the representation is to be used *for*.

We asked here about the requirements for a shape representation to be used for recognition, and we came to three main conclusions: A shape representation for recognition should (1) use an object-centered coordinate system, (2) include volumetric primitives of various sizes, and (3) have a modular organization. A representation based on a shape's natural axes (for example, the axes identified by a stick figure) follows directly from these choices. In addition, we saw that the basic process for deriving a shape description in such a representation must involve a means for identifying the natural axes of a shape in its image and a mechanism for transforming viewer-centered axis specifications to specifications in an object-centered coordinate system.

Finally, we saw how the recognition process itself involves a mixture of straightforward derivation of shape information from the image and the deployment of gradually more detailed stored 3-D models during the process of recognition–derivation. Thus critical ingredients of this process are a collection of stored shape descriptions and various indexes for the collection that allow a newly derived description to be associated with an appropriate stored description. The most important of these indexes allows shape recognition to proceed conservatively from the general to the specific based on the specificity of the information available from the image.

There are two ways in which we might try to examine empirically the relevance of these ideas to the processes of recognition in the human visual system. We can try to discover the type of information made explicit by the visual process in its representations, or we can try to find some correlate of the processes that derive and maintain them, perhaps using Shepard-like studies of mental rotation. The first approach is the more fundamental: Is a three-dimensional representation used, does it have a modular organization, and is it object centered? These questions have yet to be put to empirical test, but three observations are worth noting here. The first is that stick figure animals like those shown in Figure 5–1 are usually recognized easily despite the limited amount of shape information they portray. While this does not demonstrate that the human visual process is based on stick figures, it does suggest that the type of information carried by stick figures plays an important role in it.

Second, illusions like that shown in Figure 5–12 (due originally to Ernst Mach) provide evidence that local shape information is described relative to axes that are defined more globally. In the right row, the shapes are seen as diamonds, whereas along the diagonal they are seen as squares.

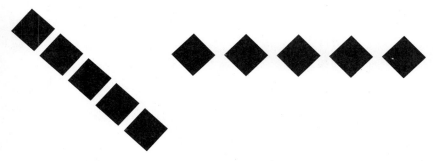

Figure 5–12. The effect of different choices of an object-centered coordinate system on the perception of shape is apparent in these diagrams. The black shapes can be seen as diamonds or squares depending on which of their several natural axes are used. (From F. Attneave, "Triangles as ambiguous figures," *Am. J. Physiol. 81,* 447–453.)

The diagonal axis is therefore being constructed during the analysis of this pattern; it influences and therefore probably precedes the description of the shapes of the local elements.

Third, Warrington and Taylor (1973) drew attention to the difficulty experienced by their patients with right parietal lesions in interpreting certain views of common objects, which Warrington and Taylor called unconventional views. For example, these patients would fail to recognize the top view of a bucket (Figure 5–9c), denying that it was a bucket even when told it was. The patients were relatively unimpaired on views like Figure 5–9(a). As Warrington and Taylor pointed out, this difference cannot be easily explained in terms of familiarity or impaired depth perception, because both views of a bucket are common and depth is just as important to the three-dimensional structure of Figure 5–9(a) as it is to that of Figure 5–9(c). However, if the internal shape representation used for recognition was based on a shape's natural axes, the second figure would be more difficult to describe correctly, since its major axis is foreshortened. If this explanation were correct, Warrington and Taylor's unconventional views would correspond to views in which an important natural axis of the shape is foreshortened in the image, making it difficult for the patient to discover or derive a description in the shape's canonical coordinate system.

CHAPTER   6

Synopsis

Our survey of this new, computational approach to vision is now complete. Although there are many gaps in the account, I hope that it is solid enough to establish a firm point of view about the subject and to prompt the reader to begin to judge its value. In this brief chapter, I shall take a very broad view of the whole approach, inquiring into its most important general features and how they relate to one another, and trying to say something about the style of research that this approach implies. It is convenient to divide the discussion into four main points.

The first point is one that we have met throughout the account—the notion of different levels of explanation. The central tenet of the approach is that to understand what vision is and how it works, an understanding at only one level is insufficient. It is not enough to be able to describe the responses of single cells, nor is it enough to be able to predict locally the results of psychophysical experiments. Nor it is enough even to be able to write computer programs that perform approximately in the desired way. One has to do all these things at once and also be very aware of the additional level of explanation that I have called the level of computational theory. The recognition of the existence and importance of this level is one of the most important aspects of this approach. Having recognized this, one can formulate the three levels of explanation explicitly (computational theory, algorithm, and implementation), and it then becomes clear how these different levels are related to the different types of empirical observation and theoretical analysis that can be conducted. I have laid particular stress on the level of computational theory, not because I regard it as inherently more important than the other two levels—the real power of the approach lies in the integration of all three levels of attack—but because it is a level of explanation that has not previously been recognized and acted upon. It is therefore probably one of the most difficult ideas for newcomers to the field to grasp, and for this reason alone its importance should not be understated in any introductory book, such as this is intended to be.

The second main point is that by taking an information-processing point of view, we have been able to formulate a rather clear overall framework for the process of vision. This framework is based on the idea that the critical issues in vision revolve around the nature of the representations used—that is, the particular characteristics of the world that are made explicit during vision—and the nature of the processes that recover these characteristics, create and maintain the representations, and eventually read them. By analyzing the spatial aspects of the problem of vision, we arrived at an overall framework for visual information processing that hinges on three principal representations: (1) the primal sketch, which is concerned with making explicit properties of the two-dimensional image, ranging from the amount and disposition of the intensity changes there to primitive representations of the local image geometry, and including at the more sophisticated end a hierarchical description of any higher-order structure present in the underlying reflectance distributions; (2) the 2½-D sketch, which is a viewer-centered representation of the depth and orientation of the visible surfaces and includes contours of discontinuities in these quantities; and (3) the 3-D model representation, whose important features are that its coordinate system is object centered, that it includes volumetric primitives (which make explicit the organization of the space occupied by an object and not just its visible surfaces), and that primitives of various size are included, arranged in a modular, hierarchical organization.

The third main point concerns the study of processes for recovering the various aspects of the physical characteristics of a scene from images of it. The critical act in formulating computational theories for such processes is the discovery of valid constraints on the way the world behaves that provide sufficient additional information to allow recovery of the desired characteristic. We saw many examples of this in Chapter 3, and they were summarized in Table 3–3. The power of this type of analysis resides in the fact that the discovery of valid, sufficiently universal constraints leads to conclusions about vision that have the same permanence as conclusions in other branches of science.

Furthermore, once a computational theory for a process has been formulated, algorithms for implementing it may be designed, and their performance compared with that of the human visual processor. This allows two kinds of results. First, if performance is essentially identical, we have good evidence that the constraints of the underlying computational theory are valid and may be implicit in the human processor; second, if a process matches human performance, it is probably sufficiently powerful to form part of a general purpose vision machine.

The final point concerns the methodology or style of this type of approach, and it involves two main observations. First, the duality between representations and processes, which is set out explicitly in Figure 6–1, often provides a useful aid to thinking how best to proceed when studying a particular problem. In the study both of representations and of processes, general problems are often suggested by everyday experience or by psychophysical or even neurophysiological findings of a quite general nature. Such general observations can often lead to the formulation of a particular process or representational theory, specific examples of which can be programmed or subjected to detailed psychophysical testing. Once we have sufficient confidence in the correctness of the process or representation at this level, we can inquire about its detailed implementation, which involves the ultimate and very difficult problems of neurophysiology and neuroanatomy.

The second observation is that there is no real recipe for this type of research—even though I have sometimes suggested that there is—any more than there is a straightforward procedure for discovering things in any other branch of science. Indeed, part of the fun is that we never really know where the next key is going to come from—a piece of daily experience, the report of a neurological deficit, a theorem about three-dimensional geometry, a psychophysical finding in hyperacuity, a neurophysiological observation, or the careful analysis of a representational problem. All these kinds of information have played important roles in establishing the framework that I have described, and they will presumably continue to contribute to its advancement in an interesting and unpredictable way. I hope only that these observations may persuade some of my readers to join in the adventures we have had and to help in the long but rewarding task of unraveling the mysteries of human visual perception.

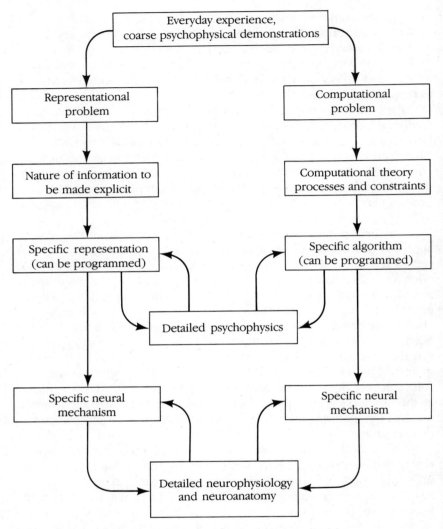

Figure 6–1. Relationships between representations and processes.

PART III

Epilogue

CHAPTER 7

In Defense
of the Approach

7.1 INTRODUCTION

In the first and second parts of this book, I have tried to set out in some
detail an approach that treats visual perception primarily as an information-
processing problem. I have incorporated answers to the objections most
commonly raised, but from my experience in trying to convey in lectures
and conversations the essence of this point of view, I expect that the reader
will still have some private difficulty or question, even if it is as simple as
thinking the scheme too farfetched or, at the opposite extreme, not imag-
inative enough.

   To have addressed all the possible objections, however, would have
disrupted the account too much, so I thought it best to attempt to answer
them separately in the form of a conversation between an imaginary skeptic
and an imaginary defender of the information-processing point of view.
The dialogue is based on lunchtime conversations at the Salk Institute
between Francis Crick, Tomaso Poggio, and myself, but it does not follow

those conversations very closely, and my imaginary objector is a combination of many real-life people. The discussion is not very structured and ranges over a variety of topics, but this seems unavoidable.

7.2   A CONVERSATION

Can we begin with the levels-of-explanation idea, since you attribute so much importance to it? How is it related to ideas about feature detectors and in particular to Horace Barlow's first dogma (1972, p. 380), which states, "A description of the activity of a single nerve cell which is transmitted to and influences other nerve cells, and of a nerve cell's response to such influences from other cells, is a complete enough description for functional understanding of the nervous system"?

Here, of course, I must disagree with Barlow's formulation, although I do agree with one of the thoughts behind this dogma, namely, that there is nothing else looking at what the cells are doing—they are the ultimate correlates of perception. However, the dogma fails to take level one analysis—the level of the computational theory—into account. You cannot understand stereopsis simply by thinking about neurons. You have to understand uniqueness, continuity, and the fundamental theorem of stereopsis. You cannot understand structure from motion without knowing a result like the structure-from-motion theorem, which shows how such a phenomenon is possible. In addition, and critically important for a researcher, the levels approach enforces a rigid intellectual discipline on one's endeavors. As long as you think in terms of mechanisms or neurons, you are liable to think too imprecisely, in similes.

Remember the moral from the early stereopsis networks discussed in section 3.3! None of them formulated the computational problem precisely at the top level, and almost all the proposed networks actually computed the wrong thing. Another example was the notion of segmentation to carve up an image into regions and objects. This wasted an enormous amount of time and led to the development of all kinds of special relaxation and hypothesize-and-test methods for agglomerating areas of the picture into useful regions (see Chapter 4). The problem again was that people became so entranced by the mechanisms for doing something that they erroneously thought they understood it well enough to build machinery for it—just as had occurred in the simpler case of stereopsis. It was only with a level-one attack—the formulation of the 2½-D sketch and its attendant and precisely stated problems—that real progress was possible.

Have I made my case strong enough yet? The levels idea is crucial, and perception cannot be understood without it—never by thinking just about synaptic vesicles or about neurons and axons, just as flight cannot be understood by studying only feathers. Aerodynamics provides the context in which to properly understand feathers. Another key point is that explanations of a given phenomenon must be sought at the appropriate level. It's no use, for example, trying to understand the fast Fourier transform in terms of transistors as it runs on an IBM 370. There's just no point—it's too difficult.

For instance, take the retina. I have argued that from a computational point of view, it signals $\nabla^2 G * I$ (the X channels) and its time derivative $\partial/\partial t\,(\nabla^2 G * I)$ (the Y channels). From a computational point of view, this is a precise specification of what the retina does. Of course, it does a lot more—it transduces the light, allows for a huge dynamic range, has a fovea with interesting characteristics, can be moved around, and so forth. What you accept as a reasonable description of what the retina does depends on your point of view. I personally accept $\nabla^2 G$ as an adequate description, though I take an unashamedly information-processing point of view. A retinal physiologist would not accept this, because he would want to know exactly *how* the retina computes this term. A receptor chemist, on the other hand, would scarcely admit that these sorts of consideration have anything at all to do with the retina! Each point of view corresponds to a different level of explanation, and all must eventually be satisfied.

Yes, I see the point. You're simply saying that, from an information-processing point of view, what is done and why assumes paramount importance— this is your top level. The implementation details don't matter so much from this perspective provided that they do the right thing.

I'd like to make that point even more strongly. Figure 7–1 shows three descriptions of essentially the same thing. At the top is the mathematical description that we're so familiar with, $\nabla^2 G * I$. Figure 7–1(b) shows a piece of the retina, which we believe does roughly this, at least in part. And Figure 7–1(c) illustrates a silicon chip, built for us by Graham Nudd of the Hughes Research Laboratories in charge-coupled device technology, which carries out the $\nabla^2 G$ convolution. So, in a real sense, all these three things— the formula, the retina, and the chip—are similar at the most general level of description of their function.

Are the different levels of explanation really independent?

Not really, though the computational theory of a process is rather independent of the algorithm or implementation levels, since it is determined solely by the information-processing task to be solved. The algorithm depends heavily on the computational theory, of course, but it also depends on the characteristics of the hardware in which it is to be implemented. For instance, biological hardware might support parallel algorithms more readily than serial ones, whereas the reverse is probably true of today's digital electronic technology.

$$\nabla^2 G * I(x,y),$$

$$\text{where } \nabla^2 G(r) = -\frac{1}{\pi\sigma^4}\left(1-\frac{r^2}{2\sigma^2}\right)\exp\left(\frac{-r^2}{2\sigma^2}\right)$$

(a)

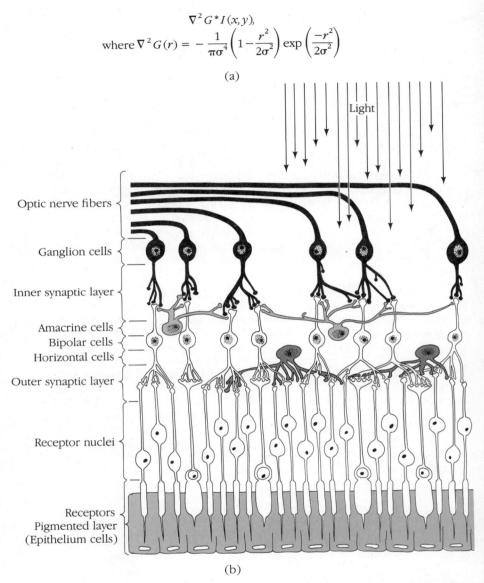

(b)

Figure 7–1. (a) The mathematical formula that describes the initial filtering of an image. ∇^2 is the Laplacian, G is a Gaussian, I (x,y) represents the image, and $*$ the operation of convolution. (b) A cross section of the retina, part of whose function is to compute (a). (c) The circuit diagram of a silicon chip, built by Graham Nudd at Hughes Research Laboratories, which is capable of computing (a) at television rates.

(c)

Figure 7–1 (continued).

I cannot really accept that the computational theory is so independent of the other levels. To be precise, I can imagine that two quite distinct theories of a process might be possible. Theory 1 might be vastly superior to theory 2, which may be only a poor man's version in some way, but it could happen that neural nets have no easy way of implementing theory 1 but can do theory 2 very well. Effort would thus be misplaced in an elaborate development of theory 1.

Yes, this could certainly happen, and I think it already has in the case of deriving shape from shading. I would not be at all surprised if it was unreasonably difficult to solve Horn's integral equations for shape from shading with neural networks, yet the equations can be solved on a computer for simple cases. Human ability to infer shape from shading is very limited, and it may be based on simplistic assumptions that are often violated—a sort of theory 2 of the kind you mentioned. Nevertheless, I doubt that the effort put into a deep study like Horn's was misplaced, even in the circumstances. Although it will not yield direct information about human shape-from-shading strategies, it probably provides indispensable background information for discovering the particular poor man's version that we ourselves use.

What about the old feature detector ideas? How did they fit in?

Historically, I think, the notion of a feature—and I would not now care to define it at all precisely—played an important role in shifting our conceptions away from Lashley's mass-action ideas (according to which the brain was a kind of thinking porridge whose only critical factor was how much was working at the time) and toward the much more specific view of single-neuron action that we now have. This movement was initiated by Barlow (1953), Kuffler (1953), Lettvin and others (1959), and, of course, Hubel and Wiesel (1962, 1968). Essentially, these findings ultimately lead to the notion that single nerve cells can have as one of their functions the job of signaling explicitly whenever a particular, very specific configuration is present in the input, and this type of thinking was formulated in terms of features.

But there are a number of fascinating points here arising mainly from the basic question, When does a specific configuration in the image imply a specific configuration in the environment? The first point, which we met in Chapter 1, has to do with how descriptions of the environment actually get made. In a true sense, for example, the frog does not detect *flies*—it detects small, moving, black spots of about the right size. Similarly, the housefly does not really represent the visual world about it—it merely computes a couple of parameters $(\psi,\dot\psi)$, which it inserts into a fast torque generator and which cause it to chase its mate with sufficiently frequent success. We, on the other hand, very definitely do compute explicit properties of the real visible surfaces out there, and one interesting aspect of the evolution of visual systems is the gradual movement toward the difficult task of representing progressively more objective aspects of the visual world. The payoff is more flexibility; the price, the complexity of the analysis and hence the time and size of brain required for it.

But wasn't there more to the features idea than that?

Yes, and that, too, is an interesting set of issues that harks back to some extent to the philosophers of perception, who thought in terms of "sense atoms" grouped into larger "molecules" of sensory experience, which were the things we could recognize. One can perhaps follow a tradition of attempts at feature-based recognition. This started with the Barlow (1953) ideas, involved Kruskal's (1964) multidimensional-scaling technique, Jardine and Sibson's (1971) excellent work on cluster analysis, my early ideas about the neocortex (Marr, 1970), and the mountainous literature on statistical decision theory.

What was the main idea?

The hope was that you looked at the image, detected features on it, and used the features you found to classify and hence recognize what you were looking at. The approach is based on an assumption which essentially says that useful classes of objects define convex or nearly convex regions in some multidimensional feature space where the dimensions correspond to the individual features measured. That is, the "same" objects—members of a common class—have more similar features than objects that are not the same.

That sounds perfectly reasonable. What went wrong?

It's just not true, unfortunately, because the visual world is so complex. Does feature refer to the image or the object? Different lighting conditions produce radically different images, as do different vantage points. Even in the very restricted world of isolated, two-dimensional, hand-printed characters, it is difficult to decide what a feature should be. Think of a 5 gradually changing into a 6—a corner disappears, a gap narrows. Almost no single feature is necessary for any numeral. The visual descriptions necessary to solve this problem have to be more complex and less directly related to what we naturally think of as their representation as a string of motor strokes.

So your main argument is that the world is just too complex to yield to the types of analysis suggested by the feature detector idea?

That is correct unless, of course, the visual environment can be rigidly constrained—the lighting, the vantage point, the domain of visible elements, and so forth. If this is done, then some progress can be made. Otherwise not, and we have to look quite carefully in the literature to see this, because people do not report negative results, even though such results can be very important in deciding whether to pursue a particular line of attack.

What are the options if the domain of study cannot be so rigidly constrained?

There are basically two: Use a more complicated decision criterion or use a better representation. Using a more complicated decision criterion means abandoning the hope that classes correspond to convex clusters of features and introducing logical ideas in the decision process so that the questions asked at a given point in the classification process may depend on the answers just obtained. It is roughly true to say that artificial intelligence grew out of this approach. It leads to a view of recognition or classification as an exercise in problem solving. Decisions and routes to the solution depend sensitively on partial results found along the way, and these in turn determine the information deployed next to allow the process to continue. We saw some examples of this type of thinking in Chapter 5. The other option is to use a representation or series of representations that are better tailored to the problem at hand. In practice, this turns out to be the more important task for the particular case of vision, although for problems like medical diagnosis the problem-solving approach may be more profitable.

Are there perhaps other ways in which we might try to think about these things? What about Winograd's (1972) procedural representation of knowledge, for example, according to which terms like pick-up *or* block *are represented by programs. If you want to pick up the block, you simply run the two programs in sequence. That sounds like a very sensible approach to me. How does that relate to your two options?*

The procedural representation idea isn't really a representation at all; it is an implementation mechanism. A representation is a much more precisely defined object. For example, there was never any result defining the scope of the procedural representation or establishing any uniqueness characteristics (in the sense of Chapter 5). It is no more a representation than is a property list! In order to define a representation, as we have seen, we must define its primitives, how they may be organized, and so on. Now the primitives in these procedural representations are simply the primitives of the underlying programming language—in Winograd's case, PLANNER or LISP. Such primitives are useless for representing what the process is actually doing in any high-level description, just as the individual instructions in a machine language program for the fast Fourier transform are useless for understanding the transform. To begin to understand and manipulate the code, one has to add comments to it. At this point it is these, not the code, that in effect provide the representation of what the code is doing from the point of view of the manipulator. G. J. Sussman's (1975) program HACKER was essentially an exercise in writing useful standard comments within a particular and restricted programming domain.

Why do you say a property list is not a way of representing knowledge? Surely it is?

I did not say that, I said it wasn't a *representation*. A property list is a programming mechanism that one may use to *implement* a representation, but it is not a representation in itself. To see this, just ask the simple question, What can and what cannot be represented in a property list, or, expressed in our earlier language, what is its scope? Is each description unique? It is meaningless to ask these questions about property lists, just as it is about procedures. Both these ideas are universal from a representational point of view, because both are in fact notions at a lower level of explanation pertaining to decisions about implementation. They are *mechanisms,* not representations. Choosing one mechanism rather than the other will affect how easy it is for the programmer to make a certain piece of information explicit, but the decision about what is to be made explicit and what is not is a decision about the representation itself and is independent of the implementing mechanism.

Ah yes, and here we come back to the feature idea again. For it was surely the notion of a feature which led eventually to the idea that a representation has as its business the making of certain information explicit, wasn't it?

Very much so. But I do think that the time has now come to abandon those older ways of thinking, it being more fruitful to think instead of systems of representations that can describe as fully as desired firstly images and then other derived aspects of the visual world. And I also think it is important not to be too anxious to relate our ideas immediately to neurons. We should first be sure that our representations and algorithms are sensible, robust, and supported by psychophysical evidence. Then we can delve into the neurophysiology.

Before leaving this topic, I feel there is one other matter we should raise. This is the question of features—well, let's call them descriptions from now on—and of measurements for getting them. What exactly is the difference between a descriptive element—perhaps we could call it an assertion— and a measurement? Is this even an important point?

There are two aspects to this. One is historical—a point I felt lay in terrible confusion back in 1974—and the second is a modern question. Let us look first at the historical question. Put most simply, people confused measurements and assertions. For example, a cell with a center–surround receptive field will respond to a blob, but it will also respond to many other things—a line, an edge, two blobs, and so forth. In fact, one can often say no more than that it signals a convolution—our old friend $\nabla^2 G * I$, for instance. Nevertheless, people did call these cells blob detectors.

Now that is not so bad in the retina, but if we were to take Hubel and Wiesel's (1962) definition of a simple cell—the simplest type of receptive field—literally, it, too, would be performing a linear convolution with one excitatory and one inhibitory strip, signaling something like a first directional derivative. I do not now believe these cells are linear convolvers (see Chapter 2), but the point is that people thought of them simultaneously as linear convolvers *and* as feature detectors, and that is criminal, intellectually. Of course, you can use the output of such convolvers to find edges, but it needs extra work. You have to find peaks in the first derivatives or zero-crossings in the second. And, of course, we now think that simple cells are in fact zero-crossing detectors. But the point is that here again, just because of imprecise thinking by computer vision people as well as by physiologists, that whole rich theory of early vision had been missed (see Chapter 2).

The second aspect is the modern one, and I have already raised it in Chapter 2. It has to do with when and how vision "goes symbolic." Most would agree that an intensity array $I(x,y)$ or even its convolution $\nabla^2 G * I$ is not a very symbolic object. It is a continuous two-dimensional array with few points of manifest interest. Yet by the time we talk about people or cars or fields or trees, we are clearly being very symbolic, and I think again that most would find suggestions of symbols in Hubel and Wiesel's (1962) recordings. Our view is that vision goes symbolic almost immediately, right at the level of zero-crossings, and the beauty of this is that the transition from the analogue arraylike representation to the discrete, oriented, sloped zero-crossing segments is probably accomplished without loss of information (Marr, Poggio, and Ullman, 1979; Nishihara, 1981).

And the use of symbols does not stop there either. Almost the whole of early vision appears to be highly symbolic in character. Terminations, discontinuities, place tokens, virtual lines, groups, boundaries—all these things are very abstract constructions, and few of their neurophysiological correlates have been found, but experiments like Stevens' (1978) tell us that such things must be there (see Chapter 2).

How else might one approach these phenomena? What about some kind of transformational or grammatical approach, like the one Chomsky used?

People have tried to write picture grammars involving rules that must be obeyed by line drawings (Narasimhan, 1970), but they have been unsuccessful in general and never successful on a real image. The best of the early approaches, was, I think, the blocks-world analysis of Guzman (1968), Mackworth (1973), and Waltz (1975). Unfortunately, this did not generalize—it suffered from the wrong choice of a miniworld, as indeed has much research in artificial intelligence. The great virtue of artificial intelligence has been that it forced people to substantiate their opinions by writing programs, and in doing so, these opinions were often found to be wrong. It forced a constructive way of thinking—disallowing, for example, Bertrand Russell's definition of the percept of an object as the set of all possible appearances of the object (Russell, 1921). But in having to program things, research was too often limited to a miniworld in which very many factors appear in only simple forms. Though the programs solved none of the individual problems, on the whole they ran just well enough to get by with luck. Winograd's (1972) blocks-world program was of this genre. The underlying conceptual fault is to ignore the modularity that must be present to help decompose the problem.

I do not follow. Why must it be there? How was it being ignored?

Once again, I think the clearest examples come from vision. An early miniworld, or domain of study if you like, was the blocks world—compositions of matte white prisms against a black background. The study of such a domain led to Waltz's (1975) careful cataloguing of the legal junctions of the various types of edges (as in Figure 1–3). Allowing for shadows, Waltz found that most line drawings of such scenes could be interpreted unambiguously. But notice that not one of the general processes listed in Chapter 3 was elucidated by this approach. The reason is that the general processes that combine to make up human vision cannot be easily studied by restricting oneself to any particular miniworld except by carefully choosing it in relation to something that one already suspects of corresponding to a genuine module, like the world of random-dot stereograms.

It is critical to appreciate the difference between these two kinds of miniworlds. One is very particular, the other general. Only the second kind has been found to be of value so far, although constraints in the spirit of

Waltz's may turn out to be useful for the 2½-D sketch (see Chapter 4). The reason is that for genuine computational modules with general and not limited abilities, we can actually prove theorems that show the modules will always work in the real world.

This is the true difference between the approach described in this book and the original conception of artificial intelligence, which, in its desperate effort to pack a whole working miniworld into a program—an endeavor that requires a huge amount of work—was forced to neglect and eventually to abandon attempts at real theory, turning instead to the development of better computer tools. This endeavor has met with little success. So although the artificial intelligence approach was necessary to haul us out of our false preconceptions about the simplicity of vision, it in turn became limited and hidebound because of its failure to recognize what a true computational theory is and how it should be deployed.

Are there any rules for doing this successfully?

I don't think so, and it's perfectly natural to get it wrong first. The example of flight that came up earlier makes a number of points in a nice way. First, it's obvious that you cannot understand how a bird flies by speculating on the fine structure of a feather. So the next natural step is to try to copy how the bird behaves—what I call the mimicry phase. So people built imitation wings and flapped them. That didn't work either. This phase is essentially copying at the lower two levels or possibly only at level two. The real advance comes only when you understand that an airfoil provides lift in accordance with Bernoulli's equation. That is the level-one part—aerodynamics. It is why a bird and a 747 are similar—and why both are dissimilar from a gnat, which keeps itself aloft not by means of an airfoil but by "treading air" in an essentially turbulent regime.

But at some stage, one has to relate one's level-one ideas directly to neural machinery, surely? You talked about the eyes—the retina and $\nabla^2 G$—but what about eye movements? I understand that from your—I should say, from an information-processing and levels point of view, they are quite trivial to deal with. But that doesn't make it any easier for me to think of compensating for them in neural machinery.

Yes, I admit that this is a thorny issue. But first, I hope I made it clear in Chapter 4 that eye movements involve much more than just a subtraction. We saw there how the representation of surface orientation, for example, is quite intimately bound up with whether you choose a retinocentric polar frame (the natural one from the point of view of imaging) or a more invariant type of retinocentric frame.

The second point is that, by delaying the transition out of a retinocentric frame, the difficulty of the arithmetic that is necessary when one at last performs the transition is correspondingly eased. In the manner of Chapter

5, we can move directly to a 3-D model representation, which is located in a stable frame around the viewer; and then all we have to check is that when the eyes move, the appropriate blob moves as expected.

Lastly, I think that here, as always, it is important not to be fooled by the apparent detail and luxury of our perception. We met this earlier in connection with the immediacy and vividness of our perception. I would be surprised if we can keep track of more than a handful of objects during eye movements, and I expect our powers are quite limited in this respect.

Yes, I see the plausibility of the argument. But this doesn't need our levels, does it? It seems a rather different kind of issue.

Absolutely true, but that is mostly because the level-one theory of eye movements is so simple that we don't notice that it's even there. In fact, general ideas along these lines were in Gibson's thinking, I suspect, and were certainly being articulated by Marvin Minsky and Seymour Papert in the late 1960s and early 1970s. But the details to these general ideas were never filled in. In a curious sense, this was because artificial intelligence remained decerebrate. It never realized that there was a level one theory to be discovered. It remained, and often still does, stuck fast in the mud of mechanistic explanations—where memory is held to be achieved by a neural net of some kind, or by a process in a computer, or by a set of procedures.

I don't know about this. These seem quite reasonable ways of explaining memory. Why do you find them so objectionable?

Well, in simple cases like eye movements, we can think in that rather direct fashion and get away with it. But it is very dangerous to hope that this type of thinking can ever give any real insight into the computational problems that the neural mechanisms are busy solving.

For example, to take a famous and elegantly expressed case, we might discuss Minsky's frames theory a little. A frame is essentially an item to which properties may be attached. For example, consider the following properties of an elephant considered as a frame:

Name	Clyde
Color	Pink
Weight	Large
Appetite	Large

Processes can also be attached to a frame and the contents of a frame may be interconnected or indexed in various ways. In his most stimulating article, Minsky (1975) describes how many "subjectively plausible" phenomena can be thought of in this way provided that the conceptual units involved are "large" enough. But I believe the approach is fundamentally flawed by its mechanism-based thinking. This harks back to our earlier point. If frames offered a representation and not just a mechanism, we would at once see what they are capable of representing and what they are not. This may still be done, but it has not yet been; until it has, we must be wary of ideas like frames or property lists. The reason is that it's really thinking in similes rather than about the actual thing—just as thinking in terms of different parts of the Fourier spectrum is a simile in vision for thinking about descriptions of an image at different scales. It is too imprecise to be useful. Real progress can only be made in such cases by precisely formulating the information-processing problems involved in the sense of our level one.

But your point isn't about just frames, is it? Doesn't it apply to almost the whole of artificial intelligence?

Yes, very true, and mechanism-based approaches are genuinely dangerous. The problem is that the goal of such studies is mimicry rather than true understanding, and these studies can easily degenerate into the writing of programs that do no more than mimic in an unenlightening way some small aspect of human performance. Weizenbaum (1976) now judges his program ELIZA to belong to this category, and I have never seen any reason to disagree. More controversially, I would also criticize on the same grounds Newell and Simon's (1972) work on production systems and some of Norman and Rumelhart's (1974) work on long-term memory.

Why, exactly?

The reason is this. If we believe that the aim of information-processing studies is to formulate and understand particular information-processing problems, then the structure of those problems is central, not the mechanisms through which their solutions are implemented. Therefore, in exploiting this fact, the first thing to do is to find problems that we can solve well, find out how to solve them, and examine our performance in the light of that understanding. The most fruitful source of such problems is operations that we perform well, fluently, and hence unconsciously, since it is difficult to see how reliability could be achieved if there was no sound, underlying method.

Unfortunately, problem-solving research has for obvious reasons tended to concentrate on problems that we understand well intellectually but perform poorly on, like mental arithmetic and cryptarithmetic* geometry-theorem proving, or the game of chess—all problems in which human skills are of doubtful quality and in which good performance seems to rest on a huge base of knowledge and expertise.

---

*For example, DONALD + GERALD = ROBERT. The object is to find the digit each letter stands for.

I argue that these are exceptionally good grounds for *not* yet studying how we carry out such tasks. I have no doubt that when we do mental arithmetic we are doing something well, but it is not arithmetic, and we seem far from understanding even one component of what that something is. I therefore feel we should concentrate on the simpler problems first, for there we have some hope of genuine advancement.

If one ignores this stricture, one is left with unlikely looking mechanisms whose only recommendation is that they cannot do something we cannot do. Production systems seem to me to fit this description quite well. Even taken on their own terms as mechanisms, they leave a lot to be desired. As programming languages, they are poorly designed and hard to use, and I cannot believe that the human brain could possibly be burdened with such poor implementation decisions at so basic a level.

This mimicry idea—is it just the business of thinking in similes that you mentioned before?

Yes, very much so. In fact, we could draw another parallel, this time between production systems for students of problem solving and Fourier analysis for visual neurophysiologists. Simple operations on a spatial-frequency representation of an image can mimic several interesting phenomena that seem to be accomplished by our visual systems. These include the detection of repetition, certain visual illusions, the notion of separate independent channels, separation of overall shape from fine local detail, and a simple expression of size invariance. The reason why the spatial-frequency domain is ignored by image analysts is that it is virtually useless for the main job of vision—building up a description of what is there from the intensity array. The intuition that visual physiologists lack, and which is so important, is for how this may be done. As a computing mechanism, a production system exhibits several interesting ideas—the absence of explicit subroutine calls, a blackboard-like communication channel, and some notion of a short-term memory.

However, just because production systems display these side effects (as a Fourier analysis "displays" some visual illusions) does not mean that they have anything to do with what is really going on. For example, I would guess that the fact that short-term memory can act as a storage register is probably the least important of its functions. I expect that there are several "intellectual reflexes" that operate on items held there about which nothing is yet known and which will eventually be held to be the crucial things about short-term memory.

Studying our performance in close relation to production systems seems to me a waste of time, because it amounts to studying a mechanism, not a problem. Once again, the mechanisms that such research is trying to penetrate will be unraveled by studying the problems that need solving, just as vision research is progressing because it is the problem of vision that is being attacked, not neural visual mechanisms.

What about human memory? You implied that the same type of misdirection was evident there. What did you mean?

I was referring to Norman and Rumelhart's work on the way information seems to be organized in long-term memory. Again the danger is that questions are not asked in relation to a clear information-processing problem. Instead, they are asked and answers proposed in terms of mechanisms—in this case the mechanism is called an "active structural network," and it is so simple and general as to be devoid of theoretical substance. Norman and Rumelhart may be able to say that such an "association" seems to exist, but they cannot say of what the association consists, nor do they say that to solve problem x (which we humans can solve) memory must be organized in a particular way; and that if this organization exists, certain apparent "associations" occur as side effects.

The phenomenological side of experimental psychology can do a valuable job in discovering facts that need explaining, including those about long-term memory, and the work of Shepard (1975), Rosch (1978), and Warrington (1975), for example, seems to me very successful at this; but like experimental neurophysiology, experimental psychology will not be able to explain those facts unless information-processing research has identified and solved the underlying information-processing problems, and I think that this is where we should be concentrating our energies.

What about Gunther Stent's work on the leech, though? Isn't that rather mechanism based, too?

Yes, but it is meant to be. It is concerned with elucidating the precise mechanism by which a leech swims. I value his work very highly, like that of the Tübingen group's on the housefly, but I think that early hopes of generalizing very far from these results have not borne fruit, and the reason is the levels story again. What higher nervous systems must do is determined by the information-processing problems that they must solve. We may have some simple leechlike oscillators inside us, and they may, to be very farfetched, eventually help us to understand some aspects of respiration. But such results will not teach us how we see.

One has a strong urge to tie explanation to structure eventually—that, of course, was the impact of molecular biology. It has to be done here, don't you think? Or do you see the endeavor as totally hopeless?

Yes, I agree it has to be done for the central nervous system, but I doubt if it can ever be done completely. The complexity barrier is just too great. But we have started to do it, don't forget! The zero-crossing detection and directional selectivity stories are very close to neurons. Don't be too impatient about the later things! As I said earlier, I bet you could never understand the fast Fourier transform as implemented in transistors on an IBM 370. I can only understand its formulas for about 10 minutes at a time—let alone understand a circuit diagram implementing them. One last word—I don't think that developmental and genetic programs will be able to be understood so directly in terms of underlying mechanisms. I would guess that some levels structure will eventually be needed to understand growth, because it is complicated.

Can we perhaps return to thinking rather specifically about visual perception and what actually happens when you see?

Well, are you happy with the primal sketch ideas?

I think so. The critical point seems to be that even very early vision is a highly symbolic activity. Assertions are actually made where lines end— yes, I've even accepted that terminology and am not too worried here about neurons!—and that objective lines and virtual lines are just as "real" as one another. Both can, for example, have their orientations detected and manipulated. Isn't this the idea?

Very much so. And if there is one more key idea, it is the idea of a place token and the ability to use crude selection criteria to group such tokens together and look for patterns, just as we saw in Figure 2–3.

I'm still a little unhappy about the representation of spatial relations—in the image, that is. I remember the discussion in Chapter 2 about coordinate systems, but was a little unconvinced. How can we be sure that important spatial information isn't lost?

Well, we have to be careful here, because I do not think much in the way of spatial relations *is* made explicit very early on. For example, certainly no intrinsic structure like the angle between two lines is. This type of information is not explicit in the full primal sketch, nor would the angle between two surfaces be in the 2½-D sketch. Such quantities do not belong to perception; their realm is that of the 3-D model representation. On the other hand, a few explicit spatial relations, like virtual lines between neighboring place tokens, often carry implicitly the entire geometry of the figure. This can be true even if the length measurements are very imprecise— perhaps only ranked by size.

A striking example of the richness of the information coming from a few clues about nearness is provided by the archaeological endeavors of Flinders Petrie. He measured the similarity of graves found along the Upper Nile by judging the number of characteristics shared by pieces of pottery found in each one. By using just this similarity information, techniques like multidimensional scaling can recover the times of burial quite accurately. The story makes fascinating reading (see Kendall, 1969), but we need note only that in two dimensions, the situation is even more constrained. I do not think there's much danger of the information being lost, but I do think only rather little spatial information is made explicit at the early stages.

So we derive the full primal sketch and then all those processes of Chapter 3 run to give us surface information? And roughly speaking, that is delivered in retinocentric polar coordinates, with perhaps slight differences for each process?

Yes, indeed, and the surface information from each process is combined in the 2½-D sketch, still in a retinocentric fashion but perhaps in a more convenient frame than the polar one. In a deep sense this is the end of pure autonomous perception. At this point the information is ready to be turned into a real 3-D model type of representation, a description that you can then remember.

I'm still unhappy about this tying-together process and the idea that from all that wealth of detail all you have left is a description. It sounds too cerebral somehow.

Well, the description can be arbitrarily rich—it's just a question of how much time and energy you spend on it. The other matter, that visual perception is just the formation of such descriptions—well, that is the conceptual leap I'm asking you to make. I personally find nothing important that this view fails to account for in general, and since we probably understand 20%–25% of the whole process already, I'm frankly ready to put my money on the rest of the process being of the same character. It's a conceptual leap, to be sure, but I think this view is worth trying to live with for a while, because thinking of visual perception in terms of the formation of particular kinds of descriptions explains so much so simply. But don't try to think about vision all the time in neurons! It's just impossible—the structure of vision is complicated enough at the top level, and outrageously so in terms of wiring.

And the result of those Chapter 3 processes, embodied in the 2½-D sketch, is the end of the immediate perception?

I think it's the right place to make the division, because up to here the processes can be influenced little or not at all by higher-order considerations. They deliver what they compute—no more, no less. The term *immediate perception* is a bit misleading, because these processes can take time—think of fusing a random-dot stereogram—but they do not involve scrutiny in Julesz's sense of an active intelligent examination of the image and comparison of its parts. This is compatible with the random-dot stereogram case, because we think that when the time to perceive one is long, most of the delay is due to random-walk-like movements of the eyes as they try to find somewhere to start fusion from.

If the 2½-D sketch changes every time you move your eyes, you lose it every time you move them (except possibly for small movements purely in depth). Isn't this a terribly wasteful thing to do?

It is wasteful, surely, but if you have the machinery there capable of recomputing the scene in real time, it doesn't matter that it's wasteful. In fact, it almost has to be this way, since the point of the 2½-D sketch is to assemble and represent incoming perceptual information, not to store it, and the alternative of economizing on computing power by using more memory is of no real use here. Just suppose, for example, that a 2½-D sketch had foveal resolution everywhere and was driven by a foveal retina in the usual way. Immediately, the memory has to contain out-of-date information (or nothing) in most of its capacity. This is not what the memory is for. Before resorting to almost any real storage, one must convert to something like the 3-D model representation, which is much more stable than the viewer-centered appearance of an object in a fleeting world. So the representation in which information from the different sources is assembled must be retinocentric and transient, it should have a foveal region where resolution is high, and it should reflect exactly and only what is coming in now.

These seem sensible distinctions, but they raise a difficulty I have in relating this to my own experience. The problem is that there seem to be so many different things going on in this model for perception, yet my perception has a unity, a oneness that I feel does not jibe with or at least is not reflected in these ideas. How is all the information tied together? How can one account for the unity of visual experience?

The basic idea is indeed that very many things are delivered through almost independent processes. At the 2½-D sketch level they are tied together, but only implicitly, whereas the next step is the creation of object-centered descriptions of the visible shapes (which is perhaps localized in a viewer-centered frame), and the description here *is* a unified object made up just by adding properties to its basic shape description, rather as a novelist adds to a description by adding qualifying adjectives.

What do you mean by being tied together "only implicitly"?

Simply that although different processes operate in different ways, there is a way of finding out when they are referring to the same visual object.

You mean if a raw primal sketch process finds an edge, and a color process finds its color, the relation between the two is implicitly available? I don't quite follow.

It's all a question of addressing. In most computers, you address information by specifying where to look for it. In some computers, you access a chunk of information by specifying pieces of the chunk. That is a content-addressable memory, and such memories are easy to build. What we might have here is a mixture of these two types of addressing—something like "the edge at roughly position (x,y) in the visual field with an orientation within, say, 30° of some given value". That would uniquely specify the edge in question both for the raw primal sketch representation and for the output for the color processes. In this way, we can tie the two things together, at least in principle.

What, dare I ask, about all those cortical areas? Isn't it natural to expect that they should each deal with a different process?

I would not be surprised.

Then what you are hinting at is, essentially, that up to this point each process runs, perhaps in a different cortical area (by now there are 10 at least, aren't there?), and that by presenting each with rough information, which could be rough position and orientation, you define precisely which visual object you are referring to.

Yes, that is the addressing problem.

And then, in addition, you get the precise information with which that particular area or process is concerned—the particular color or disparity, for example.

Exactly. And I think that the critical point about this is that the joining together of information is done symbolically.

What do you mean by that?

It's not like adding together the three impressions that a printer uses to make a printed page of color. We never see the colors of things smudged beyond their boundaries. The point is that the rough position and orientation information is used as an address. If you want the position of an item's exact boundary, you look at the raw primal sketch. If you want its color, you look at the color process.

I see. This idea means that assembling the information must be a very active process, doesn't it? Unless something specifically notices that stereo, zero-crossing x is a brown border, these two pieces of information will remain separate.

Yes, I think one has to ask for the color of x. And we must expect much of this to go on automatically as we move our eyes around. That is what the 2½-D sketch is partly for, after all—reducing information about surface geometry from many retinocentric processes to a single, more usable, viewer-centered form. At the same time, links to descriptions of other aspects of a surface are presumably made easily accessible, in preparation for the task of constructing a three-dimensional, object-centered description.

So you think it's likely that the actual combination isn't done until the 3-D model starts being constructed?

Yes.

It's as though strings are there to all the relevant information clearly marked and labeled, but you don't pull it all together unless you start making a 3-D model.

Which may be a very coarse one or parts of a very fine one. And in the same way, one might expect other properties to be coarse (for example, greenish) or quite fine (for example, a specific shade of green).

But how does this correspond to my perceptual experience? My experience appears to be complete, not at all the halfway, ill-defined, fragmented sort of thing that you describe.

Well, first remember that our visual processes can work extremely rapidly. The time between requesting information about a part of the visual field and moving the eyes there, getting it, and linking it to a 3-D model is probably usually under half a second. The second thing is, How much of a novel scene can you recall if you look at it only very briefly? Not very much! Its coarse organization, or perhaps one or two details. And once you close your eyes, the richness is gone, isn't it? I think that the richness corresponds to what is available now, at the pure perceptual level, and what you can remember immediately is much more closely related to the 3-D model description that you create for it while your eyes are open.

I begin to see more clearly the force of the idea that perception is the construction of a description.

Yes, that is the core of the thing, and a really important point to come to terms with.

But let's suppose you're right, then, that the 2½-D sketch is retinocentric and that you compute out of it little 3-D models and hang them up in a space frame centered on you. What happens when you move your eyes a lot?

One thing is that the finely detailed shape that you were just looking at—suppose it was a porcelain cat—and for which you have just built up an elaborate description is reduced to a blob in the image when you turn your eyes to study its neighbor, a porcelain dog. If the blob can be distinguished confidently in the 2½-D sketch, then I would guess that there is a process that maintains the link between it and the 3-D model you've just finished building, so that if that blob moves, you know immediately *what* has moved.

But how on earth do you do that with neurons?

Hold on there—we'll face that next. But note that basically, it's not difficult computationally.

But to tie all this up with what it feels like to see—that is difficult to swallow.

It grows on you. That first step, that vision is the computation of a description, is the crucial one. Once you have accepted that, you can go on to study exactly what description and how to make it.

And again it's not at all easy for me to allow you to talk so much about computation. The brain, after all, is made of neurons, not silicon chips. But I suppose I'll get used to it. Still, if vision is the construction of descriptions, they must be implemented neurally, mustn't they? So couldn't one hope to look for neurophysiological correlates of the 2½-D sketch or of a piece of a 3-D model? That, I would find convincing.

It would be marvelous if the implementation were that simple—close to Barlow's neural dogma! My own guess is that it is more like that than a Hebb cell assembly.

There's another more general point that is still troubling me, and it has to do with the temporal continuity of perceptual experience. I understand very well how you think continuity can be held between eye movements and so forth, but this avoids the larger question of pure continuity over time. Why, if I look at a tree, do I see it continuously as the same tree? Presumably I could at any moment start a new 3-D model for it, in which case I ought to experience it as a new tree in the same spot as the old one. Yet I don't. Do you have any comments?

The permanence of the visual world—the continuity of objects in time—is an awfully important aspect of vision, and I think it's just part of our reflexes as adults that we assume it. In fact, whole aspects of processing are based on discovering and exploiting the continuity relations—the correspondence processes of Chapter 3, for example.

Another general point. You deal only with shape here. What about the recognition as being the same thing of two objects that have different shapes but the same function—like two different kinds of chair?

This theory has nothing to say about semantic recognition, object naming or function, though that is most certainly a path almost as useful as shape determination for recognition in the external world (Warrington and Taylor, 1978). I think that the problems of understanding what we mean by the semantics of an object are fascinating, but I also think that they are very difficult indeed and at present much less accessible than the problems of visual perception.

If the overall scheme you describe is correct, would we be able to say anything about painting and drawing using this knowledge of what the visual system does with its input? Might it help to teach these skills, for example?

Perhaps, although I would hate to commit myself to a definite view yet. Nevertheless, it is interesting to think about which representations the different artists concentrate on and sometimes disrupt. The pointillists, for example, are tampering primarily with the image; the rest of the scheme is left intact, and the picture has a conventional appearance otherwise. Picasso, on the other hand, clearly disrupts most at the 3-D model level. The three-dimensionality of his figures is not realistic. An example of someone who operates primarily at the surface representation stage is a little harder—Cezanne perhaps?

With respect to other problems such as natural language, how universal is the approach you are advocating? How far can it be taken? What kind of things would it be likely to fail at?

Systems that are not modular. Things like the process by which a chain of amino acids folds to form a protein—that is to say complex, interactive systems with many influences that cannot be neglected. A burning issue in the study of natural language understanding is, of course, How modular is it, and what are the modules?

Yes, I suppose modularity is the key, but also fluency of some kind must be important, mustn't it? If a process doesn't flow well, smoothly, unattended, and without having to be patched by conscious interference, then it may have no clean theory, and that might turn it into the protein-folding class of difficult-to-understand theories. But to return to natural language, what modules have been found there?

It's not clear, and some claim it's inherently not modular and should be viewed much more heterarchically.

Doesn't that sound a little reminiscent of the early days of vision?

Yes, I'm afraid so. But there do seem to be modules and rules for modules emerging at the early level—rules for syllable formation, prosodics, and most famously Chomsky's analysis of syntax.

But how much of a module is syntax? Don't artificial intelligence workers like Schank claim that syntax is not a separable module at all?

Yes, and it is clear that the syntactical decoding of a sentence cannot proceed entirely independently of its semantical analysis. But a good case is being built up that the *amount* of interaction necessary between the two is small, and the types of questions about syntax that must be answered seem to be of a quite simple kind—for example, Should a particular clause refer to noun phrase one or to noun phrase two? Marcus (1980) was the first to explore these problems in detail; and he has shown that a very successful module can be made out of a parsing system. Above the level of syntax, however, few hints are currently available about what the modularity is, but I'm sure it must be present.

Why has artificial intelligence shown such resistance to traditional Chomskian approaches to syntactical analysis? Only Marcus seems to have embraced it.

I think there are two reasons. First, it is easy to construct examples in which syntax cannot be analyzed without some concurrent semantical analysis. Thus, syntax is not a truly isolated module, and this fact led the artificial intelligence people to jump to the opposite conclusion, that syntax is not a module at all. This is incorrect—the true situation seems to be that syntax is almost a module, requiring some interactions with semantics but only a very small number of types of interaction.

The second reason is our old friend, the levels. Noam Chomsky's transformational grammar is a level one theory, that is in no way concerned with *how* syntactical recognition should be implemented. It merely gives rules for stating *what* the decomposition of an arbitrary sentence should be. Chomsky's description of it as a competence theory was his way of saying this.

However, the levels idea has not been properly understood by computational linguists. Indeed, one of Winograd's reasons for rejecting Chomsky was that he could not invert the transformational structure and turn it into a parser! This observation could be made only by someone who failed to understand the distinction between levels one (what and why) and two (how). Winograd is not to be singled out for this error, however; everyone in artificial intelligence made it, and now that the linguists themselves are becoming computationally aware, they are falling into the same trap. The result is, I fear, that natural language computer programs have contributed rather little to natural language understanding, with the recent exception of Marcus (1980), who has begun to construct a genuine level-two theory of the parsing algorithm we use.

What do you feel are the most promising approaches to semantics?

Probably what I call the problem of multiple descriptions of objects and the resolution of the problems of reference that multiple descriptions introduce.

Could you expand on this?

Well, like many others in the field, I expect that at the heart of our understanding of intelligence will lie at least one and probably several important principles about organizing and representing knowledge that in some sense capture what is important about the general nature of our intellectual abilities. While still somewhat vague, the ideas that seem to be emerging are as follows:

1. The chunks of reasoning, language, memory, and perception ought to be larger than most recent theories in psychology have allowed (Minsky, 1975). They must also be very flexible, and incorporating this requirement precisely will not be easy.

2. The perception of an event or of an object must include the simultaneous computation of several different descriptions of it that capture diverse aspects of the use, purpose, or circumstances of the event or object.

3. The various descriptions referred to in point 2 include coarse versions as well as fine ones. These coarse descriptions are a vital link in choosing the appropriate overall scenarios demanded by point 1 and in correctly establishing the roles played by the objects and actions that caused those scenarios to be chosen.

An example will help to make these points clear. If one reads

> The fly buzzed irritatingly on the windowpane.
> John picked up the newspaper.

the immediate inference is that John's intentions toward the fly are fundamentally malicious. If he had picked up the telephone, the inference would be less secure. It is generally agreed that an "insect-damaging" scenario is somehow deployed during the reading of these sentences, being suggested in its coarsest form by the fly buzzing irritatingly. Such a scenario will contain a reference to something that can squash an insect on a brittle surface—a description that a newspaper fits, but not a telephone. We might therefore conclude that when the newspaper is mentioned (or, in the case of vision, seen) not only is it described internally as a newspaper and some rough 3-D model description of its shape and axes set up, but it is also described as a light, flexible object with area. Because the second sentence might have continued "and sat down to read," the newspaper must also be described as reading matter; similarly, it must also be described as a combustible article, as something that rustles, and so forth. Since we do not usually know in advance what aspect of an object or action is important, it follows that most of the time a given object will give rise to several different coarse internal descriptions. Similarly for actions. It may be important to note that the description of fly swatting or reading or fire lighting does not have to be attached to the newspaper—a description of the newspaper is merely available that will match its role in each scenario.

Why do you think this must be so?

Because the importance of a primitive, coarse catalogue of events and objects lies in the role that such coarse descriptions play in the ultimate access and construction of perhaps exquisitely tailored specific scenarios, rather in the way that a general 3-D animal model can finish up as a very specific Cheshire cat after due interaction between the image and information stored in the catalogue of models. What existed as little more than a malicious intent toward the innocent fly after the first sentence becomes, with the additional information about the newspaper, a very specific case of fly squashing. Exactly how this is best done and exactly what descriptions should accompany different words or perceived objects is not yet known.

What about other types of processing that the brain does, such as the planning and execution of behavior? Might not these be simpler places to start looking for modules? After all, semantics is one of the most advanced areas of human ability, so it's not unreasonable to expect that it may be complex. I would try something simpler.

I think that may be excellent advice, and it reminds me of a fascinating experiment done some time ago by Stamm (1969). He was running what is called a delayed-response task (see Figure 7–2). In this, a scrap of food

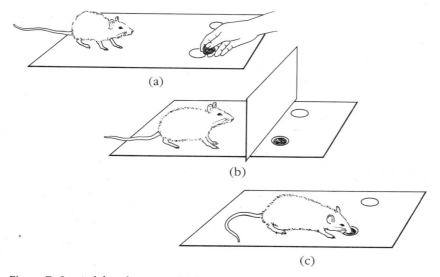

(a)

(b)

(c)

Figure 7–2.   A delayed-response task. A scrap of food is placed under one of the wells in full view of the animal. Then a screen descends for a period. When the screen is raised, the animal has to choose one of the wells. If he looks under the correct one, he is rewarded with the food.

is placed in one of two wells, a screen comes down, a delay ensues, the screen lifts, and the animal is then free to choose the well in which he thinks the food is hidden. Certain portions of the prefrontal cortex are known to be involved in this task, and the animal cannot perform it if they are removed. Stamm used a technique—depolarization—whereby he could effectively disable these areas for the precise period he desired. He asked, When must the area be operating for the task to be carried out? It turned out that the animal had to have its area working as the screen came down at the beginning of the delay; if the area was knocked out at any other time, it mattered either much less or not at all!

One possible way of thinking about this experiment is this. Any real-time computer must be able to construct plans, set them up for execution under the appropriate conditions, and set the triggers for them. One cannot recompute everything afresh each time, and indeed the structure of a human personality consists in part of thousands of such little plans, all set to run a person's behavior if the appropriate conditions arise. But something must *write* these plans, and here in Stamm's experiment maybe we are seeing a simple example of this happening. As the wells are removed from view, the animal writes into its set of plans to go to the appropriate well when it can. A simple plan, but a plan nevertheless.

If we carry this idea a little further, we see that it splits the central system into what one might call the planner and the executive. The planner writes plans and their triggers to the executive, which, when the time and conditions are ripe, executes them. Is it too absurd to suggest that during hypnosis the executive becomes externally programmable and that this is why it is possible to set up plans under hypnosis that are executed later when the assigned conditions are met? The idea bears reflection, at least.

That is an interesting idea. I have not seen any previous explanation about why it should be possible to "program" someone at all, and your suggestion is certainly plausible. But what about the stereotyped nature of the programming? We are ourselves very flexible, are we not? It's a little difficult to reconcile that with a set of programmed responses.

I think that depends entirely on how large, rich, and subtle the set of responses has grown to be. If there is wide variety of responses and considerable ability to act differently in only subtly different situations, then we would be called flexible—and freer, incidentally, since we would be taking a wider range of relevant information appropriately into account. If we take no information (random response) or only one piece (compulsive response), then we are certainly not acting flexibly or freely.

That seems a sensible distinction. But as we move closer to saying the brain is a computer, I must say I do get more and more fearful about the meaning of human values.

Well, to say the brain is a computer is correct but misleading. It's really a highly specialized information-processing device—or rather, a whole lot of them. Viewing our brains as information-processing devices is not demeaning and does not negate human values. If anything, it tends to support them and may in the end help us to understand what from an information-processing view human values actually are, why they have selective value, and how they are knitted into the capacity for social mores and organization with which our genes have endowed us.

Glossary

Action potential The self-regenerating electrical spike that propagates down an axon, thus transmitting a signal from one cell to the next via a synapse. The mechanism of the conduction of this signal was elucidated by A. L. Hodgkin and A. F. Huxley.

Adjunct relation A flexible way of specifying the relative positions of two axes in a 3-D model, usually used to relate a component axis to the model's principal axis (see Figures 5–4 and 5–5).

Area 17 The striate cortex.

Band-pass channel A filter that allows only a particular band of frequencies to pass through it.

Bit map A convenient way of representing rough position in an image. A two-dimensional array is set in correspondence with the x- and y-coordinates in an image, and the positions of items are represented by putting a 1 at the appropriate point in the array.

Blocks world The visual domain of matte white, plane-faced blocks viewed against a dark background. Much early machine vision was conducted in this domain.

Complex cells   An orientationally sensitive class of cells in the visual cortex discovered by Hubel and Wiesel. These cells are more complicated than simple cells in that their response is not a linear function of the spatial stimuli falling within their receptive fields, but they do not show any particular sensitivity to the termination of edges and bars.

Component axis   A subsidiary axis of a 3-D model, for example, the neck axis in a quadruped 3-D model.

Conjunctive eye movements   Eye movements that change the average direction of gaze of the two eyes.

Contour generator   The locus of points on a visible surface that gives rise to a contour in the surface's image.

Convolution (*)   Formally, the convolution of two functions $f(x)$ and $g(x)$ is given by $f*g(x) = \int f(x')\,g(x - x')\,dx'$. For the case of an image, its meaning may be visualized more easily in terms of receptive fields. Suppose we place at position (x,y) in an image a weighted receptive field, perhaps with a center–surround organization. This field adds up linearly the contributions from each part of the image as "seen" through the receptive field—that is, points in the center receive a strong positive weighting and those in the surround a weaker negative weighting. The result is the value of the convolution of the image with the function represented by the receptive field weights at that one particular point (x,y). Thus to calculate directly the convolution of the whole image, that is, for every point (x,y), can be a computationally expensive process.

Cooperative Algorithm   A nonlinear algorithm in which purely local operations appear to cooperate to produce order on a global scale in a well-regulated manner. So called after cooperative phenomena in physics, like the Ising model of ferromagnetism, superconductivity, and phase transitions in general. Cooperative algorithms share many characteristics with these phenomena.

$\nabla^2 G$ The Laplacian operator applied to a Gaussian distribution in two dimensions. The result has a Mexican-hat shape and can be written:

$$\nabla^2 G(r) = -1/\pi\sigma^4\,(1 - r^2/2\sigma^2)\exp(-r^2/2\sigma^2)$$

It is illustrated in Figure 2–9.

Depth   Viewer's subjective impression of the distance to the visible surface.

Description   A description is the result of applying a representation to a particular entity (see *Representation*).

Differential operators   Spatial differential operators like $\partial/\partial x$ and $\partial/\partial y$ can be realized approximately by convolution operators with appropriately shaped receptive fields. Some of these are diagrammed in Figure 2–11.

Dip   See *Slant*.

Disjunctive eye movements Eye movements that change the relative directions of gaze of the two eyes, making them more convergent or more divergent, while leaving their average direction of gaze unchanged.

Disparity If two items are positioned at different distances from the viewer, the relative positions of their images in the two eyes will differ. This difference, usually measured in minutes of arc, is called disparity. A 1-in. depth difference at a distance of 5 ft straight ahead will produce a disparity of about $1'$.

Distance Usually refers to objective three-dimensional distance from the viewer to the visible surface.

DOG A function composed of the difference of two Gaussian distributions. Such functions are thought to describe the shape of the receptive fields of the retinal ganglion cells and the shape of the receptive fields associated with Wilson's four-channel model of early visual processing. They are very close in shape to the ideal function $\nabla^2 G$ (see Figure 2–16).

Eccentricity Usually refers to the angle out from the central fovea of the retina.

Emittance angle The angle of emittance e is the angle between a ray of light reflected from a surface and the normal to the surface.

Entropy Roughly speaking, the entropy of a probability distribution measures how chaotic the distribution is. Thus the entropy is low if the distribution is concentrated around one value, and zero if it is concentrated on exactly one value. A uniform distribution has the maximum entropy. Formally, for a discrete distribution with outcomes $1, 2, \ldots, i, \ldots$ having probabilities $p_1, p_2, \ldots, p_i, \ldots$, the entropy $q(p)$ of the distribution is given by $q(p) = \sum_i - p_i \log_2 p_i$.

Fast Fourier transform A fast digital algorithm for carrying out a Fourier transform on a discrete array whose dimensions are a power of 2. It was devised by J. M. Cooley and T. W. Tukey. Recently S. Winograd devised an even faster algorithm known as the very fast Fourier transform (VFFT).

Frontal plane The plane lying perpendicular to the line of sight.

Gaussian (G) The so-called Gaussian or normal distribution has the form $G(r) = (1/2\pi\sigma^2) \exp(-r^2/2\sigma^2)$ in two dimensions.

Gradient space A way of representing three-dimensional surface orientation by a point on a two-dimensional graph, usually denoted by (p,q) (see Section 3.8 and especially Figure 3–73).

High-pass filter A filter that allows through only the high frequencies in a signal (these could be high spatial or temporal frequencies).

Horopter There are several definitions of the horopter, but in this book it refers to the zero-disparity surface for the current positions of the eyes.

Hyperacuity Humans can carry out a variety of tasks to accuracies that are more precise than the dimensions of the retinal cones from which the information originates. Foveal cones have a diameter of about 27″, yet many tasks yield accuracies of around 5″, and stereoscopic acuity may be as good as 2″. Such tasks are said to fall within the range of hyperacuity.

Incidence angle The angle of incidence i is the angle between a ray of incident light and the normal to the surface.

Isoluminance contour A reflectance map usually consists of contours of constant luminance, or isoluminance contours, plotted in (p,q) or gradient space.

Isotropic The same in all directions.

Just noticeable difference (JND) A JND experiment tests discrimination ability for a parameter over a range by measuring at each point in the range the amount the parameter has to be changed before the difference is noticed. The two test stimuli are usually juxtaposed.

Lambertian A Lambertian surface is a perfect diffuser, the reflective analogue of a blackbody radiator. Its reflectance function $\phi(i,e,g)$ is $\cos i$ and depends only on i, the angle of incidence of the illumination.

Laplacian (∇^2) Formally, $\nabla^2 = \partial^2/\partial x^2 + \partial^2/\partial y^2$. It is the lowest-order isotropic differential operator.

Lateral geniculate body (LGN) The main visual nucleus between the eye and the brain. It is fed by the optic nerve, which consists of axons of the retinal ganglion cells. The axons emerging from the LGN, called the optic radiations, project to the striate cortex in the monkey and in man.

Low-pass filter A filter that allows through only the low frequencies in a signal (these could be low spatial or temporal frequencies).

Model axis An axis, associated with a 3-D model, that defines the overall extent of the shape that the model represents.

Modulation transfer function (MTF) The amplitude of the Fourier transform of a filter or function. The MTF is useful because by looking at its graph, one can tell at a glance which frequencies are passed and which are suppressed by the filter.

Occluding contour A contour in an image that is formed by an occluding edge.

Panum's area The disparity range over which stereoscopic fusion can be achieved without eye movements.

Panum's limiting case See Figure 3–19.

Phase angle The phase angle g is the angle between the incident and emitted rays.

Place token A token that marks a point of interest in an image. Such tokens have a position, and they may possess various other properties.

They are thought to be constructed during the early analysis of the spatial arrangement of an image.

Primal sketch   A representation of the two-dimensional image that makes explicit the amount and disposition of the intensity changes there. The representation is hierarchical, the primitives at the lowest level representing raw intensity changes and their local geometrical structure, and those at the higher levels capturing groupings and alignments occurring among the lower items (see Figure 2–7).

Principal axis   The axis of a 3-D model that most component axes adjoin, for example, the torso axis of a quadruped 3-D model.

Reflectance function   Usually denoted by $\phi(i,e,g)$, the reflectance function associated with a surface specifies what fraction of the incident light is reflected under different conditions of viewing and illumination. See Figure 3–75 and Section 3.8.

Reflectance map   A graph that relates image intensities to surface orientation, not usually in a one-to-one manner. Figures 3–76 to 3–79 show some examples.

Representation   A representation of a set of entities S is a formal scheme for describing them, together with rules that specify how the scheme applies to any particular one of the entities.

Retinal ganglion cells   The final layer of cells in retinal processing. The axons of these cells leave the retina through the so-called blind spot and form the optic nerve.

Retinex   Edwin Land's term for the processing of an image by removing all gradual changes in intensity, such as might be caused by changes in illumination, while leaving all sudden changes, such as might be due to changes in reflectance.

Rhodopsin   The light-sensitive visual pigment in the rods and cones, the receptors of the eye.

Saccade   A conjunctive eye movement can either be smooth or occur in a preprogrammed ballistic jump called a saccade, which takes about 160 ms to program internally. Disjunctive eye movements, on the other hand, are always smooth and are under continuous control based on feedback about the disparity between the current vergence angle and the desired vergence angle.

Shape   The geometry of an object's physical surface.

Simple cells   A class of orientationally sensitive cells in the striate cortex, discovered by Hubel and Wiesel and defined as simple cells by the linearity of their response to stimuli falling in their receptive fields.

Slant   The angle by which a plane slants or dips away from the viewer's frontal plane. Also called *dip*.

Spatial frequency   The Fourier transform of a signal that varies in time represents that signal as the sum of sine and cosine waves, each at a different temporal frequency. If the signal varies in space rather than time, like a single image for instance, then the components by which its Fourier transform represents it are its spatial frequencies, which can be thought of as oriented sine wave gratings.

Spatial frequency channel   A channel that allows only a limited range of spatial frequencies to pass through it. The early parts of the human visual system incorporate a number of spatial frequency channels, each of which is effectively less than two octaves wide; that is, the ratio of the maximum to the minimum frequency passed is less than 4 to 1.

Striate cortex   The primary visual cortical receiving area in the monkey and in man. So called because of the stria of Genarii, a band of white matter running through only this region of the cortex.

Surface contour   The image of a contour lying on a visible surface.

Synapse   The junction between nerve cells occurring between the axon of one and the dendrite or soma (cell body) of the next. Most synapses are chemical—that is, messages are transmitted across them by release of a chemical from the axon terminal—but some synapses are electrical.

Tachistoscope   A device used in psychophysical experiments for exposing the subject to brief visual stimuli.

3-D Model   The basic building block of the 3-D model representation. It specifies a model axis, which defines the overall extent of the shape; the relative sizes and spatial arrangement of the (few) component axes of the model; and pointers to the shapes associated with these axes (see the boxes in Figure 5–3).

3-D Model representation   An object-centered representation for shapes that includes the use of volumetric primitives of various sizes, arranged in a modular, hierarchical organization (see Figure 5–3).

Tilt   The direction in which the surface slants away from the frontal plane.

2½-D Sketch   A viewer-centered representation of the depth and orientation of the visible surfaces, including contours of discontinuity in these parameters (see Figure 3–12).

Vergence eye movements   See *disjunctive eye movements.*

Volterra series   A way of representing a certain class of nonlinear systems. Provided a function is sufficiently smooth, that is, has no discontinuities or threshold or decision points, it can be expressed as a series of polynomial terms; for example,

$$f(x,y) = ax + by + cxy + dx^2y + \dots$$

In the particular case of the flight control system of the housefly, only the lower-order terms are important.

W cells, X cells, Y cells   The three classes of retinal ganglion cells. The X-cell–Y-cell distinction was originally discovered by C. Enroth-Cugell and J. D. Robson, the W cells being discovered later. These classes have been isolated anatomically and physiologically. Y cells have the largest cell bodies, the largest receptive fields and are the least frequent (about 4% of the total ganglion cells). They have a high conduction velocity and relatively transient responses, and are subject to the shift and McIlwain effects, insensitive to color, and relatively more common in the periphery. X cells are smaller than Y cells, have smaller receptive fields, and occur more frequently than Y cells (about 60% of retinal ganglion cells are X cells). They have medium conduction velocity and relatively sustained responses, and are not so subject to the shift and McIlwain effects, often color sensitive, and relatively more common toward the fovea. W cells are very small cells with slow conduction velocities, forming perhaps 40% of the ganglion cell population. These cells, which are difficult to record from, are often directionally selective and may have other rather specific properties. Many of these cells project to the superior colliculus.

Zero-crossing   Point where a function's value changes its sign.

Bibliography

Adrian, E. D. 1928. *The Basis of Sensation*. London: Christophers. (Reprint ed. New York: Hafner, 1964).

Adrian, E. D. 1947. *The Physical Background of Perception*. Oxford: Clarendon Press.

Agin, G. 1972. Representation and description of curved objects. Stanford Artificial Intelligence Project Memo AIM–173. Stanford, Ca.: Stanford University.

Anstis, S. M. 1970. Phi movement as a subtraction process. *Vision Res. 10*, 1411–1430.

Attneave, F. 1974. Apparent movement and the what-where connection. *Psychologia 17,* 108–120.

Attneave, F., and G. Block. 1973. Apparent motion in tridimensional space. *Percept. & Psychophys. 13*, 301–307.

Austin, J. L. 1962. *Sense and Sensibilia*. Oxford: Clarendon Press.

Barlow, H. B. 1953. Summation and inhibition in the frog's retina. *J. Physiol. (Lond.)* *119,* 69–88.

Barlow, H. B. 1972. Single units and sensation: a neuron doctrine for perceptual psychology? *Perception 1,* 371–394.

Barlow, H. B. 1978. The efficiency of detecting changes in random dot patterns. *Vision Res. 18,* 637–650.

Barlow, H. B. 1979. Reconstructing the visual image in space and time. *Nature 279,* 189–190.

Barlow, H. B., C. Blakemore, and J. D. Pettigrew. 1967. The neural mechanism of binocular depth discrimination. *J. Physiol. (Lond.) 193,* 327–342.

Barlow, H. B., R. M. Hill, and W. R. Levick. 1964. Retinal ganglion cells responding selectively to direction and speed of image motion in the rabbit. *J. Physiol. (Lond.) 173,* 377–407.

Barlow, H. B., W. R. Levick. 1965. The mechanism of directional selective units in rabbit's retina. *J. Physiol. (Lond.) 178,* 477–504.

Beck, J. 1972. *Surface Color Perception.* Ithaca, N.Y.: Cornell University Press.

Berry, R. N. 1948. Quantitative relations among vernier, real depth, and stereoscopic depth acuities. *J. Exp. Psychol. 38,* 708–721.

Binford, T. O. 1971. Visual perception by computer. Paper presented at the IEEE Conference on Systems and Control, December 1971, Miami.

Bishop, P.O., J. S. Coombs, and G. H. Henry. 1971. Responses to visual contours: Spatio-temporal aspects of excitation in the receptive fields of simple striate neurons. *J. Physiol. (Lond.) 219,* 625–657.

Blomfield, S. 1973. Implicit features and stereoscopy. *Nature, New Biol. 245,* 256.

Blum, H. 1973. Biological shape and visual science, part 1. *J. Theor. Biol. 38,* 205–287.

Bouguer, P. 1757. Histoire de l'Academie Royale des Sciences, Paris; and *Traite d'Optique sur la Gradation de la Lumière* (Ouvrage posthume de M. Bouguer)., l'Abbé de Lacaille, Paris, 1760.

Braddick, O. J. 1973. The masking of apparent motion in random-dot patterns. *Vision Res. 13,* 355–369.

Braddick, O. J. 1974. A short-range process in apparent motion. *Vision Res. 14,* 519–527.

Braddick, O. J. 1979. Low- and high-level processes in apparent motion. *Phil. Trans. R. Soc. Lond. B 290,* 137–151.

Brady, M. 1979. Inferring the direction of the sun from intensity values on a generalized cone. *Proc. Int. Joint Conf. Art. Intel., IJCAI–79,* 88–91.

Braunstein, M. L. 1962. Depth perception in rotation dot patterns: Effects of numerosity and perspective. *J. Exp. Psychol. 64,* 415–420.

Breitmeyer, B., and L. Ganz. 1977. Temporal studies with flashing gratings: Inferences about human transient and sustained channels. *Vision Res. 17,* 861–865.

Brindley, G. S. 1970. *Physiology of the Retina and Visual Pathway.* Physiological Society Monograph no. 6. London: Edwin Arnold.

Brodatz, P. 1966. *Textures: A Photographic Album for Artists and Designers.* New York: Dover.

Campbell, F. W. C. and J. Robson. 1968. Application of Fourier analysis to the visibility of gratings. *J. Physiol. (Lond.) 197,* 551–566.

Campbell, F. W. C. 1977. Sometimes a biologist has to make a noise like a mathematician. *Neurosciences Res. Prog. Bull. 15,* 417–424.

Carey, S., and R. Diamond. 1980. Maturational determination of the developmental course of face encoding. In *Biological Bases of Mental Processes,* D. Kaplan, ed., 1–7. Cambridge, Mass.: MIT Press.

Chomsky, N. 1965. *Aspects of the Theory of Syntax.* Cambridge, Mass.: MIT Press.

Chomsky, N., and H. Lasnik. 1977. Filters and control. *Linguistic Inquiry 8,* 425–504.

Clarke, P. G. H., I. M. L. Donaldson, and D. Whitteridge. 1976. Binocular mechanisms in cortical areas I and II of the sheep. *J. Physiol. (Lond.) 256,* 509–526.

Clocksin, W. F. 1980. Perception of surface slant and edge labels from optical flow: A computational approach. *Perception 9,* 253–269.

Corbin, H. H. 1942. The perception of grouping and apparent motion in visual space. *Arch. Psychol. Whole No. 273.*

Crick, F. H. C., D. Marr, and T. Poggio. 1980. An information processing approach to understanding the visual cortex. In *The Cerebral Cortex*, Ed. F. O. Schmitt and F. G. Worden. (The Proceedings of the Neurosciences Research Program Colloquium held in Woods Hole, Mass., May 1979.) Cambridge, Mass.: MIT Press.

Dev, P. 1975. Perception of depth surfaces in random-dot stereograms: A neural model. *Int. J. Man-Machine Stud. 7*, 511–528.

DeValois, R. L. 1965. Analysis and coding of color vision in the primate visual system. *Cold Spring Harbor Symp. Quant. Biol. 30*, 567–579.

DeValois, R. L., I, Abramov, and G. H. Jacobs. 1966. Analysis of response patterns of LGN cells. *J. Opt. Soc. Am. 56*, 966, 977.

DeValois, R. L., I. Abramov, and W. R. Mead. 1967. Single cell analysis of wavelength discrimination at the lateral geniculate nucleus in the macaque. *J. Neurophysiol. 30*, 415–433.

Dreher, B. and K. J. Sanderson. 1973. Receptive field analysis: Responses to moving visual contours by single lateral geniculate neurons in the cat. *J. Physiol. (Lond.) 234*, 95–118.

Enroth-Cugel, C. and J. D. Robson. 1966. The contrast sensitivity of retinal ganglion cells of the cat. *J. Physiol. (Lond.) 187*, 517–522.

Evans, R. M. 1974. *The Perception of Color*. New York: Wiley.

Felton, T. B., W. Richards, and R. A. Smith, Jr. 1972. Disparity processing of spatial frequencies in man. *J. Physiol. (Lond.) 225*, 349–362.

Fender, D., and B. Julesz. 1967. Extension of Panum's fusional area in binocularly stabilized vision. *J. Opt. Soc. Am. 57*, 819–830.

Forbus, K. 1977. Light source effects. MIT A.I. Lab Memo 422.

Fram, J. R., and E. S. Deutsch. 1975. On the quantitative evaluation of edge detection schemes and their comparison with human performance. *IEEE Transactions on Computers C–24*, 616–628.

Freuder, E. C. 1974. A computer vision system for visual recognition using active knowledge. MIT A.I. Lab Tech. Rep. 345.

Frisby, J. P., and J. L. Clatworthy. 1975. Learning to see complex random-dot stereograms. *Perception 4*, 173–178.

Frisby, J. P., and J. E. W. Mayhew. 1979. Does visual texture discrimination precede binocular fusion? *Perception 8*, 153–156.

Galambos, R., and H. Davis. 1943. The response of single auditory-nerve fibres to acoustic stimulation. *J. Neurophysiol.* *7*, 287–303.

Gibson, J. J. 1950. *The Perception of the Visual World.* Boston: Houghton Mifflin.

Gibson, J. J. 1958. Visually controlled locomotion and visual orientation in animals. *Brit. J. Psych. 49*, 182–194.

Gibson, J. J. 1966. *The Senses considered as Perceptual Systems.* Boston: Houghton Mifflin.

Gibson, J. J. 1979. *The Ecological Approach to Visual Perception.* Boston: Houghton Mifflin.

Gibson, J. J., and E. J. Gibson. 1957. Continuous perceptive transformations and the perception of rigid motion. *J. Exp. Psychol. 54*, 129–138.

Gibson, E. J., J. J. Gibson, O. W. Smith, and H. Flock. 1959. Motion parallax as a determinant of perceived depth. *J. Exp. Psychol. 8*, 40–51.

Gibson, J. J., P. Olum, and F. Rosenblatt. 1955. Parallax and perspective during aircraft landings. *Am. J. Psychol. 68*, 372–385.

Gilchrist, A. L. 1977. Perceived lightness depends on perceived spatial arrangement. *Sicence 195*, 185–187.

Glass, L. 1969. Moire effect from random dots. *Nature 243*, 578–580.

Glass, L., and R. Perez. 1973. Perception of random dot interference patterns. *Nature 246*, 360–362.

Glass, L., and E. Switkes. 1976. Pattern perception in humans: Correlations which cannot be perceived. *Perception 5*, 67–72.

Goodwin, A. W., G. H. Henry and P. O. Bishop. 1975. Direction selectivity of simple striate cells: Properties and mechanism. *J. Neurophysiol. 38*, 1500–1523.

Gordon, D. A. 1965. Static and dynamic visual fields in human space perception. *J. Opt. Soc. Am. 55*, 1296–1303.

Gouras, P. 1968. Identification of cone mechanisms in monkey ganglion cells. *J. Physiol. (Lond.) 199*, 533–547.

Granit, R., and G. Svaetichin. 1939. Principles and technique of the electrophysiological analysis of colour reception with the aid of microelectrodes. *Upsala Lakraef Fath. 65*, 161–177.

Green, B. F. 1961. Figure coherence in the kinetic depth effect. *J. Exp. Psychol. 62*, 272–282.

Gregory, R. L. 1970. *The Intelligent Eye*. London: Weidenfeld & Nicholson.

Grimson, W. E. L. 1979. Differential geometry, surface patches and convergence methods. MIT A.I. Lab. Memo 510. (Available as *From Images to Surfaces: A Computational Study of the Human Early Visual System*. Cambridge: MIT Press 1981.)

Grimson, W. E. L. 1980. A computer implementation of a theory of human stereo vision. MIT A.I. Lab. Memo 565. *Phil. Trans. Roy. Soc. Lond. B292*, 217–253.

Grimson, W. E. L., and D. Marr. 1979. A computer implementation of a theory of human stereo vision. In *Proceedings of ARPA Image Understanding Workshop*, L. S. Baumann, ed., SRI, 41–45.

Gross, C. G., C. E. Rocha-Miranda, and D. B. Bender. 1972. Visual properties of neurons in inferotemporal cortex of the macaque. *J. Neurophysiol. 35*, 96–111.

Guzman, A. 1968. Decomposition of a visual scene into three-dimensional bodies. In *AFIPS Conf. Proc. 33*, 291–304. Washington, D.C.: Thompson.

Harmon, L. D., and B. Julesz. 1973. Masking in visual recognition: Effects of two-dimensional filtered noise. *Science 180*, 1194–1197.

Hartline, H. K. 1938. The response of single optic nerve fibres of the vertebrate eye to illumination of the retina. *Am. J. Physiol. 121*, 400–415.

Hartline, H. K. 1940. The receptive fields of optic nerve fibers. *Am. J. Physiol. 130*, 690–699.

Hassenstein, B., and W. Reichardt. 1956. Systemtheoretische Analyse der Zeit-, Reihenfolgen- and Vorzeichenauswertung bei der Bewegungsperzeption des Russelkafers. *Chlorophanus. Z. Naturf. 11b*, 513–524.

Hay, C. J. 1966. Optical motions and space perception—An extension of Gibson's analysis. *Psychol. Rev. 73*, 550–565.

Helmholtz, H. L. F. von. 1910. *Treatise on Physiological Optics*. Translated by J. P. Southall, 1925. New York: Dover.

Helson, H. 1938. Fundamental principles in color vision. I. The principle governing changes in hue, saturation, and lightness of non-selective samples in chromatic illumination. *J. Exp. Psychol. 23*, 439–471.

Hershberger, W. A., and J. J. Starzec. 1974. Motion parallax cues in one dimensional polar and parallel projections: Differential velocity and acceleration/displacement change. *J. Exp. Psycol. 103,* 717–723.

Hildreth, E. 1980. A computer implementation of a theory of edge detection. MIT A.I. Lab Tech. Rep. 579.

Harai, Y., and K. Fukushima. 1978. An inference upon the neural network finding binocular correspondence. *Biol. Cybernetics, 31,* 209–217.

Hochstein, S., and R. M. Shapley. 1976a. Linear and non-linear spatial subunits in Y cat retinal ganglion cells. *J. Physiol. (Lond.) 262,* 265–284.

Hochstein, S., and R. M. Shapley. 1976b. Quantitative analysis of retinal ganglion cell classification. *J. Physiol. (Lond.) 262,* 237–264.

Hollerbach, J. M. 1975. Hierarchical shape description of objects by selection and modification of prototypes. MIT A.I. Lab. Tech. Rep. 346.

Horn, B. K. P. 1973. The Binford-Horn LINEFINDER. MIT A.I. Lab. Memo 285.

Horn, B. K. P. 1974. Determining lightness from an image. *Computer Graphics and Image Processing 3,* 277–299.

Horn, B. K. P. 1975. Obtaining shape from shading information. In *The Psychology of Computer Vision,* P. H. Winston, ed., 115–155. New York: McGraw-Hill.

Horn, B. K. P. 1977. Understanding image intensities. *Artificial Intelligence 8,* 201–231.

Horn, B. K. P., R. J. Woodham, and W. M. Silver. 1978. Determining shape and reflectance using multiple images. MIT A.I. Lab. Memo 490.

Hubel, D. H., and T. N. Wiesel. 1961. Integrative action in the cat's lateral geniculate body. *J. Physiol. (Lond.) 155,* 385–398.

Hubel, D. H., and T. N. Wiesel. 1962. Receptive fields, binocular interaction and functional architecture in the cat's visual cortex. *J. Physiol. (Lond.) 166,* 106–154.

Hubel, D. H., and T. N. Wiesel. 1968. Receptive fields and functional architecture of monkey striate cortex. *J. Physiol. (Lond.) 195,* 215-243.

Hubel, D. H., and T. N. Wiesel. 1970. Cells sensitive to binocular depth in area 18 of the macaque monkey cortex. *Nature 225,* 41–42.

Hueckel, M. H. 1973. An operator which recognizes edges and lines. *J. Assoc. Comput. Mach. 20,* 634–647.

Huffman, D. A. 1971. Impossible objects as nonsence sentences. *Machine Intelligence 6*, 295–323.

Ikeda, H., and M. J. Wright. 1972. Receptive field organization of "sustained" and "transient" retinal ganglion cells which subserve different functional roles. *J. Physiol. (Lond.) 227*, 769–800.

Ikeda, H., and M. J. Wright. 1975. Spatial and temporal properties of "sustained" and "transient" neurons in area 27 of the cat's visual cortex. *Exp. Brain Res. 22*, 363–383.

Ikeuchi, K. Personal communication.

Ito, M. 1978. Recent advances in cerebellar physiology and pathology. In *Advances in Neurology,* R. A. P. Kark, R. N. Rosenberg, and L. J. Shut, eds., 59–84. New York: Raven Press.

Ittelson, W. H. 1960. *Visual Space Perception*, New York: Springer.

Jardine, N., and R. Sibson. 1971. *Mathematical Taxonomy*. New York: Wiley.

Johansson, G. 1964. Perception of motion and changing form. *Scand. J. Psychol. 5*, 181–208.

Johansson, G. 1975. Visual motion perception. *Sci. Am. 232*, 76–88.

Johnston, I. R., G. R. White, and R. W. Cumming. 1973. The role of optical expansion patterns in locomotor control. *Am. J. Psychol. 86*, 311–324.

Judd, D. B. 1940. Hue saturation and lightness of surface colors with chromatic illumination. *J. Opt. Soc. Am. 30*, 2–32.

Judd, D. B. 1960. Appraisal of Land's work on two-primary color projections. *J. Opt. Soc. Am. 50*, 254–268.

Julesz, B. 1960. Binocular depth perception of computer generated patterns. *Bell Syst. Tech. J. 39*, 1125–1162.

Julesz, B. 1963. Towards the automation of binocular depth perception (AUTO-MAP-1). In *Proceedings of the IFIPS Congress,* C. M. Popplewell, ed. Amsterdam: North Holland.

Julesz, B. 1971. *Foundations of Cyclopean Perception*. Chicago: University of Chicago Press.

Julesz, B. 1975. Experiments in the visual perception of texture. *Sci. Am. 232*, 34–43.

Julesz, B., and J. J. Chang, 1976. Interaction between pools of binocular disparity detectors tuned to different disparities. *Biol. Cybernetics 22*, 107–120.

Julesz, B., and J. E. Miller. 1975. Independent spatial-frequency-tuned channels in binocular fusion and rivalry. *Perception 4*, 125–143.

Kaufman, L. 1964. On the nature of binocular disparity. *Am. J. Psychol. 77*, 393–402.

Kelly, D. H. 1979. Motion and vision. II. Stabilized spatio-temporal threshold surface. *J. Opt. Soc. Am. 69*, 1340–1349.

Kendall, D. G. 1969. Some problems and methods in statistical archaeology. *World Archaeology 1*, 68–76.

Kidd, A. L., J. P. Frisby, and J. E. W. Mayhew. 1979. Texture contours can facilitate stereopsis by initiating appropriate vergence eye movements. *Nature 280*, 829–832.

Koenderick, J. J., and A. J. van Doorn. 1976. Local structure of movement parallax of the plane. *J. Opt. Soc. Am. 66*, 717–723.

Koffka, K. 1935. *Principles of Gestalt Psychology*. New York: Harcourt, Brace & World.

Kolers, P. A. 1972. *Aspects of Motion Perception*. New York: Pergamon Press.

Kruskal, J. B. 1964. Multidimensional scaling. *Psychometrika 29*, 1–42.

Kuffler, S. W. 1953. Discharge patterns and functional organization of mammalian retina. *J. Neurophysiol. 16*, 37–68.

Kulikowski, J. J., and D. J. Tolhurst. 1973. Psychophysical evidence for sustained and transient detectors in human vision. *J. Physiol. (Lond.) 232*, 149–162.

Land, E. H. 1959a. Color vision and the natural image. *Proc. Natl. Acad. Sci. 45*, 115–129, 636–645.

Land, E. H. 1959b. Experiments in color vision. *Sci. Am. 200*, 84–94, 96–99.

Land, E. H., and J. J. McCann. 1971. Lightness and retinex theory. *J. Opt. Soc. Am. 61*, 1–11.

Leadbetter, M. R. 1969. On the distributions of times between events in a stationary stream of events. *J. R. Statist. Soc. B 31*, 295–302.

Lee, D. N., 1974. Visual information during locomotion. In *Perception: Essays in Honor of James J. Gibson*, I. D. G. MacLed and O. Pick, eds. Ithaca, N.Y.: Cornell University Press.

Legge, G. E. 1978. Sustained and transient mechanisms in human vision: Temporal and spatial properties. *Vision Res. 18*, 69–81.

Lettvin, J. Y., R. R. Maturana, W. S. McCulloch, and W. H. Pitts. 1959. What the frog's eye tells the frog's brain. *Proc. Inst. Rad. Eng. 47*, 1940–1951.

Logan, B. F., Jr. 1977. Information in the zero-crossings of bandpass signals. *Bell Syst. Tech. J. 56*, 487–510.

Longuet-Higgins, H. C., and K. Prazdny. 1980. The interpretation of moving retinal images. *Proc. R. Soc. Lond. B 208,* 385–387.

Longuet-Higgins, M. S. 1962. The distribution of intervals between zeros of a stationary random function. *Phil. Trans. R. Soc. Lond. A 254*, 557–599.

McCann, J. J., S. P. McKee, and T. H. Taylor. 1976. Quantitative studies in retinex theory: a comparison between theoretical predictions and observer responses to the color Mondrian experiments. *Vison Res. 16*, 445–458.

McCulloch, W. S., and W. Pitts. 1943. A logical calculus of ideas immanent in neural nets. *Bull. Math. Biophys. 5*, 115–137.

Mackworth, A. K. 1973. Interpreting pictures of polyhedral scenes. *Art. Intel. 4*, 121–137.

Marcus, M. P. 1980. *A Theory of Syntactic Recognition for Natural Language.* Cambridge, Mass.: MIT Press.

Marr, D. 1969. A theory of cerebellar cortex. *J. Physiol. (Lond.) 202*, 437–470.

Marr, D. 1970. A theory for cerebral neocortex. *Proc. R. Soc. Lond. B 176*, 161–234.

Marr, D. 1974a. The computation of lightness by the primate retina. *Vision Res. 14*, 1377–1388.

Marr, D. 1974b. A note on the computation of binocular disparity in a symbolic, low-level visual processor. MIT A.I. Lab. Memo 327.

Marr, D. 1976. Early processing of visual information. *Phil. Trans. R. Soc. Lond. B 275*, 483–524.

Marr, D. 1977a. Analysis of occluding contour. *Proc. R. Soc. Lond. B 197*, 441–475.

Marr, D. 1977b. Artificial intelligence—a personal view. *Artificial Intelligence 9*, 37–48.

Marr, D. 1978. Representing visual information. *Lectures on Mathematics in the Life Sciences 10,* 101–180. Reprinted in *Computer Vision Systems,* A. R. Hanson and E. M. Riseman, eds., 1979, 61–80. New York: Academic Press.

Marr, D. 1980. Visual information processing: the structure and creation of visual representations. *Phil. Trans. R. Soc. Lond. B 290,* 199–218.

Marr, D., and E. Hildreth. 1980. Theory of edge detection. *Proc. R. Soc. Lond. B 207,* 187–217.

Marr, D., and H. K. Nishihara. 1978. Representation and recognition of the spatial organization of three-dimensional shapes. *Proc. R. Soc. Lond. B 200,* 269–294.

Marr, D., G. Palm, and T. Poggio. 1978. Analysis of a cooperative stereo algorithm. *Biol. Cybernetics 28,* 223–229.

Marr, D., and T. Poggio. 1976. Cooperative computation of stereo disparity. *Science 194,* 283–287.

Marr, D., and T. Poggio. 1977. From understanding computation to understanding neural circuitry. *Neurosciences Res. Prog. Bull. 15,* 470–488.

Marr, D., and T. Poggio. 1979. A computational theory of human stereo vision. *Proc. R. Soc. Lond. B 204,* 301–328.

Marr, D., T. Poggio, and E. Hildreth. 1980. The smallest channel in early human vision. *J. Opt. Soc. Am. 70,* 868–870.

Marr, D., T. Poggio, and S. Ullman. 1979. Bandpass channels, zero-crossings, and early visual information processing. *J. Opt. Soc. Am. 69,* 914–916.

Marr, D., and S. Ullman. 1981. Directional selectivity and its use in early visual processing. *Proc. R. Soc. Lond. B 211,* 151–180.

Marroquin, J. L. 1976. Human visual perception of structure. Master's thesis, MIT.

Maturana, H. R., and S. Frenk. 1963. Directional movement and horizontal edge detectors in pigeon retina. *Science 142,* 977–979.

Maturana, H. R., J. Y. Lettvin, W. S. McCulloch, and W. H. Pitts. 1960. Anatomy and physiology of vision in the frog *(Rana pipiens). J. Gen. Physiol. 43* (suppl. no. 2, Mechanisms of Vision), 129–171.

Mayhew, J. E. W., and J. P. Frisby. 1976. Rivalrous texture stereograms. *Nature 264,* 53–56.

Mayhew, J. E. W., and J. P. Frisby. 1978a. Stereopsis masking in humans is not orientationally tuned. *Perception 7,* 431–436.

Mayhew, J. E. W., and J. P. Frisby. 1978b. Texture discrimination and Fourier analysis in human vision. *Nature 275*, 438–439.

Mayhew, J. E. W., and J. P. Frisby. 1979. Convergent disparity discriminations in narrow-band-filtered random-dot stereograms. *Vision Res. 19*, 63–71.

Metelli, F. 1974. The perception of transparency. *Sci. Am. 230*, 91–98.

Miles, W. R. 1931. Movement in interpretations of the silhouette of a revolving fan. *Am. J. Psychol. 43*, 392–404.

Minsky, M. 1975. A framework for representing knowledge. In *The Psychology of Computer Vision*, P. H. Winston, ed., 211–277. New York: McGraw-Hill.

Mitchell, D. E. 1966. Retinal disparity and diplopia. *Vision Res. 6*, 441–451.

Monasterio, F. M. de, and P. Gouras. 1975. Functional properties of ganglion cells of the rhesus monkey retina. *J. Physiol. (Lond.) 251,* 167–195.

Movshon, J. A., I. D. Thompson, and D. J. Tolhurst. 1978. Spatial and temporal contrast sensitivity of neurones in areas 17 and 18 of the cat's visual cortex. *J. Physiol. (Lond.) 283,* 101–120.

Nakayama, K., and J. M. Loomis. 1974. Optical velocity patterns, velocity sensitive neurons, and space perception: A hypothesis. *Perception 3*, 63–80.

Narasimhan, R. 1970. Picture languages. In *Picture Language Machines*, S. Kaneff, ed., 1–25. New York: Academic Press.

Nelson, J. I. 1975. Globality and stereoscopic fusion in binocular vision. *J. Theor. Biol. 49*, 1–88.

Neuhaus, W. 1930. Experimentelle Untersuchung der Scheinbewegung. *Arch. Ges. Psychol. 75,* 315–458.

Newell, A., and H. A. Simon. 1972. *Human Problem Solving*. Englewood Cliffs, N.J.: Prentice-Hall.

Newton, I. 1704. *Optics*. London.

Nishihara, H. K. 1978. Representation of the spatial organization of three-dimensional shapes for visual recognition. Ph.D. dissertation, MIT.

Nishihara, H. K. 1981. Reconstruction of $\nabla^2 G$ filtered images from gradients at zero-crossings. (In preparation.)

Norman, D. A., and D. E. Rumelhart. 1974. *Explorations in Cognition*. San Francisco: W. H. Freeman and Company. See esp. 35–64.

Pearson, D. E., C. B. Rubinstein and G. J. Spivack. 1969. Comparison of perceived color in two-primary computer generated artificial images with predictions based on the Helsen-Judd formulation. *J. Opt. Soc. Am. 59*, 644–658.

Pettigrew, J. D., and M. Konishi. 1976. Neurons selective for orientation and binocular disparity in the visual wulst of the barn owl (*Tyto alba*). *Science 193*, 675–678.

Poggio, G. F., and B. Fischer. 1978. Binocular interaction and depth sensitivity of striate and prestriate cortical neurons of the behaving rhesus monkey. *J. Neurophysiol. 40*, 1392–1405.

Poggio, T., and W. Reichardt. 1976. Visual control of orientation behavior in the fly. Part II. Towards the underlying neural interactions. *Quart. Rev. Biophys. 9*, 377–438.

Poggio, T., and V. Torre. 1978. A new approach to synaptic interactions. In *Approaches to Complex Systems*, R. Heim and G. Palm, eds. 89–115. Berlin: Springer-Verlag.

Potter, J. 1974. The extraction and utilization of motion in scene description. Ph.D. dissertation, University of Wisconsin.

Prazdny, K. 1980. Egomotion and relative depth from optical flow. *Biol. Cybernetics, 36*, 87–102.

Ramachandran, V. S., and R. L. Gregory. 1978. Does colour provide an input to human motion perception? *Nature 275*, 55–56.

Ramachandran, V. S., V. R. Madhusudhan, and T. R. Vidyasagar. 1973. Apparent movement with subjective contours. *Vision Res. 13*, 1399–1401.

Rashbass, C., and G. Westheimer. 1961a. Disjunctive eye movements. *J. Physiol. (Lond.) 159*, 339–360.

Rashbass, C., and G. Westheimer. 1961b. Independence of conjunctive and disjunctive eye movements. *J. Physiol. (Lond.) 159*, 361–364.

Regan, D., K. I. Beverley, and M. Cynader. 1979. Stereoscopic subsystems for position in depth and for motion in depth. *Proc. R. Soc. Lond. B 204*, 485–501.

Reichardt, W., and T. Poggio. 1976. Visual control of orientation behavior in the fly. Part I. A quantitative analysis. *Quart. Rev. Biophys. 9*, 311–375.

Reichardt, W., and T. Poggio. 1979. Visual control of flight in flies. In *Recent Theoretical Developments in Neurobiology*, W. E. Reichardt, V. B. Mountcastle, and T. Poggio, eds.

Rice, S. O. 1945. Mathematical analysis of random noise. *Bell Syst. Tech. J. 24*, 46–156.

Richards, W. 1970. Stereopsis and stereoblindness. *Exp. Brain Res. 10*, 380–388.

Richards, W. 1971. Anomalous stereoscopic depth perception. *J. Opt. Soc. Am. 61*, 410–414.

Richards, W., and E. A. Parks. 1971. Model for color conversion. *J. Opt. Soc. Am. 61*, 971–976.

Richards, W. 1977. Stereopsis with and without monocular cues. *Vision Res. 17*, 967–969.

Richards, W., and D. Regan. 1973. A stereo field map with implications for disparity processing. *Invest. Opthal. 12*, 904–909.

Riggs, L. A., and E. W. Niehl. 1960. Eye movements recorded during convergence and divergence. *J. Opt. Soc. Am. 50*, 913–920.

Roberts, L. G. 1965. Machine perception of three-dimensional solids. In *Optical and electro optical information processing*, ed. J. T. Tippett et al., 159–197. Cambridge, Mass.: MIT Press.

Rock, I., and S. Ebenholtz. 1962. Stroboscopic movement based on change of phenomenal rather than retinal location. *Am. J. Psychol. 72*, 221–229.

Rodieck, R. W., and J. Stone. 1965. Analysis of receptive fields of cat retinal ganglion cells. *J. Neurophysiol. 28*, 833–849.

Rosch, E. 1978. Principles of categorization. In *Cognition and categorization*, E. Rosch and B. Lloyd, eds., 27–48. Hillsdale, N.J.: Lawrence Erlbaum Associates.

Rosenfeld, A., R. A. Hummel, and S. W. Zucker. 1976. Scene labelling by relaxation operations. *IEEE Trans. Man Machine and Cybernetics SMC–6*, 420–433.

Rosenfeld, A., and M. Thurston. 1971. Edge and curve detection for visual scene analysis. *IEEE Trans. Comput. C–20*, 562–569.

Russell, B. 1921. *Analysis of Mind*. London: Allen & Unwin.

Saye, A., and J. P. Frisby. 1975. The role of monocularly conspicuous features in facilitating stereopsis from random-dot stereograms. *Perception 4*, 159–171.

Schatz, B. R. 1977. The computation of immediate texture discrimination. MIT A.I. Lab Memo 426.

Schiller, P. H., B. L. Finlay, and S. F. Volman. 1976a. Quantitative studies of single-cell properties in monkey striate cortex. I. Spatiotemporal organization of receptive fields. *J. Neurophysiol. 39*, 1288–1319.

Schiller, P. H., B. L. Finlay, and S. F. Volman. 1976b. Quantitative studies of single-cell properties in monkey striate cortex. II. Orientation specificity and ocular dominance. *J. Neurophysiol. 39*, 1320–1333.

Shepard, R. N. 1975. Form, formation and transformation of internal representations. In *Information Processing and Cognition: The Loyola Symposium*, R. Solso, ed., 87–122. Hillsdale, N.J.: Lawrence Erlbaum Associates.

Shepard, R. N. 1981. Psychophysical complementarity. In *Perceptual Organization*, M. Kubovy and J. R. Pomerantz, eds. Hillsdale, N.J.: Lawrence Erlbaum Associates.

Shepard, R. N., and J. Metzler. 1971. Mental rotation of three-dimensional objects. *Science 171*, 701–703.

Shipley, W. G., F. A. Kenney, and M. E. King. 1945. Beta-apparent movement under binocular, monocular and interocular stimulation. *Amer. J. Psychol. 58*, 545–549.

Shirai, Y. 1973. A context-sensitive line finder for recognition of polyhedra. *Artificial Intelligence 4*, 95–120.

Sperling, G. 1970. Binocular vision: A physical and neural theory. *Am. J. Psychol. 83*, 461–534.

Stamm, J. S. 1969. Electrical stimulation of monkey's prefrontal cortex during delayed response performance. *J. Comp. Phys. Psych. 67*, 535–546.

Stevens, K. A. 1978. Computation of locally parallel structure. *Biol. Cybernetics 29*, 19–28.

Stevens, K. A. 1979. Surface perception from local analysis of texture and contour. Ph.D. dissertation, MIT. (Available as The information content of texture gradients. *Biol. Cybernetics 42* (1981), 95–105; also, The visual interpretation of surface contours. *Artificial Intelligence 17* (1981), 47–74.)

Sugie, N., and M. Suwa. 1977. A scheme for binocular depth perception suggested by neurophysiological evidence. *Biol. Cybernetics 26*, 1–15.

Sussman, G. J. 1975. *A Computer Model of Skill Acquisition*. New York: American Elsevier.

Sutherland, N. S. 1979. The representation of three-dimensional objects. *Nature 278*, 395–398.

Szentagothai, J. 1973. Synaptology of the visual cortex. In *Handbook of Sensory Physiology*, vol. 7/3B, R. Jung, ed., 269–324. Berlin: Springer-Verlag.

Tenenbaum, J. M., and H. G. Barrow. 1976. Experiments in interpretation-guided segmentation. Stanford Research Institute Tech. Note 123.

Tolhurst, D. J. 1973. Separate channels for the analysis of the shape and the movement of a moving visual stimulus. *J. Physiol. (Lond.) 231*, 385–402.

Tolhurst, D. J. 1975. Sustained and transient channels in human vision. *Vision Res. 15*, 1151–1555.

Torre, V., and T. Poggio. 1978. A synaptic mechanism possibly underlying directional selectivity to motion. *Proc. R. Soc. Lond. B 202,* 409–416.

Trowbridge, T. S., and K. P. Reitz. 1975. Average irregularity representation of a rough surface for ray reflection. *J. Opt. Soc. Am. 65*, 531–536.

Tyler, C. W. 1973. Stereoscopic vision: cortical limitations and a disparity scaling effect. *Science 181*, 276–278.

Tyler, C. W., and B. Julesz. 1980. On the depth of the cyclopean retina. *Exp. Brain Re., 40*, 196–202.

Ullman, S. 1976a. Filling-in the gaps: The shape of subjective contours and a model for their generation. *Biol. Cybernetics 25*, 1–6.

Ullman, S. 1976b. On visual detection of light sources. *Biol. Cybernetics 21*, 205–212.

Ullman, S. 1977. Transformability and object identity. *Percept. Psychophys. 22*, 414–415.

Ullman, S. 1978. Two dimensionality of the correspondence process in apparent motion. *Perception 7*, 683–693.

Ullman, S. 1979a. The interpretation of structure from motion. *Proc. R. Soc. Lond. B 203*, 405–426.

Ullman, S. 1979b. *The Interpretation of Visual Motion*. Cambridge, Mass.: MIT Press.

von der Heydt, R., Cs. Adorjani, P. Hanny, and G. Baumgartner. 1978. Disparity sensitivity and receptive field incongruity of units in the cat striate cortex. *Exp. Brain Res. 31*, 523–545.

Wallach, H., and D. N. O'Connell. 1953. The kinetic depth effect. *J. Exp. Psychol. 45*, 205–217.

Waltz, D. 1975. Understanding line drawings of scenes with shadows. In *The Psychology of Computer Vision*, P. H. Winston, ed., pp. 19–91. New York: McGraw-Hill.

Warrington, E. K. 1975. The selective impairment of semantic memory. *Quart. J. Exp. Psychol. 27*, 635–657.

Warrington, E. K., and A. M. Taylor. 1973. The contribution of the right parietal lobe to object recognition. *Cortex 9*, 152–164.

Warrington, E. K., and A. M. Taylor. 1978. Two categorical stages of object recognition. *Perception 7*, 695–705.

Watson, B. A., and J. Nachmias. 1977. Patterns of temporal interaction on the detection of gratings. *Vision Res. 17*, 893–902.

Weisstein, N. 1973. Beyond the yellow Volkswagen detector and the grandmother cell: A general strategy for the exploration of operations in human pattern recognition. In *Contemporary Issues in Cognitive Psychology: The Loyola Symposium*, R. Solso, ed. Washington, D.C.: W. H. Winston & Sons.

Weizenbaum, J. 1976. *Computer Thought and Human Reason*. San Francisco: W. H. Freeman and Company.

Wertheimer, M. 1912. Experimentelle Studien uber das Sehen von Bewegung. Zeitschrift f. Psychol. 61, 161–265.

Wertheimer, M. 1938. Laws of Organization in Perceptual Forms. Harcourt, Brace & Co., London. 71–88.

Westheimer, G., and S. P. McKee. 1977. Spatial configurations for visual hyperacuity. *Vision Res. 17*, 941–947.

Westheimer, G., and D. E. Mitchell. 1969. The sensory stimulus for disjunctive eye movements. *Vision Res. 9*, 749–755.

White, B. W. 1962. Stimulus-conditions affecting a recently discovered stereoscopic effect. *Am. J. Psychol. 75*, 411–420.

Williams, R. H., and D. H. Fender. 1977. The synchrony of binocular saccadic eye movements. *Vision Res. 17*, 303–306.

Wilson, H. R. 1979. Spatiotemporal characterization of a transient mechanism in the human visual system. Unpublished manuscript.

Wilson, H. R., and J. R. Bergen. 1979. A four mechanism model for spatial vision. *Vision Res. 19*, 19–32.

Wilson, H. R., and S. C. Giese. 1977. Threshold visibility of frequency gradient patterns. *Vision Res. 17*, 1177–1190.

Winograd, T. 1972. *Understanding Natural Language*. New York: Academic Press.

Woodham, R. J. 1977. A cooperative algorithm for determining surface orientations from a single view. *Proc. Int. Joint Conf. Art. Intel., IJCAI-77*, 635–641.

Woodham, R. J. 1978. Photometric stereo: A reflectance map technique for determining surface orientation from image intensity. *Image Understanding Systems and Industrial Applications, Proc. S.P.I.E. 155.* Also available as MIT A.I. Lab Memo 479.

Zeeman, W. P. C., and C. O. Roelofs. 1953. Some aspects of apparent motion. *Acta Psychol. 9,* 159–181.

Zucker, S. 1976. Relaxation labelling and the reduction of local ambiguities. University of Maryland Computer Science Center, Tech. Rep. 451.

Index